BIOCATALYSIS

VAN NOSTRAND REINHOLD
CATALYSIS SERIES

Burtron Davis, Series Editor

Metal-Support Interactions in Catalysis, Sintering, and Redispersion, edited by Scott A. Stevenson, R.T.K. Baker, J.A. Dumesic, and Eli Ruckenstein

Molecular Sieves: Principles of Synthesis and Identification, R. Szostak

Raman Spectroscopy for Catalysis, John M. Stencel

Theoretical Aspects of Heterogeneous Catalysis, John B. Moffat

Biocatalysis, edited by Daniel A. Abramowicz

BIOCATALYSIS

Edited by

Daniel A. Abramowicz

VAN NOSTRAND REINHOLD CATALYSIS SERIES

VNR VAN NOSTRAND REINHOLD
New York

Copyright © 1990 by Van Nostrand Reinhold

Library of Congress Catalog Card Number: 89-24847
ISBN 0-442-23848-7

Printed in the United States of America

Van Nostrand Reinhold
115 Fifth Avenue
New York, New York 10003

Van Nostrand Reinhold International Company Limited
11 New Fetter Lane
London EC4P 4EE, England

Van Nostrand Reinhold
480 La Trobe Street
Melbourne, Victoria 3000, Australia

Nelson Canada
1120 Birchmount Road
Scarborough, Ontario M1K 5G4, Canada

16 15 14 13 12 11 10 9 8 7 6 5 4 3 2 1

Library of Congress Cataloging-in-Publication Data

Biocatalysis / edited by Daniel A. Abramowicz.
    p.   cm.—(Van Nostrand Reinhold catalysis series)
   ISBN 0-442-23848-7
   1. Enzymes—Biotechnology. 2. Enzymes—Industrial applications.
3. Organic compounds—Synthesis. I. Abramowicz, Daniel A.
II. Series.
TP248.65.E59B543   1990
660′.2995—dc20   89-24847

Dedicated to my parents, Albert and Veda Abramowitz
and to my wife, Alice Abramowicz

Contents

Chapter Summaries

cluded chemical synthesis, immobilized enzyme synthesis, and whole cell fermentations. The microbial whole cell synthesis using an *E. coli aroB* strain proved to be the most convenient synthesis on a larger scale.

The application of transaminase enzymes to the preparation of chiral amino acids was investigated. Enzymes were immobilized onto silica-based supports that remained active for >1,000 h. These systems could catalyze the conversion of 2-ketoacids to the corresponding L-amino acids at high yields. In addition, the genes encoding this activity were cloned to facilitate the production of highly active enzyme preparations.

The microbial synthesis of 2-keto-L-gulonic acid (2-KLG), a key intermediate in the synthesis of ascorbic acid, is discussed. This was achieved by metabolic pathway engineering, combining portions of the catalytic pathway from two diverse organisms. By cloning the 2,5-diketo-D-gluconic acid (2,5-DKG) reductase gene from a *Corynebacter* species and expressing this gene in *Erwinia herbicola*, a strain was created that produces 2-KLG from glucose in a single fermentative step.

The application of certain pesticides as their single active stereoisomer rather than as racemates provides a real opportunity to utilize biocatalysis to the synthesis of key optically active intermediates. The production of phenoxyproponate herbicides as active stereoisomers requires S-2-chloropropanoic acid. This paper describes the application of a novel dehalogenase enzyme to affect kinetic resolution of racemic chloropropanoic acid to produce the desired isomer in high enantiomeric excess and in high yield.

The use of enzymes to affect resolutions, via enzymatic hydrolysis of a racemic mixture of esters, has created engineering challenges in the area of bioreactor design. A multiphase membrane reactor can be used to overcome the limitations of such immobilized enzyme systems with water-insoluble substrates. The application of this bioreactor for the enzymatic resolution of ibuprofen is described, along with process design improvements.

Microbial Reduction of Carbonyl Compounds: A Way to Pheromone Synthesis   178

In an extension of previous efforts developed for the enzymatic reduction of carbonyl compounds, the chiral syntheses of natural insect pheromones has been demonstrated. These whole cell processes illustrate that fermentation conditions can dramatically alter the observed stereoisomer ratio. In all cases, the biosynthetic routes described are more efficient and convenient than previously reported chemical methods.

Resolution of Binaphthols and Spirobiindanols Using Pancreas Extracts   195

The enantiospecific hydrolysis of binapthol and spirobiindanol diesters has been achieved with the enzyme cholesterol esterase. Synthetic scale (200 g) resolution of 1,1'-binaphthalene-2,2'-diol yielded each enantioner at greater than or equal to 99% enantiomeric excess (ee). Hydrolysis of the diesters to diols involves two enzyme-catalyzed steps, and calculations show that such multistep resolutions can yield higher purity stereoisomers.

Chiral Synthons by New Oxidoreductases and Methodologies   217

The stereospecific microbial reduction of carboxyl compounds was investigated. Applications include the reduction of a range of 2-enolates with *Clostridium tyrobutyricum* and the reduction of a range of 2-oxo-carboxylates to (2R)-hydroxy-carboxylates with *Proteus vulgaris*. In addition, a newly detected tungsten containing enzyme has been shown to reduce carboxylic acids to aldehydes.

Enzymes from Extreme Environments   243

Very stable enzymes are readily available from organisms growing in extreme environments, especially extreme thermophiles. Such thermostable enzymes isolated from extremely thermophilic archaebacteria display significant half-lives above 100°C. These enzymes should enable biosynthetic applications under harsh conditions. In addition, it has been shown that when cloned into mesophiles, heat treatment enables rapid, large-scale purification of the thermostable cloned enzymes.

Biocatalysis in Anaerobic Extremophiles 255
Certain anaerobic microorganisms have developed unique biocatalytic mechanisms for adaptation to extreme environments (including temperature, pH, and salt concentration). This chapter describes these organisms and discusses the application of enzymes isolated from these organisms in biosynthesis. The Haloanaerobes and Acidoanaerobes appear to have evolved intracellular enzymes that function under high salt or acidic conditions. In addition, unique enzymes from Thermoanaerobes may have potential industrial utility due to their high physicochemical stability and broad substrate specificity.

Large-Scale Bioconversion of Nitriles into Useful Amides and Acids 277
Nitrile-hydrolyzing enzymes, such as nitrile hydratase and nitrilase, have demonstrated great potential as catalysts for the conversion of nitriles into higher-value amides or acids. Recently, bacterial nitrile hydratase has been utilized for the production of the important chemical commodity acrylamide on an industrial scale. This work describes this enzymatic process as well as recent progress in the microbial transformation of nitriles.

Aldolases in Organic Synthesis 319
The synthetic utility of adolases has been demonstrated for the synthesis of common and uncommon sugars with bacterial fructose-1,6-diphosphate aldolase and N-acetylneuraminic acid aldolase. Both thermodynamically and kinetically controlled C–C bond formations have been developed for the synthesis of C-alkyl and N-containing sugars. In addition, the use of enol esters in enzymatic transformations of sugar-related compounds has been shown.

Two-Liquid Phase Biocatalysis: Reactor Design 337
The application of a two-liquid phase bioreactor to reactions involving compounds of low aqueous solubility is investigated. In this chapter,

the elucidation of the engineering parameters required for reactor design are discussed. The relevant design criteria are identified and methods suitable for obtaining such data are presented. The hydrolysis of benzyl acetate by pig liver esterase serves as a quantitative example of this methodology.

**Enzymes That Do Not Work in Organic Solvents: Too Polar
Substrates Give Too Tight Enzyme-Product Complexes   357**
The replacement of water by an apolar organic solvent largely effects the association between the enzyme and either substrate or product. This effect is important for enzyme complexes that demonstrate strong hydrogen bonding in aqueous solutions. It is believed that carbohydrate-converting enzymes are not catalytically active in organic solvents for this reason. This is supported by results that such enzymes are indeed active in organic solvents when the product formed is much less polar than the substrate.

Series Introduction

The action of enzymes fascinated mankind long before they were recognized for the complex chemicals that they are. The first application of these remarkable compounds to produce ethanol by fermentation is lost to antiquity. Payer and Persoz (*Ann. Chim. Phys., 53,* 73 (1833ii)) appear to have provided the first step toward understanding this complex area when they reported the isolation of diastase in 1833. These workers showed that diastase could catalyze the hydrolysis of starches to sugars. Somewhat earlier Kirchhoff (*Schwigger's Journal, 4,* 108 (1812)) had shown that a small amount of dilute acid could hydrolyze a seemingly endless amount of starch to sugars. The genius of Berzelius recognized the commonality of these two observations in connection with a few other isolated observations and in 1834 coined the term catalysis to describe such actions.

Professor Leibig was one of the giants of the chemical world in 1840. In addition to his own work, Liebig was training the world's next generation of chemists in his laboratory in Giessen. This cadre of chemists were very impressed by the master teacher so that is it only natural that Liebig's views should dominate with this next generation of chemists. Leibig was, in the 1830s and 1840s, developing his mastery of agricultural chemistry. The mechanism of putrefaction was of great concern to Leibig, and he turned to the newly defined area of catalysis for an explanation. However, Liebig did not adopt Berzelius' explanation of catalysis involving a new force—a catalytic force. Instead, Liebig viewed catalysis to be an induction of activity in an inactive body by the actions of a nearby active body. Thus, an unreacting mass could be made to react by placing it so that the motions of a highly active reacting mass could be transmitted to the unreacting body, and thereby activate it. Simply stated: place a rotting apple into a barrel of good apples and you soon have a barrel of rotten apples—to Liebig the rotting apple was in a highly active vibrating state and thus induced vibrations in the good apples so that they also became activated and rotten. This induced-vibration theory spread just as the graduates of Liebig's school spread throughout the scientific world.

Some 50 years later Ostwald, a future Nobel Prize winner, would refute this vibration theory as one that is worthless since it could not be subjected to experimental verification. Ostwald offered instead a kinetic definition of catalysis (*Z. Physik Chem., 15,* 705 (1894)). In Ostwald's view, a catalyst acted only to increase the rate of a reaction that was already occurring at a slow rate. Ostwald's view worked very well where the catalytic species could be described on a quantitative basis as, for example, the hydrogen ion. Thus, in the early 1900s the future Nobel Prize winner Langmuir was establishing the kinetic laws for heterogeneous catalysis (*Trans. Faraday Soc., 17,* 621 (1921)). Even earlier, Michaelis and Menton (*Biochem. Z., 49,* 333 (1913)) were making a definition of the kinetic law for enzyme action that was identical to the form deduced independently by Langmuir, but their law did not immediately attract the attention of the catalytic world.

Thus, in the early 1900s, a description of the kinetics in catalytic and enzymatic reactions was at about the same stage. However, the understanding to enzyme action was destined to progress at a slower rate, and this was primarily due to the complexity of the enzyme. Difficulties in purification and structure identification ensured that the progress would be slow, and that advances could only come with great effort. But in recent years the situation has changed rapidly. Sophisticated instrumentation now allow a determination of composition, both chemical sequence and three-dimensional configuration, that early workers could not even imagine in their wildest dreams. With this rapid advance in defining the structure of the enzyme, and even the catalytic site, understanding of enzyme action has speedily advanced. This volume defines much of these recent advances as it applies to the synthesis of organic products, both at the fundamental level and at the level important for commercial applications.

BURTRON H. DAVIS

Preface

This volume contains the applications of enzymes (biocatalysts) to the synthesis of specialty chemicals. These efforts involve whole cell as well as isolated and immobilized enzyme systems. By definition this work excludes the synthesis of low value bulk chemicals (methane, ethanol, biomass, etc.) and of hormone and protein products (human growth hormone, insulin, etc.). Instead, this volume focuses on the relatively new application of enzymes to the synthesis of organic chemicals with intermediate value (~$1–20/lb), an area of great commercial importance. The major themes presented in this work include:

- History of biocatalysis
- Applications in plastics (monomer and polymer synthesis)
- Biosynthesis of biotic metabolites (carbohydrates, peptides, vitamins)
- Chiral resolutions (herbicides, pharmaceuticals, chiral auxiliaries)
- Enzymes from extremophiles (thermophiles, acidophiles, halophiles)
- Applications of aldolases, enzymes in organic solvents, bioengineering concepts, and large-scale applications.

The volume is an outgrowth of the first international conference on the biocatalysis of organics and selected papers are included. The conference "Biocatalytic Synthesis of Organic Compounds" was held at Skidmore College in Saratoga Springs, New York, August 8–12, 1988. The conference was organized by an Executive Committee consisting of:

Conference Co-chairman   Daniel A. Abramowicz (GE)
                                 Alexander M. Klibanov (MIT)

Organizing Committee   David L. Anton (duPont)
Arnold Demain (MIT)
Charles R. Keese (GE)
Saul Neidleman (Cetus)

I would also like to take this opportunity to acknowledge the assistance of Herman L. Finkbeiner (GE) who conceived of the Biocatalysis Conference and was therefore responsible for its success and, indirectly, for this volume.

Contributors

Daniel A. Abramowicz, Ph.D., Manager of Environmental Technology, Biological Sciences Laboratory, General Electric Company, Schenectady, New York

Stephen Anderson, Ph.D., Senior Scientist, Department of Biomolecular Chemistry, Genentech, Inc., South San Francisco, California

Judy Bragger, Ph.D., Thermophile Research Group, School of Science, University of Waikato, Hamilton, New Zealand

D. L. Conley, Ph.D., Department of Chemistry, Purdue University, West Lafayette, Indiana

Steven P. Crump, Assistant Manager, Coors Biotech Products, Winchester, Kentucky

Roy M. Daniel, Ph.D., Thermophile Research Group, School of Science, University of Waikato, Hamilton, New Zealand

Mark S. Dennis, Research Associate, Department of Biomolecular Chemistry, Genentech Inc., South San Francisco, California

K. M. Draths, Ph.D., Department of Chemistry, Purdue University, West Lafayette, Indiana

Annie Fauve, Docteurés Sciences, Laboratoire de Chimie Organique Biologique, Université Blaise Pascal, Aubiére, France

James W. Frost, Ph.D., Department of Chemistry, Purdue University, West Lafayette, Indiana

Jeffery S. Heier, Research Associate, The Genetics Institute, Inc., Cambridge, Massachusetts

Romas J. Kazlauskas, Ph.D., Assistant Professor, Department of Chemistry, McGill University, Montreal, Quebec, Canada

Charles R. Keese, Ph.D., Biological Sciences Laboratory, General Electric Company, Schenectady, New York (Currently at Rensselaer Polytechnic Institute, Troy, New York)

Antonius P. G. Kieboom, Ph.D., Laboratory of Organic Chemistry, Delft University of Technology, Delft, The Netherlands

Margery G. Lazarus, Research Assistant, Department of Fermentation Research and Development, Genentech Inc., South San Francisco, California

Robert A. Lazarus, Ph.D., Senior Scientist, Department of Biomolecular Chemistry, Genentech Inc., South San Francisco, California

Suzanne H. Lockwood, Associate Staff, Biological Sciences Laboratory, General Electric Company, Schenectady, New York

J. L. Lopez, Ph.D., Sepracor Inc., Marlborough, Massachusetts

Susan E. Lowe, Ph.D., Visiting Assistant Professor, Department of Biochemistry, Michigan State University, East Lansing, Michigan

Cara B. Marks, Research Associate, Department of Biomolecular Chemistry, Genentech, Inc., South San Francisco, California

F. X. McConville, Ph.D., Sepracor Inc., Marlborough, Massachusetts

Hugh W. Morgan, Ph.D., Thermophile Research Group, School of Science, University of Waikato, Hamilton, New Zealand

Cary J. Morrow, Ph.D., Associate Professor, Department of Chemistry, University of New Mexico, Albuquerque, New Mexico

Toru Nagasawa, Ph.D., Department of Agricultural Chemistry, Kyoto University, Kyoto, Japan

Saul L. Neidleman, Ph.D., Cetus Corporation, Emeryville, California

D. L. Pompliano, Ph.D., Department of Chemistry, Purdue University, West Lafayette, Indiana

L. M. Reimer, Ph.D., Department of Chemistry, Purdue University, West Lafayette, Indiana

J. David Rozzell, Ph.D., Director, Industrial Chemicals, Celgene Corp., Warren, New Jersey

Badal C. Saha, Ph.D., Research Scientist, Michigan Biotechnology Institute, Lansing, Michigan, and the Department of Food Science and Human Nutrition, Michigan State University, East Lansing, Michigan

Jana L. Seymour, Research Associate, Department of Biomolecular Chemistry, Genentech Inc., South San Francisco, California

Helmut Simon, Ph.D., Institute for Organic Chemistry, Technical University, Munich, Federal Republic of Germany

R. Kevin Stafford, Scientist, Department of Fermentation Research and Development, Genentech Inc., South San Francisco, California

Stephen C. Taylor, Ph.D., ICI Biological Products, Cleveland, United Kingdom

J. Shield Wallace, Ph.D., Research Associate, Department of Chemistry, University of New Mexico, Albuquerque, New Mexico

S. A. Wald, Ph.D., Sepracor Inc., Marlborough, Massachusetts

Chi-Huey Wong, Ph.D., Professor, Department of Chemistry, Texas A&M University, College Station, Texas (Currently at the Department of Chemistry, The Research Institute of Scripps Clinic, La Jolla, California)

J. M. Woodley, Ph.D., Department of Chemical and Biochemical Engineering, SERC Centre for Biochemical Engineering, University College, London, United Kingdom

Hideaki Yamada, Ph.D., Department of Agricultural Chemistry, Kyoto University, Kyoto, Japan

J. Gregory Zeikus, Ph.D., President, Distinguished Senior Scientist, Michigan Biotechnology Institute, Lansing, Michigan, and the Departments of Biochemistry and Microbiology and Public Health, Michigan State University, East Lansing, Michigan

BIOCATALYSIS

1
The Archeology of Enzymology

SAUL L. NEIDLEMAN

From the past, we may understand the present and anticipate the future. The science of enzymology is no exception. Further, it seems comforting and encouraging to realize that a horde of predecessors have helped to get us where we are and, with perhaps a few exceptions, are cheering us on to new horizons. Archeology is an investigative and interpretative science and requires time and patience. In this chapter, as at the early stages of a particular quest, only a small proportion of the obscured material is revealed and only preliminary conclusions may be drawn. Bits and pieces will be offered so that an incipient friendship can be established between the reader and some of the players and concepts of the past. From these dispersed artifacts, it is hoped that a beginning perspective will result that will clarify where we are and where we may be going.

EARLY HIGHSPOTS

In an initial survey of the excavation site, some obvious highlight events may be recognized and noted (Table 1.1) (Waksman and Davidson 1926; Haldane 1930; Tauber 1937; Sumner and Somers 1947; Gortner and Gortner 1949; Fruton and Simmonds 1958; Dixon and Webb 1979; Fruton 1981). Because this table cannot show the fun, turmoil, and intellectuality of scientific discovery a few highlights will be examined in greater detail.

CONFESSION OF A PEROXIDASE ADDICT

For this author, studies on peroxidases have constituted a deep, long-standing interest. As suggested in the introductory section, a sense of the historical background can give added dimensions to research routine.

Table 1.1. Selected great moments in early enzymology.

DATES	INVESTIGATORS	DISCOVERIES
1752	Reamur	Chemical aspect to gastric digestion
1783	Spallanzani	The same
1820	Planche	Plant extracts cause guaiacum blueing
1830	Robiquet	Amygdalin hydrolysis by bitter almond extract
1831	Leuchs	Ptyalin activity
1833	Payen and Persoz	Diastase (amylase) activity
1833	Beaumont	Food solvent other than HCl
1835	Fauré	Siningrinase activity
1835	Berzelius	Defined catalysis
1836	Schwann	Pepsin activity
1837	Liebig and Wöhler	Emulsin activity
1846	Dubonfaut	Invertase activity
1856	Corvisart	Trypsin activity
1867	Kühne	Proposed term *enzymes* for unorganized ferments
1894	Fischer	Enzyme stereoselectivity
1894	Takamine	Diastase patent
1897	Buchner	Unorganized ferments convert glucose to ethanol
1898	Duclaux	—*ASE* to indicate enzymes
1911–13	Bourquelot, Bridel, and Verdon	Glucoside synthesis or degradation by enzymes in >80% ethanol, acetone
1915	Röhm	Tryptic enzyme patent
1922	Willstätter	Träger theory
1926	Sumner	Urease crystals
1930	Northrop	Pepsin crystals
1932	Northrop and Kunitz	Trypsin crystals

There is a close historical connection between guaiacum and peroxidases. In 1820, Planche observed that an extract of guaiacum turned blue in the presence of horseradish or milk. In 1855, Schönbein reported that the blueing reaction required three components: an extract of some plant or animal tissues, guaiacum, and air or "oxygenated water" (Paul 1987).

In 1494, Europe, in the throes of a syphilis epidemic, was blessed with the appearance of guaiacum, obtained from chips from the Caribbean trees *G. sanctum* and *G. officinale*, which "cured" syphilis when added to water in a steam bath (Munger 1949). The "cure" retained its credibility for nearly two centuries.

You can see how running an assay for peroxidase with guaicol (present in guaiacum) is elevated to a higher level of enjoyment by such an anecdote of the past.

More specifically, a personal focus on the halogenating capacity of peroxidases has been a favorite area of concern. These enzymes can oxidize chloride, bromide, and iodide. A book has derived from this interest (Neidleman and Geigert 1986). Unfortunately, the authors were not aware of a wayward dogma expressed by Harvey (1918) in a paper on bioluminescence wherein he begins a sentence by proclaiming: "Because NaCl could not possibly be oxidized by photogenin (=luciferase)—or any other substance—. . .". Thus a whole family of enzymes was not considered to be conceivable. This phrase would have had a prominent place in the book's introduction, but this information was not accessible to the authors.

GASTRIC JUICE: A HEARTY BREW

Having just illustrated the charm of detail available in enzyme archeology, one is compelled to reveal more instances with the same appealing quality.

Gastric digestion has attracted the attention of a number of talented experimentalists (Carlson 1923). Reaumur in 1752 and Spallanzani in 1783 introduced food and sponges, contained in perforated metal or wooden capsules, into the stomachs of fish, birds, and man. These capsules were subsequently recovered by means of attached strings, vomiting, or rectal passage. It was demonstrated that gastric digestion involved chemical solution rather than physical rupture. In 1822, Alexis St. Martin endured an accidental shotgun blast that resulted in the development of a gastric fistula (Young 1985). In 1833, Beaumont did experiments on St. Martin and determined that gastric juice contained a food solvent other than hydrochloric acid. In 1839, Wassman showed this to be pepsin.

Arrhenius, the famous Swedish chemist and physicist, attempted to demonstrate that there was a mathematical relationship between the quantity of ingesta and the quantity of gastric juice produced. However, it remained for Pavlov's laboratory to report, in 1914, that the situation was more complex. It was concluded that, dependent on dietary intake, there were meat, bread, and milk juices and these varied in their secretion period, volume, acidity, and pepsin content. It is not clear (Carlson 1923) that any of Pavlov's work was true, but it is comforting to realize that even 70 years ago gastric digestion gave rise to a bewildering and complex biochemical network.

WHAT IS CATALYSIS?

Early pronouncements on catalysis and then biocatalysis were a polyphonic mix of several themes, including: (1) what is a catalyst, (2) what is a biocatalyst, and (3) what is the structure of a biocatalyst.

A survey of the more prominent visions will indicate the focus of the intellectual exertion exercised through a 100-year period.

In 1836, Berzelius defined a catalyst as a substance capable of wakening energies dormant at particular temperatures, merely by its presence (Waksman and Davidson 1926; Fruton and Simmonds 1961). Further, Berzelius prophetically recognized the similarity of catalysis in the chemical laboratory and the living cell.

Such speculations did not, however, go unchallenged. Liebig stated that the assumption of this new force was detrimental to scientific progress, satisfying, rather, the human spirit. He suggested that any catalytic agent was itself unstable and during its decomposition caused unreactive substances to undergo chemical change.

The challenger was then challenged. In 1878, Traube retorted that ferments were not unstable substances transmitting chemical vibrations to unreactive materials, but were chemicals, related to proteins, possessing a definite chemical structure that evoked changes in other chemicals through specific chemical affinities (Fruton and Simmonds 1958).

In the 18th and 19th centuries, and even into the 20th century, a dominant concept was that referred to as "Lebenskraft" or "Spiritus Vitae." Among its tenets was the idea that minerals (inorganics) were products of ordinary physical forces, but organic compounds owed their synthesis to an organic vital force associated with the living cell. Even Berzelius in 1827 believed that organic synthesis in the laboratory was impossible. Organic chemists struggled with this belief for well over 100 years.

Levene (1931) wrote that the story of the rise and fall of biochemistry in the esteem of the scientific community is connected to its revolt against the concept of the vital force. The biocatalysts were clearly enmeshed in this controversy. There was the distinction between formed or organized ferments such as yeast and the unorganized ferments such as pepsin and diastase. In 1867, Kühne suggested the name enzyme for the latter. Mechanists, such as Berthelot and Hoppe-Seyler, considered active agents that performed their catalytic function within the cells as ferments and when these agents were excreted outside the cell, they were called enzymes. Vitalists asserted that fermentations were necessarily intracellular and enzyme reactions were extracellular. The work of Buchner and Hahn on cell-free alcohol fermentation by yeast extracts dissipated this differentiation. The conclusion was that living protoplasm produced and carried enzymes, and they could be separated from the living tissue and retain catalytic activity (Waksman and Davidson 1926), this in the face of the belief by some researchers that various enzymes separated from the cell still retained a residue of vital force from the living material (Haldane 1930).

Vital forces aside, our present definition of enzymes still invokes the dictum that they are produced by living cells. A direct corollary of this concept was succinctly expressed by Tauber (1943), who noted that Ryshkov and Sukhov in 1928 had analyzed tobacco mosaic virus for the enzymatic activity of amylase, asparaginase, catalase, chlorophyllase, oxidase, peroxidase, phosphatase, and protease. Their results were negative and Tauber concluded, therefore, that the virus is not a living thing. Enzymes are produced by living things; living things must produce enzymes.

One of the other major areas of speculation, agreement, and dissent was concerned with the structure of an enzyme. Even a sketchy review of this subject indicates that the primary difficulty was related to the nature and relevance of proteins. They were ill understood and this fact led to conceptual problems.

In 1922, Willstätter proposed his "Trager" or carrier theory. This viewpoint held that enzymes contain a smaller reactive group possessing a particular affinity for certain groupings in the substrate, thus accounting for the specificity of enzyme behavior. The reactive group, or enzyme proper, is considered to be attached to a colloidal carrier, and enzyme action is determined by the affinity of the active group for the substrate, and by the colloidality of the entire aggregate. When the colloidal properties of the aggregate are destroyed, the activity of the enzyme disappears.

A particular colloidal carrier did not appear to be essential, but any suitable colloidal carrier could act as a protective colloid for the active group (Gortner and Gortner 1949).

In a refinement of Willstätter's concept, Northrop (1930) stated that it is possible that enzymes are similar to hemoglobin, containing an active group combined with an inert group. The active group may be too unstable to exist alone. Further, it is quite conceivable that a series of compounds may exist containing varying numbers of active groups combined with the protein, and that the activity of a compound depends on the number of active groups. It might be possible to attach more active groups to the inert group, thus increasing the activity above that of the original compound. This hypothetical complex does not differ much from that proposed by Willstätter and his co-workers, except that it supposes a definite chemical compound with a stabilizing moiety in place of an adsorption complex.

The interesting observation can be made that even Northrop's concept does not indicate a specific role for the protein in a catalytic sense. Despite this, a step had been taken in a positive direction since Kunitz and Northrop (1935) in a related speculation indicated that if the native protein is merely a carrier for an active group, it is necessary to assume that the active group will become inactive when the protein is denatured and then will become active again when the protein reverts to the native condition. This is an early description of the reversible denaturation of an enzyme. It is also pertinent to point out that neither of these theories considered that parts of the protein (or colloid) itself might constitute the active site.

Quastel and Wooldridge (1928) published an incisive set of thoughts on the relationship of adsorption, activation, and specificity at the active center of an enzyme. They said that it was to be expected that a center possessing certain groups would adsorb a particular type of compound and another center with different groups would adsorb a different type of compound, i.e., the active center would evince a specificity of adsorption. But of the total number of molecules capable of being adsorbed at a particular center, only a few would be activated.

The number of these would depend on the strength and nature of the polarizing field and the structure of the substrate molecules.

The mention of a "polarizing field" recalls that in 1919–1921, Barendrecht offered a radiation theory for enzyme activity: a molecule of urea absorbs radiation from urease and is hydrolyzed (Fearon 1923).

We return to the currently accepted notion that (almost all) enzymes are proteins. Even in the 1930s and 1940s this fact rested uneasily in the minds of many scientists. J. B. S. Haldane (1930) observed that preparations of gastric, pancreatic, and hepatic esterases, and of yeast saccharase and pancreatic amylase were obtained by Willstätter and Bamann that were free from protein reactions (the biuret, Millon, ninhydrin, and tryptophan reactions). Haldane concluded that the amount of protein in these preparations must have been small, and if, as many workers believed, the enzymes are all proteins, it was remarkable that the majority of the successful attempts to purify them led to the isolation of substances that were at least predominantly nonproteins, although the original material from which they were derived consisted largely of protein. Haldane also remarked that the purest preparations, which give no protein reactions, were still nondialyzable and, when analyzed, contained C, H, O, and N. Even more drastic and dramatic was the claim of Rao et al. (1941) that renin, a proteolytic enzyme, was probably less complex a structure than supposed, because it contained no detectable nitrogen, sulfur, or phosphorus. Their purest enzyme preparation contained carbon, hydrogen, and oxygen along with some metals. Berridge (1945) proposed that further work was required to clarify the contradictions inherent in the work of Rao et al. and other investigators who stated the renin contained nitrogen.

Theorell (1976) added more to the story by indicating that in 1926, Willstätter described his work on peroxidase purification, and it came to the point where ordinary analyses for protein, sugar, or iron indicated nothing: but enzyme activity remained. Willstätter concluded that enzymes might not contain any of these and did not belong to a known chemical class. He was included to believe that enzyme activity derived from a natural force. Even nothing has an explanation! The problem, of course, was that analytical methods were not sensitive enough to deal with these situations.

Levene (1931) poignantly commented on the state of knowledge relevant to protein structure. He noted that the history of proteins, known from earliest time, was most discouraging. The term protein was introduced by Mulder in 1860, yet how little was known about the details of the structure of even a single protein, although the number of them in nature was endless.

On the more optimistic side, Haldane (1930) commented that our definition of enzymes, produced by living cells, would be out of date when an enzyme was prepared synthetically. He tempered this prophetic remark by adding that his book would be out of date long before this occurred.

ANTICIPATING AN ENZYME

Often, in the early days of enzymology, logic anticipated the existence of a biocatalyst before its existence was, in fact, confirmed. Two illustrations, out of many, will be considered: carbonic anhydrase and lipoxygenase.

The evolution of carbon dioxide from bicarbonate in the presence of blood solutions was an area of study in the 1920s and early 1930s. It had been tentatively concluded that the hemoglobin was the catalytic agent. However, Van Slyke and Hawkins (1930) concluded that this was not the case and that some other catalyst was responsible. The problem centered about the fact that hemoglobin preparations, which showed carbon dioxide evolution, were contaminated with another substance that was subsequently identified and named carbonic anhydrase (Meldrum and Roughton 1933). To insert some sense of humanity into these academic flurries, Davenport (1984), a later worker with carbonic anhydrase, made the following pithy analysis of Roughton: "Roughton was not, to put it gently, an accomplished technician in the laboratory. My guess is that after Meldrum's death, when Roughton had to finish the work himself, he missed finding carbonic anhydrase in other tissues by sheer ineptitude."

The second example, that of lipoxygenase, is of a different sort because it graphically illustrates the marvelous but unsuspected anticipation of an enzyme and its active site by studies in chemical catalysis. Warburg (1925) was studying the effect of iron in oxidation reactions of living cells. A segment of his research was devoted to a detailed investigation of model systems, involving iron, for the specific oxidation of amino acids, fructose, and fatty acids. With regard to the fatty acids, Warburg noted that iron, combined with the sulfhydryl (—SH) group of cysteine, did not oxidize amino acids or sugar, but did specifically oxidize unsaturated fatty acids such as linoleic acid.

One train of thought that might be developed from such a finding is that these two entities, iron and —SH groups, might constitute important functionalities at the active site of an enzyme devoted to the oxidation of unsaturated fatty acids. Such an enzyme family is, in fact, that of the lipoxygenases. It has been demonstrated that iron and —SH groups are involved in their oxidative activity (Grossman et al. 1984; Feiters, Veldink, and Vliegenthart 1986). This is a case, then, in which the model system anticipated the enzyme, in contrast to many modern examples where the enzyme anticipates the model system or synzyme (artificial enzyme).

SOME EARLY ENZYME PATENTS

In the midst of discussing the intellectual travail inherent in the work of pioneer enzymologists, it adds a touch of perspective to realize that there were those who had practical thoughts. Even if the basic understanding of enzymatic catalysis was still struggling, there were profitable applications to be developed. Table 1.2 gives a very scant indication of where some of the action was. Granting that all of these patents, as well as many that are uncited, symbolize progress in the application of biocatalysis, one must be singled out for an accolade, that of O. Röhm in 1908.

Workers in the leather industry must have given three cheers for Röhm. In a close reading of the patent, it is realized that Röhm's new process, using an aqueous extract of mammalian pancreas, replaced the previous commercial item of choice for bating of hides: dog manure.

WATER-INSOLUBLE ENZYMES CAN WORK TOO

The interactions of enzymes and water-insoluble matrices in enzyme immobilization is an area of active research at present. The past also had its moments. In 1910, Starkenstein reported on the adsorption of amylase on its substrate starch (Hais 1986). He demonstrated that

Table 1.2. Selected early enzyme patents.

INVENTORS	YEAR	PATENT NUMBER	TITLE	ENZYME
J. Takamine	1894	US 525,823	Process of Making Diastatic Enzyme	Amylase
O. Röhm	1908	US 886,411	Preparation of Hides for the Manufacture of Leather	Trypsin and steapsin
L. Wallerstein	1911	US 995,820	Beer and Method of Preparing Same	Malt protease
L. Wallerstein	1911	US 995,824	Method of Treating Beer or Ale	Pepsin
L. Wallerstein	1911	US 995,825	Method of Treating Beer or Ale	Papain
L. Wallerstein	1911	US 995,826	Method of Treating Beer or Ale	Bromelin
L. Wallerstein	1911	US 997,873	Method of Treating Beer or Ale	Yeast protease
O. Röhm	1915	GER 283,923	Process for Cleaning Laundry of All Types	Pancreatin
V. G. Bloede	1918	US 1,257,307	Process of Manufacturing Vegetable Glue	Amylase
H. S. Paine and J. Hamilton	1922	US 1,437,816	Process for Preparing Fondant or Chocolate Soft Cream Centers	Invertase
A. Boidin and J. Effront	1924	US 1,505,534	Treatment of Textile Fabrics or Fibres	Amylase
M. Wallerstein	1932	US 1,854,355	Method of Making Chocolate Syrups	Amylase or papain
R. Douglas	1932	US 1,858,820	Process of Preparing Pectin	Amylase

dialyzed amylase could be adsorbed on water-insoluble starch. The complex could be recovered, washed, and then activated to cause starch degradation in the presence of chloride ion. He ascribed the formation of the amylase–starch complex to purely physical forces. The enzyme activity did not depend on the presence of a soluble enzyme.

In a continuation of this theme of enzyme immobilization by adsorption, Nelson and Griffin (1916) demonstrated that yeast invertase could be absorbed on charcoal and retain its catalytic activity. Levene and Weber (1924) showed that nucleosidase adsorbed on kaolin retained its activity and could not be extracted from the matrix by a variety of aqueous solutions.

CHEMICAL MODIFICATION OF ENZYMES

Early workers were not only aware that enzyme activity did not necessarily depend on water solubility but that, in addition, enzyme structure could be modified without total elimination of catalytic activity. These were studies designed to investigate the importance of particular functionalities in determining enzyme activity. Herriott and Northrop (1934) and Tracy and Ross (1942) worked with derivatives of pepsin using ketene and carbon suboxide to acetylate and malonylate, respectively, the enzyme. Both amino and phenolic hydroxyl groups could be altered. It was determined that, whereas loss of amino groups did not appear to affect enzyme activity, loss of tyrosyl hydroxyl groups inactivated the enzyme. It was further claimed that malonylation of the lysyl amino moieties did not alter pepsin specificity: that amino groups could be replaced by carboxyl groups without effect.

As a final illustration of studies related to modification of enzyme structure, Gjessing and Sumner (1942) showed that the "natural" iron porphyrin present in horseradish peroxidase could be replaced by a manganese porphyrin and the resultant, regenerated catalyst had 20–30% of the activity obtained when iron porphyrin was used in its place.

The work noted in this and the preceding section illustrates early enzymologists realized that enzymes, as isolated from nature, could be chemically and physically altered without a major loss in activity.

ENZYMES IN ORGANIC SOLVENTS CAN GO EITHER WAY

Research related to the catalytic activity of enzymes in the presence of high concentrations of organic solvents has reached explosive intensity in the past decade. In reporting on this, the statement has been made that, in the early 1980s, Klibanov found that lipase activities were retained in nonaqueous solutions (Gillis 1988). The inference that this constituted a pristine discovery is inaccurate by a margin of over 50 years, and work with another enzyme in organic solvents was carried out over 75 years ago.

In an extraordinary series of papers, Bourquelot and co-workers (Bourquelot and Bridel 1911, 1912a–d; Bourquelot and Verdon 1913a,b) studied the action of emulsin (containing β-glucosidase) in organic solvents. It was reported that this enzyme from almonds, when placed in suspension in 85% alcohol containing glucose in solution, caused at ordinary temperature the combination of sugar and alcohol with formation of β-ethylglucoside. Furthermore, it was found that emulsin as a powder had a hydrolytic activity on dissolved glucosides in neutral liquids such as acetone and ethyl ether, although these liquids did not dissolve a trace of enzyme.

These studies were early demonstrations that enzymes were involved in equilibrium reactions and could go either way depending on conditions. Further, it established that enzyme activity in organic solvents was feasible.

With regard to investigations on lipase activity in the presence of organic solvents, the following may be noted:

Sym (1936) showed that in the case of pig pancreatic lipase, the application of organic solvents, practically insoluble in water, gave much higher yields of esters (often >95%) than the use of aqueous systems, owing to the fact that the products of reaction were removed from the medium where the reaction proceeds. In addition, the nonaqueous phase may be regarded as a reservoir of the components of the reaction, supplying substrates to the aqueous phase. The action of the enzymic preparations was studied, in most cases, in systems containing 1 M n-butanol alcohol, 0.43 M butyric acid, and benzene. In a number of studies, the enzyme preparation was an acetone powder containing 8–12% water. Sperry and Brand (1941) showed cholesterol esterification using a similar enzyme preparation in $\geq 90\%$ carbon tetrachloride. With butyric acid, 65.2% esterification occurred; with palmitic acid, 30%; and with oleic acid, 40%.

It is evident that the explosive expansion of research on enzyme reactions in organic solvents began building energy over 75 years ago. The literature awaits further excavation.

PUBLICATION VELOCITY: A SURVIVAL FACTOR

The insatiable drive to establish scientific priority survives unabated in these times. It is, however, not a product of our times. Two examples from the past will illustrate this with blinding clarity. The first might be entitled: "At 4:01 p.m. on August 22, 1888, I knew I had a ferment." Green (1890) reported an experiment designed to demonstrate that a ferment existed in extracts of seeds of *Ricinus communis* (Castor oil plant) that could produce fatty acids from castor oil:

> Tube F was prepared by mixing the extract and the emulsion of castor oil in the proportions given above, and was put into an incubator at 12:30 o'clock on August 22, 1888. A boiled control was put with it, labelled F_1. Both were carefully made neutral. At 4 p.m. 10 drops litmus solution were added to each. F was acid, F_1 neutral.

The experiment was successful and priority was ensured.

The second example is by Dakin (1910) who had submitted a paper on the conversion of acetoacetic acid to β-oxybutyric acid in cats and dogs. He comments:

> This fact was recorded in a short paper sent on March 16 to the *Journal of the American Medical Association*, which, however, not withstanding a promise of prompt publication, did not appear until April 29. In the meantime, a paper appeared by Blum (March 29) in which the same fact was established by slightly different methods.

The Blum paper was published in a different journal and even more irritating was the fact, noted by Dakin, that Blum made reference to similar experiments by someone named Maase, the publication of which was not yet accessible to Dakin.

ENZYME SPECIFICITY: AN INTELLECTUAL CALDRON

In a landmark paper on the effect of configuration on enzyme activity, Fischer (1894) expressed a number of opinions that ring true after almost 100 years:

". . . bezweisle ich ebensowenig wie die Brauchbarkeit der Enzyme für die Ermittlung der Configuration asymmetrischer Substanzen.''

[I have little doubt about the usefulness of enzymes for the determination of the configuration of asymmetric substances.]

"Noch wichtiger für dieselbe aber scheint mir der Nachweis zu sein, dass der früher vielfach angenommene Unterschied zwischen der chemischen Thätigkeit der lebenden Zelle und der Wirkung der chemischen Agentien in Bezug auf moleculare Asymmetrie thatsächlich nicht besteht.''

[Also important it appears to me is that the evidence shows that the earlier repeated and accepted difference between the chemical activity of the living cell and that of chemicals with regard to asymmetric molecules does not in fact exist.]

"Aber schon genügen die Beobachtungen, um principiell zu beweisen, dass die Enzyme bezüglich der Configuration ihrer Angriffsobjecte ebenso wählerisch sind, . . .''

[But the results already suffice to prove the principle that enzymes are fussy about the configuration of their object of attack.]

"Um ein Bild ze gebrauchen, will ich sagen, dass Enzym and Glucosid wie Schloss und Schlüssel zu einander passen müssen, um eine chemische Wirkung auf einander ausüben zu können.''

[To use an image, I would say that the enzyme and glucoside must fit each other like a lock and key to be able to carry on a chemical reaction on each other.]

The future confirmed the genius of Fischer.

It is to be emphasized that the lock-and-key metaphor is invaluable as an image for those learning to think of enzyme–substrate interactions, but it should also be emphasized that modern work has built on this rigid portrait and added dimensions of conformational flutter. Enzyme and substrates stand still for no reaction.

Thunberg (1920) presented another indication that enzymes have specificity by observing that individual enzymes in a mixture could be differentiated by their different temperature responses, especially their thermostability. Therefore, he reasoned there are a variety of distinct enzymes catalyzing a variety of reactions. In 1921, Dakin commented that Thunberg's argument was rather unconvincing but Bernheim (1928) offered an analogous suggestion to that of Thunberg's by saying that one type of evidence for enzyme specificity was in the separation of the enzyme from the tissue:

This evidence is fairly conclusive for the separation depends on differences in physical properties of the enzymes, i.e., the centres responsible for the activations are attached to different colloids. It would be difficult to explain these facts on the assumption of one enzyme in the tissue which was originally capable of effecting all the activations but which during the process of extraction has been so altered as to appear specific for one substrate. Four dehydrases have been separated in this way: the succinic, lactic, citric and xanthine.

Thunberg argued for specific enzymes on the basis of thermostability and Bernheim because of separation individuality.

It is more and more common to read in the present literature of "unexpected" catalytic activity of known enzymes (Klibanov 1983). From this viewpoint, it is enlightening to consider how the "expected" catalytic activity of two specific enzymes metamorphosed over a period of years. The two enzymes are xanthine oxidase and urease.

Morgan, Stewart, and Hopkins (1922) stated that the enzyme xanthine oxidase showed marked specificity, accepting xanthine and hyperoxanthine as substrates but not guanine, caffeine, uracil, thymine, or cytosine, for example. The authors were, however, skeptical about reports that the enzyme could also oxidize aldehydes such as acetaldehyde. They thought it unlikely that the enzyme would extend its activity to such compounds. Dixon (1926) was equally unsure that xanthine oxidase could oxidize both purines and aldehydes:

> With regard to the case of aldehyde, it cannot be maintained that it is definitely established that the oxidations of aldehyde and purines are due to the same enzyme. If they are not, then *both* enzymes must be remarkably specific. On the other hand, there appears to be a fairly strong balance of evidence in favor of identity, although a certain part of this can be explained away by supposing that we have two different enzymes adsorbed on the same colloid.

This is, indeed, a classic case of inconclusiveness.

Keilen and Hartree (1935) concluded, on the basis of the work of others, that the oxidation of purines and aldehydes was definitely due to a single enzyme, xanthine oxidase. Coming nearly full circle after nearly 50 years, Dixon and Webb (1979) stated that xanthine oxidase had a rare, dual specificity, oxidizing purines and aldehydes. It is delightful to report that these authors actually wrote *duel* specificity, as if to indicate subconsciously that arguments over enzyme specificity can be barbed on occasion, as in this case wherein two camps existed: the single enzyme and two enzyme forces.

The background related to the specificity of urease is simpler in the sense that only urea and closely related analogues were studied rather than diverse substances such as purines and aldehydes. In the case of xanthine oxidase, Armstrong and Horton (1912) concluded, quite correctly, based on the range of substrates examined, that urease was specific for urea. Other substrates examined were biuret and various methyl and ethyl derivatives of urea. The authors also suggested that the enzyme must "correspond very closely in structure with the hydrolyte urea with which alone it is in correlation."

Sumner (1926) succeeded in preparing the first crystalline enzyme, urease. One can sense the stamp of priority in his opening paragraph:

> After work both by myself and in collaboration with Dr. V. A. Graham and Dr. C. V. Noback that extends over a period of a little less than 9 years, I discovered on the 29th of April a means of obtaining

from the jack bean a new protein which crystallizes beautifully and whose solutions possess to an extraordinary degree the ability to decompose urea into ammonium carbonate.

Thus, it is established that Sumner had a vested interest in urease and so it is of more than passing interest to quote the following from Sumner and Somers (1947): "Urease is an enzyme that is absolutely specific. It acts only upon urea and nothing else." Therefore, it might be assumed that this is an example of a known, purified enzyme with no hope for unexpected substrates. There is always hope! Gazzola, Blakely, and Zerner (1973) reported other compounds hydrolyzed by urease were N-hydroxyurea, Semicarbazide (N-aminourea), and N,N'-dihydroxyurea, while carbazide (N,N'-diaminourea), N-methylurea, N-butylurea, N-methoxyurea, N-(2-aminoethyl)urea, 2-imidazolidinone, biuret, 1-phenylsemicarbazide, ethyl carbamate, and carbamoyl azide were among many derivatives that were not hydrolyzed.

Sumner, as the first crystallizer of an enzyme, deserves a biographical sketch. Laskowski (1982) offered some first-hand insights into the person. He was a millionaire descendent of a Mayflower family. He lost an arm at 15, and won a Nobel prize at 59. His lectures were considered dull and bloated with detail. His mind was a living encyclopedia of biochemistry.

The above brief and superficial discussion of enzyme specificity suggests that unexpected substrates may be unexpected but not a surprise because we still know relatively little about the total potential of any given enzyme under a variety of reaction conditions, given an exception now and again.

DEVELOPMENTS OVER TIME

One of the obvious advances made in enzymology during the past 50 years is the development of high technology analytical capabilities. When the pioneers of enzymology were building their proteinaceous pyramids, they had only a vague idea of what the building material was. Today, with our molecular microscopes, we nearly know it all. One brief illustrative example is in the conversion of chymotrypsinogen to chromotrypsin.

Kunitz and Northrop (1935) reported that:

The transformation of chymo-trypsinogen into chymo-trypsin is accompanied by a change in optical activity and a slight increase in amino nitrogen. There is no detectable non-protein nitrogen fraction formed nor is there any significant change in molecular weight. The reaction, therefore, is probably an internal rearrangement, possibly due to the splitting of a ring.

In a subsequent publication, Herriott and Northrop (1936) refined this interpretation by suggesting that the rupture of a peptide bond was involved. Given the state of their art, this was an incisive conclusion. However, the transformation was later summarized in more graphic detail by Stryer (1975):

Some conformational changes that have been elucidated include:

1. Cleavage of peptide bond at $Arg^{15}Ile^{16}$.
2. $-NH_2$ of Ile^{16} interacts with $-COOH$ of Asp^{194}.
3. Other changes follow:
   A. Met^{192} moves from interior to surface.
   B. Residues 187 and 193 become extended.
   C. Substrate specificity pocket is thus created.

The contrast is shocking when it is realized that the best the early workers could do was to implicate the cleavage of an unidentified peptide bond, whereas later workers could deal with the transformation of specific molecular domains. The intellect was willing but information and technique were wanting.

AN EXAMPLE OF SERENDIPITY IN SCIENTIFIC DISCOVERY

Fleming had a prepared mind in 1929, when he reported the discovery of penicillin. Earlier in his career, he had another incident of the prepared mind: his discovery of lysozyme. Fleming (1922) described the event:

In this communication I wish to draw attention to a substance present in the tissues and secretions of the body, which is capable of rapidly dissolving certain bacteria. As this substance has properties akin to those of ferments I have called it a "Lysozyme," . . .

In the first experiment nasal mucus from the patient, with coryza, was shaken up with five times its volume of normal salt solution, and the mixture was centrifuged. A drop of the clear supernatant fluid was placed on an agar plate, which had previously been thickly planted with *M. lysodeikticus*, and the plate was incubated at 37°C for 24 hours, when it showed a copious growth of the coccus, except in the region where the nasal mucus had been placed. Here there was complete inhibition of growth, and this inhibition extended for a distance of about 1 cm. beyond the limits of the mucus."

What makes this occurrence even more intriguing is the claim that the patient was Fleming (Stryer 1975):

In 1922, Alexander Fleming, a bacteriologist in London, had a cold. He was not one to waste a moment, and consequently used his cold as an opportunity to do an experiment. He allowed a few drops of his nasal mucus to fall on a culture plate containing bacteria. He was excited to find some time later that the bacteria near the mucus had been dissolved away and thought that the mucus might contain the universal antibiotic he was seeking.

TIME AND THE CYTOCHROMES: TERMINATION AND RESUSCITATION

Destructive criticism wanders through the pages of enzymology. A particularly devastating instance involves the cytochromes. MacMunn (1887) described what he believed to be novel respiratory pigments in muscle and other tissues of animals. He used the terms myohaematin and histohaematin to identify them. Levy (1889) repeated the work of MacMunn and regarded the substances of MacMunn as haemochromogen, derived from hemoglobin. An exchange of disagreements was initiated by MacMunn (1889) and Hoppe-Seyler (1890) who sided with

Levy. MacMunn (1890), made one last attempt to salvage his work. His effort was discredited when Hoppe-Seyler (1890) appended a terse editorial note at the end of this publication saying that MacMunn argued with no new facts and further discussion was not to occur: there was no reason for pigeon muscle to contain a special pigment.

MacMunn's new respiratory pigment was forgotten. Fortunately, Keilen (1925) revived the subject. He renamed MacMunn's pigment cytochrome and demonstrated that cytochrome was distinct from muscle hemoglobin and that both pigments could be readily detected in the same muscle of a bird or mammal and that cytochrome also occurred in yeast, bacteria, and higher plants.

EFFECT OF TEMPERATURE ON FAT UNSATURATION

Neidleman (1987) published a broad review of the effects of temperature on lipid unsaturation. An oversimplified, single sentence summary would be as follows:

The colder the living system, the more unsaturated are its lipids; the warmer the living system, the less unsaturated are its lipids and enzymatic unsaturation is intimately involved in these effects.

In preparing the review, a wonderful experiment described by Henriques and Hanson (1901) was not rediscovered. These investigators acquired three young pigs from the same litter. Each was given special and individualized attention. One pig was kept in a room at 30–35°C, while the others were kept in pens at about 0°C (the experiment was completed during the winter). Of the latter two pigs, one animal had a sheepskin, the wool side turned inward, placed on its back, stomach, and sides. All three animals were fed only a corn diet. After 2 months all three animals were slaughtered and the skin fat, as well as the kidney and omental fat were examined.

The results indicated that the 20–35°C pig had the least unsaturated fat, the unadorned pig at 0°C had the most unsaturated fat, and the pig in a blanket was intermediate in lipid unsaturation. Enzymes involved in determining lipid unsaturation are affected by temperature.

CONCLUSIONS

Studying the archeology of enzymology adds new dimensions to participation in the catalytic discipline. There is an increase in the joy inherent in the process of creative science that flows from an involvement with past: its brilliance, its conflict, its humanity, and its satisfaction. Drama abounds. Working or thinking about the cytochromes must be more compelling when the scenario of MacMunn, Hoppe-Seyler, Levy, and Keilen is recalled. There is, in addition, a real chance that, by steeping oneself in the history of science, new insights will ignite under the stimulus of the work and thought of our predecessors, who may be physically gone but are intellectually in the seat next to each of us.

REFERENCES

Armstrong, H. E., and E. Horton. 1912. Studies on enzyme action. XV. Urease: a selective enzyme. *Proc. Roy. Soc. Lond. Ser. B. Biol.* **85:** 109–27.

Bernheim, F. 1928. CXLVII. The specificity of the dehydrases. The separation of the citric acid dehydrase from liver and of the lactic acid dehydrase from yeast. *Biochem. J.* **22:** 1178–1192.

Berridge, N. J. 1945. The purification and crystallization of rennin. *Biochem. J.* **39:** 179–186.

Bourquelot, E., and M. Bridel. 1911. Action de l'émulsine sur la gentiopicrine, en milieu alcoolique. *J. Pharm. Chim.* **4:** 385–390.

Bourquelot, E., and M. Bridel. 1912a. Action de l'émulsine sur la salacine en milieu alcoolique. *Compt. Rend. Acad. Sci.* **154:** 944–946.

Bourquelot, E., and M. Bridel. 1912b. Sur une action synthétisante de l'émulsine. *Compt. Rend. Acad. Sci.* **154:** 1375–1378.

Bourquelot, E., and M. Bridel. 1912c. Synthèse de glucosides d'alcools à l'aide de l'émulsine: méthylglucoside β et propylglucoside β. *Compt. Rend. Acad. Sci.* **155:** 86–88.

Bourquelot, E., and M. Bridel. 1912d. Nouvelles synthèses de glucosides d'alcools à l'aide de l'émulsine: butylglucoside β, isobutylglucoside β et allylglucoside β. *Compt. Rend. Acad. Sci.* **155:** 437–439.

Bourquelot, E., and E. Verdon. 1913a. Recherches sur la synthèse biochimique du méthylglucoside β dans un liquide neutre, étranger à la reaction. *Compt. Rend. Acad. Sci.* **156:** 1264–1266.

Bourquelot, E., and E. Verdon. 1913b. De l'emploi de proportions croissantes de glucose dans la synthèse biochimique du méthylglucoside β. Influence du glucoside formé sur l'arrêt la réaction. *Compt. Rend. Acad. Sci.* **156:** 1638–1640.

Carlson, A. J. 1923. The secretion of gastric juice in health and disease. *Physiol. Rev.* **3:** 1–40.

Dakin, H. D. 1910. The formation in the animal body of *l-β* oxybutyric acid by the reduction of aceto-acetic acid. *J. Biol. Chem.* **8:** 97–104.

Davenport, H. W. 1984. The early days of research on carbonic anhydrase. *Ann. NY Acad. Sci.* **429:** 4–9.

Dixon, M. 1926. XCIII. Studies on xanthine oxidase. VII. The specificity of the system. *Biochem. J.* **20:** 703–718.

Dixon, M., and E. C. Webb. 1979. *Enzymes.* New York: Academic Press.

Fearon, W. R. 1923. XII. Urease. Part I. The chemical changes involved in the zymolysis of urea. *Biochem. J.* **17:** 84–93.

Feiters, M. C., G. A. Veldink, and J. F. G. Vliegenthart. 1986. Heterogeneity of soybean lipoxygenase 2. *Biochim. Biophys. Acta* **870:** 367–371.

Fischer, E. 1894. Einfluss der configuration auf die wirkung der enzyme. *Chem. Ber.* **27:** 2985–2993.

Fleming, A. 1922. On a remarkable bacteriolytic element found in tissues and secretions. *Proc. Roy. Soc. Lond. Ser. B. Biol.* **93:** 306–317.

Fruton, J. S. 1981. Enzyme. In *The Encyclopedia Americana. Vol. 10*, pp. 489–493. Danbury: Grolier.

Fruton, J. S., and S. Simmonds. 1958. *General Biochemistry*, 2nd Edition. New York: John Wiley & Sons.

Gazzola, C., R. L. Blakely, and B. Zerner. 1973. On the substrate specificity of Jack Bean urease (urea aminohydrolase, EC 3.5.1.5). *Can. J. Biochem.* **51:** 1325–1330.

Gillis, A. 1988. Research discovers new roles for lipases. *J. Am. Oil Chem. Soc.* **65:** 846–850.

Gjessing, E. C., and J. B. Sumner. 1942. Synthetic peroxidases. *Arch. Biochem. Biophys.* **1:** 1–8.

Gortner, R. A., Jr., and W. A. Gortner. 1949. *Outlines of Biochemistry*, 3rd Edition. New York: John Wiley & Sons.

Green, J. R. 1890. On the germination of the seed of the castor-oil plant (*Ricinus communis*). *Proc. Roy. Soc. Lond. Ser. B. Biol.* **48:** 370–392.

Grossman, S., B. P. Klein, B. Cohen, D. King, and A. Pinsky. 1984. Methylmercuric iodide modification of lipoxygenase-1. Effects on the anaerobic reaction and pigment bleaching. *Biochim. Biophys. Acta* **793:** 455–462.

Hais, I. M. 1986. Biospecific or non-specific adsorption of amylase on starch in Starkenstein's experiments (1910)? *J. Chromatogr.* **373:** 265–269.

Haldane, J. B. S. 1930. *Enzymes.* New York: Longmans, Green and Co.; Reprint. 1965. Cambridge: M.I.T. Press.

Harvey, E. N. 1918. Studies on bioluminescence. VII. Reversibility of the photogenic reaction in *Cypridinia*. *J. Gen. Physiol.* **1:** 133–145 (1918).

Henriques, V., and C. Hansen. 1901. Vergleichende untersuchungen über die chemische zusammensetzung des thierischen fettes. *Skand. Arch. Physiol.* **11**: 151–165.

Herriott, R. M., and J. H. Northrop. 1934. Crystalline acetyl derivatives of pepsin. *J. Gen. Physiol.* **18**: 35–67.

Herriott, R. M., and J. H. Northrop. 1936. Isolation of crystalline pepsinogen from swine gastric mucosae and its autocatalytic conversion into pepsin. *Science* **83**: 469–470.

Hoppe-Seyler, F. 1890. Ueber muskelfarbotoffe. *Z. Physiol. Chem.* **14**: 106–108.

Keilen, D. 1925. On cytochrome, a respiratory pigment, common to animals, yeast, and higher plants. *Proc. Roy. Soc. Lond. Ser. B. Biol.* **98**: 312–339.

Keilen, D., and E. F. Hartree. 1935. Uricase, amino acid oxidase, and xanthine oxidase. *Proc. Roy. Acad. Sci. Lond. Ser. B. Biol.* **119**: 114–140.

Klibanov, A. M. 1983. Unconventional catalytic properties of conventional enzymes: applications in organic chemistry. *Basic Life Sci.* **8**: 497–518.

Kunitz, M., and J. H. Northrop. 1935. Crystalline chymo-trypsin and chymotrypsinogen. I. Isolation, crystallization, and general properties of a new proteolytic enzyme and its precursor. *J. Gen. Physiol.* **18**: 433–458.

Laskowski, M., Sr. 1982. Nucleases: historical perspectives. *Cold Spring Harbor Monograph Ser.* **14**: 1–21.

Levene, P. A. 1931. The revolt of the biochemists. *Science* **74** (1906): 23–27.

Levene, P. A., and I. Weber. 1924. On nucleosidases. II. Purification of the enzyme. *J. Biol. Chem.* **60**: 707–715.

Levy, L. 1889. Ueber farbstoffe in den muskeln. *Z. Physiol. Chem.* **13**: 309–325.

MacMunn, C. A. 1887. Further observations on myohaematin and the histohaematins. *J. Physiol. (Lond.)* **8**: 57–65.

MacMunn, C. A. 1889. Ueber das myohämatin. *Z. Physiol. Chem.* **13**: 497–499.

MacMunn, C. A. 1890. Ueber das Myohämatin. *Z. Physiol. Chem.* **14**: 328–329.

Meldrum, N. U., and F. J. W. Roughton. 1933. Carbonic anhydrase. Its preparation and properties. *J. Physiol. (Lond.)* **80**: 113–142.

Morgan, E. J., C. P. Stewart, and F. G. Hopkins. 1922. On the anaerobic and aerobic oxidation of xanthin and hypoxanthin by tissues and by milk. *Proc. Roy. Acad. Sci. Lond. Ser. B. Biol.* **94**: 109–131.

Munger, R. S. 1949. Guaiacum, the holy wood from the New World. *J. History Med.* **4**: 196–229.

Neidleman, S. L. 1987. Effects of temperature on lipid unsaturation. *Biotechnol. Genetic Eng. Rev.* **5**: 245–267.

Neidleman, S. L., and J. Geigert. 1986. *Biohalogenation: Principles. Basic Roles and Applications.* Chichester: Ellis Horwood Ltd.

Nelson, J. M., and E. G. Griffin. 1916. Adsorption of invertase. *J. Chem. Soc.* **38**: 1109–1115.

Northrop, J. 1930. Crystalline pepsin. I. Isolation and tests of purity. *J. Gen. Physiol.* **13**: 739–766.

Paul, K. G. 1987. Peroxidases: past and present. *J. Oral Pathol.* **16**: 409–411.

Quastel, J. H., and W. R. Wooldridge. 1928. LXXXIV. Some properties of the dehydrogenating enzymes of bacteria. *Biochem. J.* **22**: 689–702.

Rao, C. N. B., M. V. L. Rao, M. S. Ramaswamy, and V. Subrahmanyan. 1941. Purification and chemical nature of rennin. *Curr. Sci.* **4**: 179–186.

Sperry, W. M., and F. C. Brand. 1941. A study of cholesterol esterase in liver and brain. *J. Biol. Chem.* **137**: 377–387.

Stryer, L. 1975. *Biochemistry.* San Francisco: W. H. Freeman and Co.

Sumner, J. B. 1926. The isolation and crystallization of the enzyme urease. Preliminary paper. *J. Biol. Chem.* **69**: 435–441.

Sumner, J. B., and G. F. Somers. 1947. *Chemistry and Methods of Enzymes.* New York: Academic Press, Inc.

Sym, E. A. 1936. LXXXVII. Action of esterase in the presence of organic solvents. *Biochem. J.* **30**: 609–617.

Tauber, H. 1937. *Enzyme Chemistry.* New York: John Wiley & Sons.

Tauber, H. 1943. *Enzyme Technology.* New York: John Wiley & Sons.

Thunberg, T. 1920. Zur kenntnis des intermediären stoffwechsels und der dabei wirksamen enzyme. *Skand. Arch. Physiol.* **40**: 1–91.

Theorell, H. 1976. Enzymes revisited. In *Reflections on Biochemistry in Honor of Severo Ochoa*, pp. 57–63. (A. Kornberg, B. L. Horecker, L. Cornudella, and J. Oro, eds.). New York: Pergamon Press.

Tracy, A. H., and W. F. Ross. 1942. Carbon suboxide and proteins. VII. Malonyl pepsin. *J. Biol. Chem.* **146**: 63–68.

Van Slyke, D. D., and J. A. Hawkins. 1930. Studies of gas and electrolyte equilibria in blood. XVI. The evolution of carbon dioxide from blood and buffer solutions. *J. Biol. Chem.* **87**: 265–279.

Waksman, S. A., and W. C. Davison. 1926. *Enzymes. Properties, Distribution, Methods and Applications.* Baltimore: Williams & Wilkins.

Warburg, O. 1925. Iron, the oxygen-carrier of respiration-ferment. *Science* **61**: 575–582.

Young, S. 1985. A Christmas digest. *New Scientist* **108**: 24–27.

2
Synthesis of Polyesters by Lipase-Catalyzed Polycondensation in Organic Media

CARY J. MORROW AND J. SHIELD WALLACE

OLIGOMER AND POLYMER SYNTHESIS WITH HYDROLASES

During the past 5 years, research arising from the pioneering work of Klibanov and co-workers (Zaks and Klibanov 1984; Klibanov 1986) has led to enzyme-catalyzed reactions in anhydrous, organic media taking their place as valuable tools for synthetic organic chemistry. Among the useful processes involving hydrolytic enzymes are esterifications (Gatfield 1984; Cambou and Klibanov 1984b; Zaks and Klibanov 1985; Kirchner, Scollar, and Klibanov, 1985; Langrand et al. 1985, 1986; Chen et al. 1987; Hemmerle and Gais 1987; Gil et al. 1987; Sonnet 1987; Bianchi, Cesti, and Battistel 1988; Yamamoto, et al., 1988), transesterifications (Cambou and Klibanov 1984a,b; Zaks and Klibanov 1985; Kirchner, Scollar, and Klibanov 1985; Ramos Tombo et al. 1986; Belan et al. 1987; Degueil-Castaing et al. 1987; Njar and Caspi 1987; Francalanci et al. 1987; Stokes and Oehlschlager 1987; Riva and Klibanov 1988; Theisen and Heathcock 1988; Wang and Wong 1988; Wang et al. 1988; Abramowicz and Keese 1989), aminolyses (Zaks and Klibanov 1985; Margolin and Klibanov 1987; West and Wong 1987; Kitaguchi et al. 1989), and lactonizations (Makita, Nihira, and Yamada 1987; Gutman, Zuobi, and Boltansky 1987; Guo and Sih 1988). Most of these transformations have been shown to occur with remarkable regio- and/or enantioselectivity. Our interest in the use of enzymes for organic synthesis (Wilson et al. 1983) and in polymer chemistry (Wallace,

25

Arnold, and Tan 1989) has led us to explore the possibility of preparing polyesters by enzyme-catalyzed transesterification.

Reasons for Developing Enzyme-Catalyzed Polymerizations

There are many rationales for developing enzyme-catalyzed methods of polymerization (Lipinsky 1985). Among the most compelling are the following:

1. The polymerizations are carried out at ambient temperature or slightly higher. This should allow reactive functional groups, such as epoxides, to be present in the monomers which would probably not survive normal polycondensation conditions.
2. The stereo- and regioselectivity of the enzyme should permit the construction of polymers having highly regular structures. Obviously the regularity associated with optically active polymers is one example. A second example is that the regioselectivity of enzymes (Hennen et al. 1988; Riva et al. 1988; Therisod and Klibanov 1986, 1987) should allow construction of polymers from monomers such as polyols with the primary alcohols becoming incorporated in the polymer backbone while secondary alcohols are left open as sites for grafting. Similarly, our results suggest that an aromatic ester may be left open for cross-linking or grafting while aliphatic esters in the same monomer should be incorporated into the polymer.
3. Our results (Wallace and Morrow 1989a,b), as well as those reported by Kitazume, Sato, and Kobayashi (1988), suggest that an enzyme can polymerize a monomer or pair of monomers very rapidly but the condensation stops when the polymeric product attains a certain size. Thus, enzymatic polymerizations may offer an approach to synthesis of polymers having a very narrow, and, perhaps easily controlled, molecular weight distribution.
4. There should be several environmental advantages in using enzymes as polymerization catalysts. First, the catalysts themselves are nontoxic, or, at worst, minor irritants. Second, because the polymers have been formed biologically, they should also be susceptible to biodegradation.

5. A careful study of the mechanisms operating in enzyme-catalyzed polymerizations may provide a model for the development of new, highly specific synthetic catalysts.

Previous Attempts to Prepare Polyesters Enzymatically

Except for naturally occurring polyesters such as poly(β-hydroxybutyrate) (Vergara and Figini 1977; Peoples et al. 1987; Walsh 1987; Bloembergen et al. 1987; Brandl et al. 1988; Marchessault 1988; Marchessault et al. 1988; Gross et al. 1989; Gross et al. 1989), there appears to have been limited previous effort to prepare either achiral or optically active polyesters using enzymes (Okumura, Iwai, and Tominaga 1984; Ajima et al. 1985; Matsumura and Takahashi 1986; Margolin, Crenne and Klibanov 1987; Gutman et al. 1987; Kitazume, Sato, and Kobayashi 1988; Abramowicz and Keese 1989; Gutman and Bravdo 1989). In the report of one earlier study of an enzymatic polycondensation, which is summarized in Figure 2.1, Okumura, Iwai, and Tominaga (1984) describe the *Aspergillus niger* lipase (ANL)-catalyzed oligomerization of 1,2-ethanediol and 1,3-propanediol with the diacids from 1,6-hexanedioic acid through 1,14-tetradecanedioic acid using either excess diol or excess diol with a small amount of water added as the solvent system. The only products examined in detail proved to be a "trimer," a "pentamer," and a "heptamer" of the forms AA–BB–AA, AA–BB–AA–BB–AA, and AA–BB–AA–BB–AA–BB–AA, re-

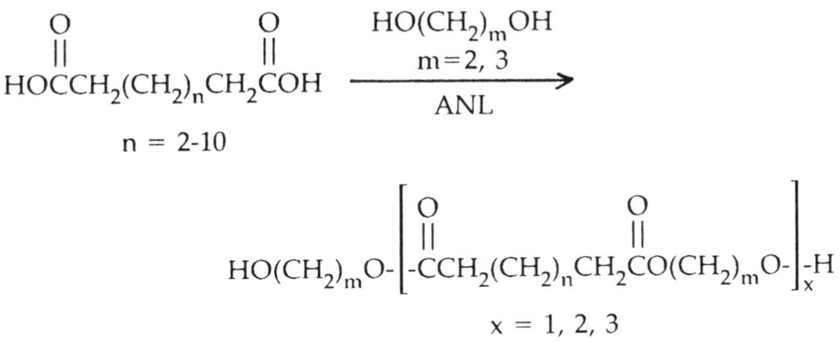

$$\underset{\substack{\text{n = 2-10}}}{\overset{O}{\underset{\|}{HOCCH_2(CH_2)_nCH_2COH}}} \quad \xrightarrow[\text{ANL}]{\overset{HO(CH_2)_mOH}{m=2,\ 3}}$$

$$HO(CH_2)_mO-\left[\overset{O}{\underset{\|}{-CCH_2(CH_2)_nCH_2CO(CH_2)_mO-}}\right]_x-H$$

$$x = 1, 2, 3$$

Figure 2.1. Formation of [AA–BB]$_x$ oligomers by *Aspergillus niger*-catalyzed esterification of diacids in excess diol. (Okamura, Iwai, and Tominaga 1984.)

spectively, which formed from 1,3-propanediol (AA) and 1,13-tridecanedioic acid (BB) in a ratio of $1:8:4.5$ after 24 h. (The words "dimer," "trimer," etc., when placed in quotes reflect that the total number of monomer units rather than the number of repeat units in an $[AA\text{-}BB]_x$ oligomer is being shown. Thus, a "dimer" is really a repeat unit, a "trimer" is really a 1.5 mer, a "pentamer" is really a 2.5 mer, etc.) Separation of the higher oligomers from the reaction mixture seems to limit the degree of polymerization possible, but, at the same time, protects the oligomers from enzymatic hydrolysis or transesterification by the large excess of diol present. Apparently the "heptamer" is either too insoluble to have a favorable rate of conversion to higher oligomers, or its rate for acylating the enzyme at other than a terminal ester is rapid, and it is converted back to lower oligomers. For reasons we have discussed elsewhere (Wallace and Morrow 1986b), it seems unlikely that these authors' explanation for the absence of even "mer" oligomers is supported by the experimental results.

Ajima and co-workers (1985) have described the first attempted enzyme-catalyzed polymerization of an A–B type monomer, 10-hydroxydecanoic acid, **1**, $n = 7$. The reaction was performed in benzene using a poly(ethyleneglycol)-solubilized lipoprotein lipase from *Pseudomonas fluorescens*. The major product was found to have a longer retention on gel permeation chromatography (GPC) than did the monomer, suggesting a structure having a smaller molecular volume. In the absence of data to the contrary, it seems likely that the product was a lactone (Makita, Nihira, and Yamada 1987; Gutman, Zuobi, and Boltansky 1987; Guo and Sih 1988) or, possibly, a bislactone (dilide) (Kruzinga and Kellog 1981; Wallace and Morrow, unpublished results) rather than the proposed high-molecular-weight polymer. Matsumura and Takahashi (1986) have also described attempted polymerizations of ω-hydroxycarboxylic acids in water and in organic solvents using lipases from *Candida rugosa* and *Chromobacterium viscosum* as the catalysts. The substrates chosen are shown in Figure 2.2. They include the primary alcohols 12-hydroxydodecanoic acid, **1**, $n = 9$, and 16-hydroxyhexadecanoic acid, **1**, $n = 13$, and the secondary alcohols 12-hydroxyhexadecanoic acid, **2**, and 12-hydroxy-*cis*-9-octadecenoic acid, **3**. While most of the substrate was consumed, the products, even with the primary alcohols, were principally trimers and tetramers. Somewhat higher degrees of polymerization, but comparable molecu-

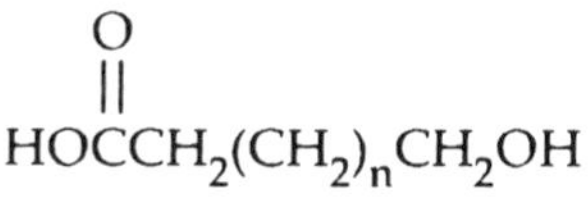

$$\underset{\text{1, n=7, 9, 13}}{\overset{\overset{\displaystyle O}{\|}}{HOCCH_2(CH_2)_nCH_2OH}}$$

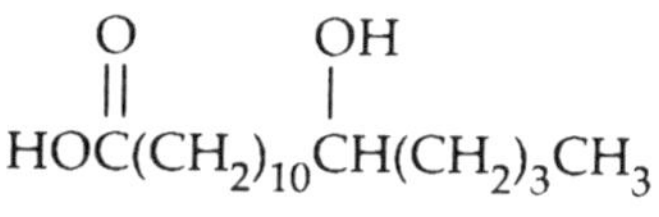

$$\underset{\textbf{2}}{\overset{\overset{\displaystyle O}{\|}\quad\overset{\displaystyle OH}{|}}{HOC(CH_2)_{10}CH(CH_2)_3CH_3}}$$

$$\overset{\overset{\displaystyle O}{\|}}{HOC(CH_2)_6CH_2} \diagdown \quad \diagup \overset{\overset{\displaystyle OH}{|}}{CH_2CH(CH_2)_5CH_3}$$
$$\underset{H \quad\quad H}{C=C}$$

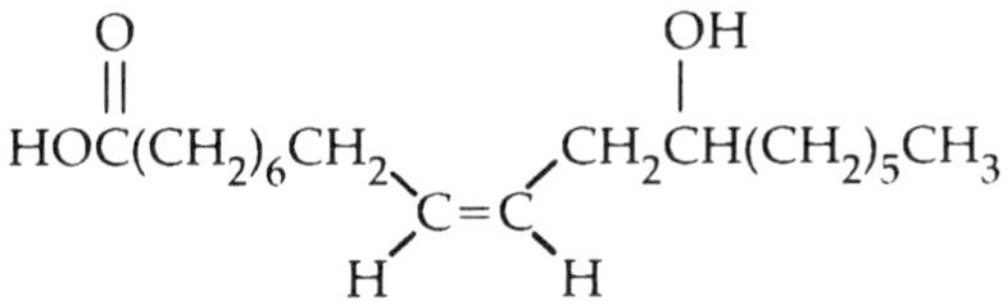

3

Figure 2.2. Hydroxy acid monomers oligomerized using lipases from *Chromobacterium viscosum* and *Candida rugosa* in water and organic media. (Matsumura and Takahashi 1986.)

lar weights, were reported by Gutman et al. (1987) for the porcine pancreas lipase (PPL)-catalyzed transesterification oligomerization of the methyl esters of primary β-, δ-, and ε-hydroxyalkanoic acids, **1**, $n = 0, 2, 3$. In contrast with γ-hydroxy esters, which give only lactones (Gutman, Zuobi, and Boltansky 1987), these monomers were found to undergo oligomerization exclusively in ether giving a degree of polymerization (DP) averaging near 7. This group has recently succeeded in increasing the DP to >100 (A. L. Gutman, personal communication).

The first attempt to prepare a polycarbonate by an enzyme-catalyzed process has been described by Abramowicz and Keese (1989). These workers explored the transesterification of diphenyl carbonate with the well-known monomer bisphenol A, as shown in Figure 2.3, as well as with a range of simple alcohols, under a variety of conditions. The catalysts chosen were porcine liver esterase (PLE) and lipase from the yeast *Candida cylindracea* (CCL). The maximum transesterification rate using bisphenol A was achieved with CCL that had been immobilized on fumed silica using water-saturated ether as the solvent. Oligomers having molecular weights of up to 900 daltons were formed under some conditions. Assuming each chain is terminated by one phenoxycarbonyl from the diphenyl carbonate and by one bisphenol A, as would be expected from the 1 : 1 stoichiometry used, this weight corresponds most closely to a "hexamer."

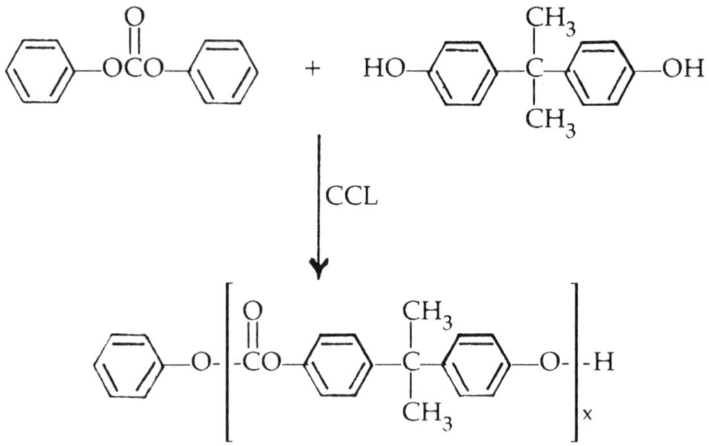

Figure 2.3. *Candida cylindricea* lipase (CCL)-catalyzed oligomerization of diphenyl carbonate with bisphenol A. (Abramowicz and Keese 1989.)

Previous Attempts to Prepare Optically Active Polyesters Enzymatically

Optically active polyesters have generally been prepared by polymerizing synthetic or naturally occurring monomers that have already been resolved or are naturally optically active (Selegny 1979; Vogl and Jaycox 1986). Kitazume, Sato, and Kobayashi (1988) have described a successful enzyme-catalyzed polymerization of an optically active A–B monomer. As shown in Figure 2.4, the prochiral, fluorine-substituted diester, **4**, was chiroselectively hydrolyzed to the corresponding half-ester, **5**, with CCL. The resulting carboxylic acid was coupled nonenzymatically with the amine in 4-(*t*-butyldimethylsilyloxy)aniline or the unprotected amine in *N*-acetyl-1,4-benzenediamine to provide, after deprotection and hydrolysis of the remaining ester, an hydroxy acid or an amino acid A–B monomer, **6**. (In Figure 2.4, ℗ represents the protecting groups.) The monomer was polymerized using a modified cellulase from *Trichoderma viride* (TVC) in benzene, hexane, or Cl_2CFCF_2Cl. Molecular weights (M_W) as high as 18,500 daltons and low polydispersities (1.09–1.51) were reported for the polyester and polyamide products, **7**. The source of this unique example's significant success, in view of the unusual catalyst chosen and the very limited success of the related reactions described previously, is not yet clear.

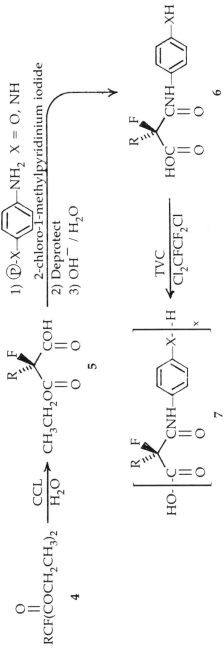

Figure 2.4. Preparation of an optically active, fluorine-containing polymer by a combination of enzymatic and chemical processes. (Kitazume, Sato, and Kobayashi 1988.)

31

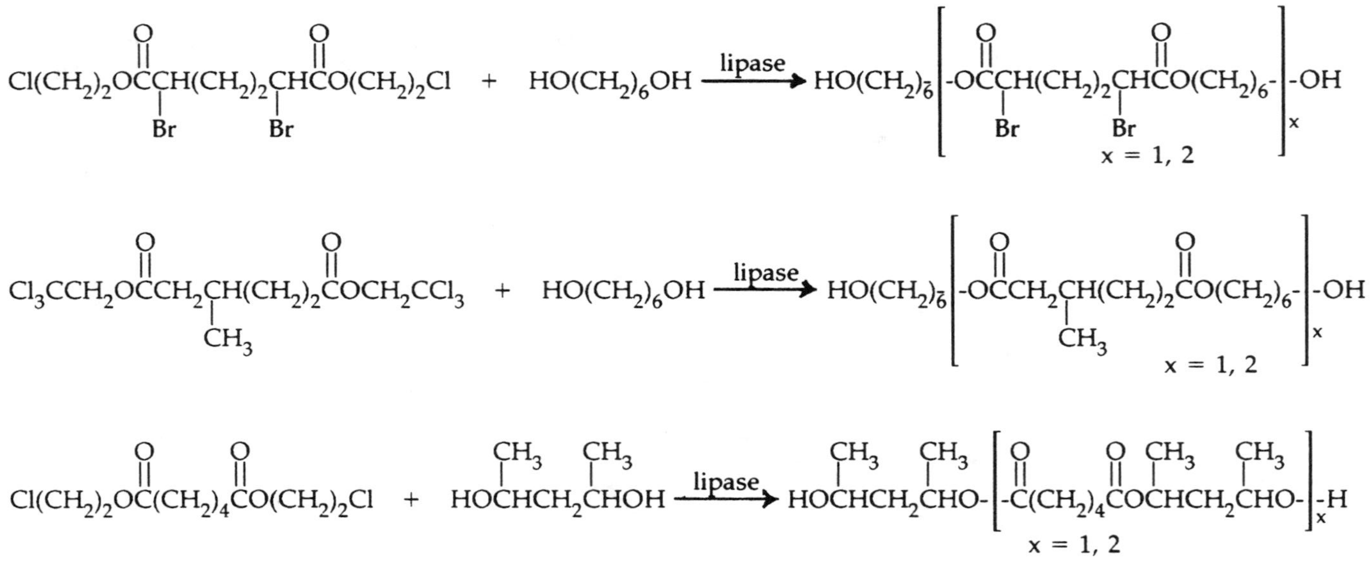

Figure 2.5. Preparation of optically active oligomers by lipase-catalyzed transesterification. (Margolin, Crenne, and Klibanov 1987.)

In the only report other than our own (Wallace and Morrow 1989a) of an attempt to carry out an AA + BB polycondensation enantioelectively, Margolin, Crenne, and Klibanov (1987) have described the stereoselective reaction of bis(2-chlorethyl) (±)-2,5-dibromoadipate with 1,6-hexanediol, the reaction of bis(2,2,2-trichloroethyl) (±)-3-methyladipate with 1,6-hexanediol, and the reaction of bis(2-chloroethyl) adipate with (±)-2,4-pentanediol as summarized in Figure 2.5. Each reaction was carried out in toluene using a commercially available lipase as the catalyst. In each case, though the reaction was allowed to continue for an extended period, it provided, principally, a mixture of a "trimer" and a "pentamer" having the forms AA–BB–AA and AA–BB–AA–BB–AA where AA represents the diol and BB represents the diacid moiety of the oligomer. It is not surprising that the two reactions involving 1,6-hexanediol resulted in products terminated by alcohols, for the diol was used in excess in each case. However, termination by an alcohol when (±)-2,4-pentanediol was used is unexpected, for equimolar quantities of diol and diester were used and it is reported that only one enantiomer of the diol was consumed, meaning that the diester was in twofold excess.

FORMATION OF [AA–BB]ₓ POLYESTERS BY LIPASE-CATALYZED TRANSESTERIFICATION

Our own research has led to the development of a general preparation of $[AA–BB]_x$ type polyesters retaining reactive end groups and having weight average molecular weights (M_W) of approximately 5,000–15,000 daltons (Wallace and Morrow 1989b). As shown in Figure 2.6, the polymerizations were achieved using PPL-catalyzed transesterification of simple, achiral bis(2,2,2-trichloroethyl) alkanedioates, **8**, by simple primary diols, **9**, in low to intermediate polarity organic solvents, such as ether, tetrahydrofuran (THF), and mixtures of methylene chloride with hexane and ether.

Rationale for the Monomers and Reaction Conditions Chosen

In our studies, we attempted to avoid some of the problems that arose during many of the previously reported efforts by careful selection of

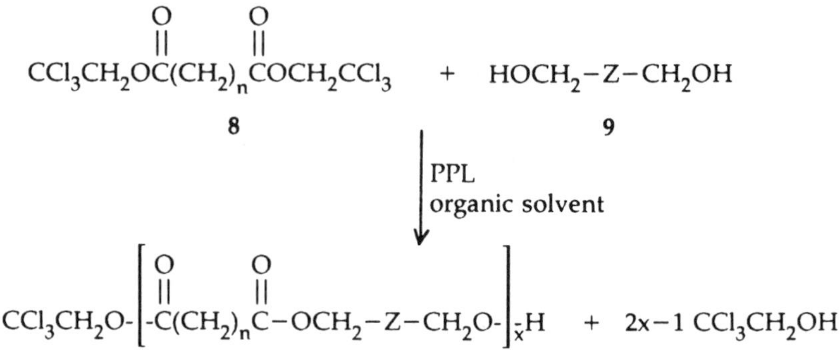

$$CCl_3CH_2OC(CH_2)_nCOCH_2CCl_3 \quad + \quad HOCH_2-Z-CH_2OH$$

$$\overset{O}{\overset{\|}{\quad}}\overset{O}{\overset{\|}{\quad}}$$

8   **9**

$$\downarrow \begin{array}{l} \text{PPL} \\ \text{organic solvent} \end{array}$$

$$CCl_3CH_2O-\left[-C(CH_2)_nC-OCH_2-Z-CH_2O-\right]_x H \quad + \quad 2x-1\ CCl_3CH_2OH$$

Figure 2.6. Formation of $[AA-BB]_x$ polyesters by PPL-catalyzed transesterification in organic media. (Wallace and Morrow 1989b.)

both the monomers to be studied and the polymerization conditions to be used. Among the choices made were the following:

1. Precisely measured, equimolar quantities of highly purified monomers were used to avoid early termination of the polymerization by an excess of one monomer.

2. Displacement of an alcohol having low nucleophilicity and the use of anhydrous aprotic organic solvents were chosen to avoid hydrolysis or alcoholysis of any oligomeric or polymeric product formed.

3. The initial experiments were carried out using AA + BB type polycondensations at room temperature and high concentrations, rather than using AB type self-condensations, to limit the opportunity for lactonization (Makita, Nihira, and Yamada 1987; Gutman, Zuobi, and Boltansky 1987, Guo and Sih 1988) or dilide formation (Kruzinga and Kellog 1981; Wallace and Morrow, unpublished results). Subsequent to our decision to work at room temperature, it was reported that lipase-catalyzed reaction of diacids with diols gives oligomers at room temperature whereas significant yields of lactonized products are formed when the reaction is performed at an elevated temperature (45°C) (Guo and Sih 1988). A similar temperature effect is not yet certain in transesterifications, but the structure of the acyl enzyme intermediate

undergoing these two competing reactions should be independent of its formation from acid or ester.

4. The alcohol to be displaced was chosen to increase the electrophilicity of the acyl carbonyl as a means of increasing the rate of the transesterification, if the rate-limiting process proved to be transfer of an acyl group to the enzyme.

5. Only primary alcohols were chosen among the initially studied diols to ensure that any rate reduction arising from steric hindrance in the second step of the transesterification mechanism was minimized.

A Survey of AA + BB Polycondensations

The results of 10 experiments carried out in the course of a survey of AA + BB polycondensations based on the reaction given in Figure 2.6 are summarized in Table 2.1. In a typical experiment, an ether solution was prepared that contained precisely equimolar quantities of bis(2,2,2-trichloroethyl) adipate and 1,4-butanediol and that was approximately 1 M in each. To this was added (0.3 g/mmol of diester) an inexpensive, but very crude commercial preparation of PPL. The resulting mixture was stirred in a nitrogen atmosphere for 5 days, about four times as long as was required for the starting materials to disappear, then the catalyst was removed from the viscous mixture by filtration, the solvents and trichloroethanol were evaporated, and the residue was reprecipitated from dichloromethane with methanol. For the example case, end-group analysis of the polymer by high-field nuclear magnetic resonance (NMR) provided a number-average molecular weight (M_N) of 4,900 daltons, while analysis of the polymer by GPC provided a weight-average molecular weight (M_W) for the polymer of 5,200 daltons by comparison with polystyrene standards.

Among the conclusions that are supported by the data from this series of experiments are the following: The choice of diol and solvent are both broad, and the diester may be a glutarate or adipate. Preliminary results not included in Table 2.1 indicate that two other simple aliphatic diesters, succinate, **8**, $n = 2$, and sebacate, **8**, $n = 8$, are also candidates for the polymerization, but, in a single experiment using the aromatic diester bis(2,2,2-trichloroethyl) terephthalate as a monomer, no reaction with 1,4-cyclohexanedimethanol was observed using PPL

Table 2.1. Enzymatic synthesis of AA–BB polymers by transesterification.

						MOLECULAR WEIGHT		
ENTRY	DIESTER 8, $n =$	DIOL 9, $Z =$	SOLVENT	REACTION TIME (h)	YIELD (%)	M_W (GPC)	M_N (^{1}H NMR)	M_W/M_N
1	3	$(CH_2)_2$	Ether	122	89	11,800	8,200	1.44
2	3	$c\text{-}C_6H_{10}$	THF	74	92	14,900	—	—
3	4	$(CH_2)_2$	Ether	120	84	5,200	4,900	1.06
4	4	$(CH_2)_2$	Ether	135	89	—	5,600	—
5	4	$(CH_2)_4$	Ether	120	86	—	6,500	—
6	4	$(CH_2)_{10}$	THF	92	87	5,900	2,100	2.80
7	4	$c\text{-}C_6H_{10}$	Ether	96	82	5,100	2,100	2.40
8	4	$c\text{-}C_6H_{10}$	THF	112	81	5,300	2,350	2.25
9	4	$c\text{-}C_6H_{10}$	Hexane/ CH_2Cl_2 2:1	144	81	2,800	1,300	2.15
10	4	$c\text{-}C_6H_{10}$	Ether/ CH_2Cl_2 2.5:1	96	84	4,900	3,200	1.53

as the catalyst. The molecular weights presented in Table 2.1 are only qualitatively comparable with each other because there is considerable variation in how these survey experiments were carried out. However, it may prove to be significant that higher molecular weights are found with glutarates, for more recent experiments continue to support that conclusion. It is also possible that the fairly low polydispersities (M_W/M_N) found in most cases will be characteristic of enzyme-catalyzed polycondensations. This same observation was made by Kitazume and co-workers in the very different polycondensation discussed in the introduction to this chapter (Kitazume, Sato, and Kobayashi 1988). The yield of polymeric products that can be precipitated by methanol appears to be nearly independent of the polycondensation carried out. Determining the effect of the solvent, the reaction time, and the precise combination of monomers that is chosen on the outcome of the polymerization will require a more systematic evaluation than is provided by the data in Table 2.1.

The Course of the Early Steps in a PPL-Catalyzed Polycondensation

To learn something about the polymerization process, the progress of the reaction of bis(2,2,2-trichloroethyl) glutarate with 1,4-butanediol in ether solution (entry 1, Table 2.1) was monitored by proton NMR spectrometry. Unfortunately, the diol is only partially soluble in ether in the absence of the enzyme, so the heterogeneous reaction mixture was first sampled 1 h after the enzyme was added. The region of the NMR spectrum displaying the most significant changes is that lying between 1.8 and 3.8 ppm. This region of the spectra obtained after reaction times of 1, 6, 24, and 72 h is shown as Figure 2.7(a), 2.7(b), 2.7(c), and 2.7(d), respectively. The series of quintets lying between 1.85 and 2.12 ppm arises from changes in the environment of the single methylene at the center of the glutarate moiety as the reaction progresses. Although this region was not interpreted in detail, its total area must remain constantly proportional to two protons and may be used to confirm that the areas of other absorption bands have been assigned the correct number of protons. It is also apparent that the absorption shifts upfield as first one, then both of the trichloroethyl groups are replaced. The most informative peaks in the spectra are four triplets

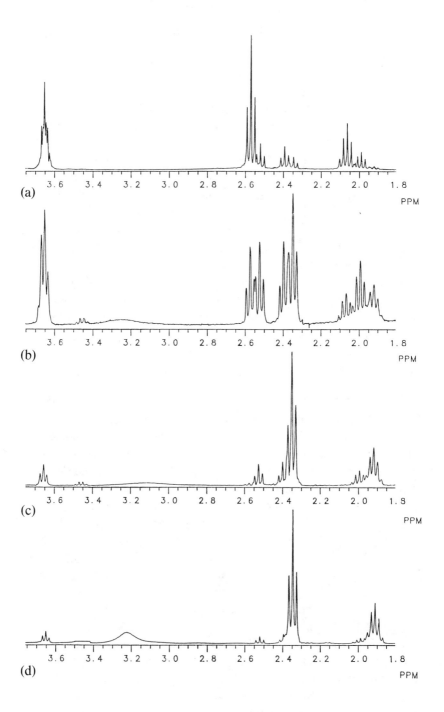

centered at δ 2.57, 2.52, 2.39, and 2.35 ppm assigned to the emboldened protons in the structures shown in the first four entries of Figure 2.8. The combined areas of these four absorptions must remain constantly proportional to four protons. A second useful region is that near 3.65 ppm. As is shown in the last entry of Figure 2.8, the triplet at δ 3.67 ppm has been assigned to the protons on a methylene bearing a free hydroxyl group in the "dimer" structure where only one end of the diester monomer has been transesterified. This triplet is apparent in the spectra recorded after 1 and 6 h but is very weak in the spectrum recorded after 24 h. The second, slightly upfield triplet at δ 3.65 ppm has been tentatively assigned to the indicated protons in 1,4-butanediol (Fig. 2.8) and to the protons of a methylene bearing a free hydroxyl group in any structure other than the dimer.

It is apparent from Figure 2.7(a) that most of the starting diester is unchanged after 1 h, for the triplet at δ 2.57 is still large. Integration of this absorption shows that 63% of the starting diester has not yet reacted with the diol. It is also apparent from the absorption at δ 2.35 that, after only 1 h, in a significant fraction (approximately 7%) of the starting bis(2,2,2-trichloroethyl) diester both trichloroethyl groups have been displaced by diol. The most probable structure for diester having no trichloroethyl groups this early in the reaction is the "trimer," 11 shown in Figure 2.9. Though it is not always possible to integrate the peaks accurately, it appears visually that the triplets at δ 2.39 and 2.52 remain equal in size throughout the polymerization as is required by their assignments. From the sum of the integrals of these two absorptions, it appears that, after 1 h, about 30% of the starting diester has been converted to products in which one acid group still retains the trichloroethyl group while the other acid group has been transesterified.

Figure 2.7. Appearance of the ^{1}H NMR absorptions assigned to the six methylene protons α (between δ 2.3 and 2.6 ppm, 4H) and β (between δ 1.85 and 2.15 ppm, 2H) to the carbonyls of the diester moieties and to methylene protons on carbons bearing an unesterified hydroxyl group in the diol monomer and in diol monomer that has reacted at only one end (near δ 3.65 ppm) during the course of lipase-catalyzed transesterification polymerization of bis(2,2,2-trichloroethyl) glutarate with 1,4-butanediol after the reaction had progressed for the following times: (a) 1 h; (b) 6 h; (c) 24 h; (d) 72 h. See Figure 2.8 for assignments of the absorptions. (Wallace and Morrow 1989b.)

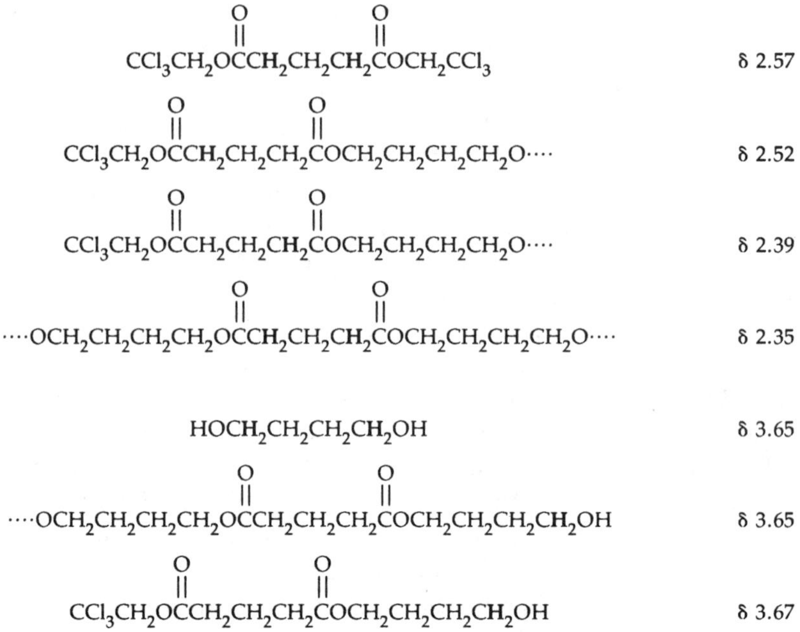

$$\overset{\displaystyle O}{\overset{\displaystyle \|}{\text{CCl}_3\text{CH}_2\text{OCCH}_2\text{CH}_2\text{CH}_2}}\overset{\displaystyle O}{\overset{\displaystyle \|}{\text{COCH}_2\text{CCl}_3}} \qquad \delta\ 2.57$$

$$\overset{\displaystyle O}{\overset{\displaystyle \|}{\text{CCl}_3\text{CH}_2\text{OCCH}_2\text{CH}_2\text{CH}_2}}\overset{\displaystyle O}{\overset{\displaystyle \|}{\text{COCH}_2\text{CH}_2\text{CH}_2\text{O}}}\cdots \qquad \delta\ 2.52$$

$$\overset{\displaystyle O}{\overset{\displaystyle \|}{\text{CCl}_3\text{CH}_2\text{OCCH}_2\text{CH}_2\text{CH}_2}}\overset{\displaystyle O}{\overset{\displaystyle \|}{\text{COCH}_2\text{CH}_2\text{CH}_2\text{O}}}\cdots \qquad \delta\ 2.39$$

$$\cdots\text{OCH}_2\text{CH}_2\text{CH}_2\text{CH}_2\overset{\displaystyle O}{\overset{\displaystyle \|}{\text{OCCH}_2\text{CH}_2\text{CH}_2}}\overset{\displaystyle O}{\overset{\displaystyle \|}{\text{COCH}_2\text{CH}_2\text{CH}_2\text{O}}}\cdots \qquad \delta\ 2.35$$

$$\text{HOCH}_2\text{CH}_2\text{CH}_2\text{CH}_2\text{OH} \qquad \delta\ 3.65$$

$$\cdots\text{OCH}_2\text{CH}_2\text{CH}_2\text{CH}_2\overset{\displaystyle O}{\overset{\displaystyle \|}{\text{OCCH}_2\text{CH}_2\text{CH}_2}}\overset{\displaystyle O}{\overset{\displaystyle \|}{\text{COCH}_2\text{CH}_2\text{CH}_2\text{CH}_2\text{OH}}} \qquad \delta\ 3.65$$

$$\overset{\displaystyle O}{\overset{\displaystyle \|}{\text{CCl}_3\text{CH}_2\text{OCCH}_2\text{CH}_2\text{CH}_2}}\overset{\displaystyle O}{\overset{\displaystyle \|}{\text{COCH}_2\text{CH}_2\text{CH}_2\text{CH}_2\text{OH}}} \qquad \delta\ 3.67$$

Figure 2.8. Assignments of the absorptions appearing in the proton NMR spectra shown in Figure 2.7.

Integration of the absorption at δ 2.57 ppm in Figure 2.7(b) indicates that, after 6 h, most of the bis(trichloroethyl) ester, **8**, has reacted. As expected, the areas of the absorptions at δ 2.52 and 2.39 ppm appear equal. Together, they account for about 43% of the diester moieties now present. The appearance of the absorption near 3.65 ppm in Figure 2.7(b) suggests that a significant fraction of the molecules contributing to the triplets at 2.39 and 2.52 ppm are the "dimer" **10**, in Figure 2.9, rather than the "trimer" **12**, in which both ends of a butanediol molecule have reacted with molecules of diester. The remaining 48% of the diester moieties are found to be contributing to the absorption at δ 2.35 ppm. From the peak heights in two different sets of peaks in the 3.65

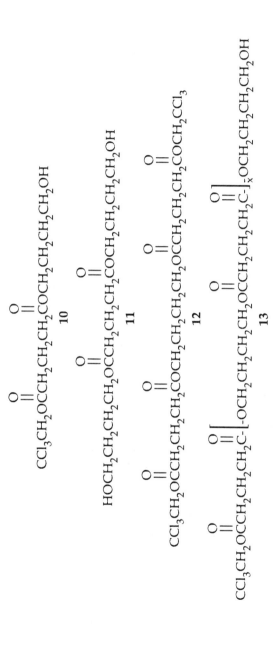

Figure 2.9. Structures of intermediates formed during the early stages of the PPL-catalyzed polycondensation of bis(2,2,2-trichloroethyl) glutarate with 1,4-butanediol. (Wallace and Morrow 1989b.)

41

and the 2.35–2.57 ppm regions of the spectrum in Figure 2.7(b), rough, but consistent, values for the relative amounts of **10**, **11**, **12**, and **13** present after 6 h could be estimated. When these results were combined with those obtained from integrating the areas under the four triplets, some interesting conclusions concerning the status of the poly-condensation at the 6-h point were suggested:

1. Consumption of the starting diol and diester progresses at different rates as indicated by the fact that nearly all of the diol has been consumed, but almost 20% of the diester is still present.
2. Between one-quarter and one-third of the product mixture comprises the "dimer" **10**.
3. Only about one-tenth of the starting diester has been converted on to oligomers having structure **13**.
4. The quantity of "trimer" **11** present in the product mixture after 6 h is about 1.75 times the quantity of "trimer" **12**.

The fourth of these conclusions is the most interesting, for it may be interpreted as indicating that the enzyme reacts more rapidly with the second trichloroethyl ester group in **10**, than with the starting bis(2,2,2-trichloroethyl) ester, **8**, despite the facts that there are two reactive sites in each of the latter molecules and the two trichloroethyl groups should make it more electrophilic than the remaining trichloroethyl ester in **10**. This unexpected result suggests that the enzyme does not always release **11** to the reaction medium immediately after its formation, but rather moves to the second activated ester group and transesterifies it with butanediol as well. Proving this point conclusively will require the development of a method for verifying and refining the results suggested by the NMR peak heights.

As we have described elsewhere (Wallace and Morrow 1989b), the appearance and disappearance of the various forms of the diester moiety can be seen more easily in the presentation of these same data provided by Figure 2.10. Each plot shows the fraction of the starting diesters that is unchanged, that has been transesterified once, or that has been transesterified twice as the reaction progresses. The fraction of unchanged bis(2,2,2-trichloroethyl) glutarate present at each time was determined by dividing the area of the absorption at δ 2.57 ppm by 4. The fraction of diester having one trichloroethyl group and one

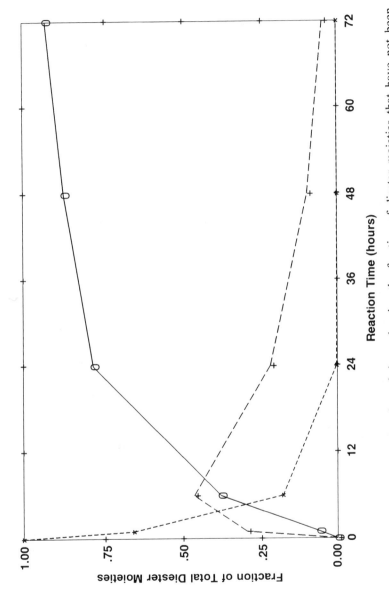

Figure 2.10. Plots, as a function of time, showing the fraction of diester moieties that have not been transesterified (– – –), that have been transesterified at one end (– – –), and that have been transesterified at both ends (——), during the course of lipase-catalyzed polymerization of bis(2,2,2-trichloroethyl) glutarate with 1,4-butanediol. (Wallace and Morrow 1989b.)

43

group from butanediol is found by adding the areas of the absorptions at δ 2.52 and 2.39 ppm together, then dividing the sum by 4. The fraction of diesters in which both trichloroethyl groups have been replaced with groups from butanediol was found by dividing the area of the absorption at δ 2.35 ppm by 4. The fraction of the starting diester must be one at time zero, whereas the fractions of mono- and disubstituted diesters must be zero.

The plot drawn with short dashes shows that the fraction of bis(trichloroethyl) ester **8** drops rapidly at the beginning of the reaction. About 80% has been consumed after 6 h and it has completely disappeared after 24 h. The plot drawn with long dashes shows that the number of diester moieties that have been transesterified at one end rises rapidly at the beginning of the reaction. Most of the product having this structural feature is probably the "dimer," **10**, containing one diester and one diol. The number of diesters in this form reaches a maximum at about 6 h after the beginning of the reaction and then begins to decrease, reflecting that the rate of the enzyme's reaction with the remaining trichloroethyl ester group in "dimer" **10** becomes faster than the formation of **10** from diester **8**. The plot drawn with a solid line has the shape expected for a curve representing the rate of appearance of diester groups esterified at both ends with butanediol. It rises rapidly early in the reaction when the concentration of 1,4-butanediol hydroxyls is high and the formation of "trimer" **11** is underway, but its slope decreases markedly after 6 h when nearly all of the butanediol has reacted at one end.

Figure 2.7(c) shows the bis(trichloroethyl) ester **8** to be nearly consumed after 24 h, while Figure 2.7(c) and 2.7(d) shows the oligomers **12** and **13**, which are terminated by a trichloroethyl ester moiety, disappear much more slowly. We have hypothesized (Wallace and Morrow 1989b) that this phase of the polymerization involves enzymatic activation of a trichloroethyl ester-terminated oligomer with subsequent transfer of the acyl group from the enzyme to an alcohol-terminated oligomer to give a higher oligomer. It is also apparent from Figure 2.10 that a plot of the sums of the areas of the absorptions at δ 2.52 and δ 2.39 decreases linearly with time after the diester, **8**, has been consumed, while a similar plot of the area of the absorption at δ 2.35 ppm increases linearly with time. Because the slope of the latter plot has an absolute value equal to that of the former, consumption of the trichloroethyl esters apparently is associated with an increase in the polymer size, rather than nonproductive processes. After 168 h, the reaction was stopped and the DP was estimated to be 30 by dividing the area of the absorption at δ 2.52 into one-half the area of the absorption at δ 2.35 ppm. This method of calculating the DP assumes that the end groups of the polymer molecules are equally divided between trichloroethyl esters, which may be estimated directly from the area of the absorption at δ 2.52 ppm, and hydroxymethyl groups, which give an absorption near δ 3.65 ppm that appears visually to be the same size, but that cannot be integrated accurately because of interferences from another absorption.

ENZYMATIC SYNTHESIS OF AN OPTICALLY ACTIVE POLYESTER

Optically active polymers have been studied fairly extensively (Selegny 1979; Vogl and Jaycox 1986) and are of interest as reagents in enantioselective syntheses, as adsorbents for the chromatographic resolution of enantiomers, and in the synthesis of ferroelectric liquid crystals, among other potential applications. However, with the exception of naturally occurring polyesters such as poly(β-hydroxybutyrate) (Vergara and Figini 1977; Peoples et al. 1987; Walsh 1987; Bloembergen et al. 1987; Brandl et al. 1988; Marchessault 1988; Marchessault et al. 1988; Gross et al. 1989; Gross et al. 1989), there appears to have been only limited effort to prepare optically active polyesters using enzymes (Margolin, Crenne, and Klibanov 1987; Kitazume, Sato, and Kobayashi 1988; Wallace and Morrow 1989a). Although the oligomers of A–B monomers **2** and **3** reported by Matsumura and Takahashi (1986) are potentially optically active, apparently they were not analyzed for stereochemical enrichment. The only successful attempt to prepare an optically active [AA–BB]$_x$ type polyester having more than a few repeat units from a racemic monomer is detailed in our recent report on the enantioselective polymerization of bis(2,2,2-trichloroethyl) ($\pm$)-3,4-epoxyadipate with 1,4-butanediol (Wallace and Morrow 1989a), and will be discussed further below. In contrast, the optically active polyamide **7**, x = NH (Fig. 2.4), reported by Kitazume,

Sato, and Kobayashi (1988), is an [AA–BB]$_x$ polymer, but the enzymatic polycondensation was of an A–B repeat unit, the previously prepared, and already optically active, AA–BB "dimer," **6**.

Monomer Selection, Preparation, and Polymerization

In devising a system for demonstrating stereoselective enzyme-catalyzed polycondensation, we attempted to ensure success by selecting monomers that would unambiguously demonstrate stereo- and chemoselectivity, and by performing the reaction under the traditional condensation polymerization conditions of a strict equimolar ratio of highly purified reactants. The following six factors were considered in choosing the chiral substrate (Wallace and Morrow 1989a):

1. The chiral centers should bear groups previously shown to lead to high steroselectivity during lipase-catalyzed transesterification in organic media.
2. The chiral substrate should have rotational symmetry (C$_2$) to remove any requirement that the enzyme display regioselectivity between two identical functional groups in different environments.
3. It should be possible to prepare the substrate free of any of the corresponding *meso* diastereomer, for incorporation of a group having both configurations of a chiral center into the polymer chain would be expected to terminate the polymerization if, as hoped, the enzyme proved to be strongly enantioselective and, therefore, able to perform a transesterification on only one configuration.
4. The leaving group in the ester substrate should be highly activated to ensure both as rapid a rate and as favorable an equilibrium position as possible for the polymerization.
5. The chirality should appear in the diacid moiety rather than the diol moiety of the polyester, if possible, to avoid the slow transesterification rate which was observed when a secondary alcohol, 2,4-pentanediol, was used to provide the chirality (Margolin, Crenne, and Klibanov 1987).
6. Finally, to demonstrate further the synthetic potential of enzymatic polymerization, a substrate was sought that would intro-

duce reactive functionality into the polymer that would probably not survive the usual conditions for polyester formation.

The chiral substrate chosen, bis(2,2,2-trichloroethyl) ($\pm$)-3,4-epoxyadipate, **14** in Figure 2.11, has the required symmetry, the reactive epoxy functional group, and leaving groups in the ester that were expected to give the desired rate enhancement and reaction driving force. Moreover, β,γ-epoxy esters have been shown to undergo highly stereoselective reactions with hydrolase enzymes (Ladner and Whitesides 1984; Mohr, Rosslein, and Tamm 1987; Bianchi, Cabri et al. 1988). This diester monomer was synthesized via the two steps summarized in Figure 2.11, in an overall yield of 86% from the commercially available *trans*-3-hexendioic acid (hydromuconic acid). Conversion of hydromuconic acid to bis(2,2,2-trichloroethyl) ester **15** was achieved by treating the diacid with 2,2,2-trichloroethanol in the presence of N,N'-dicyclohexylcarbodiimide (DCC) and a catalytic amount of 4-(dimethylamino)pyridine (DMAP) as described by Hassner and Alexanian (1978). Diester **15** was then converted to **14** by epoxidation with *m*-chloroperoxybenzoic acid (mCPBA) following the general procedure described by Dahill, Dorsky, and Easter (1970).

The polycondensation was carried out as summarized in Figure 2.12 using diester, which had been purified to monomer grade and equally high purity, commercial 1,4-butanediol. Because it was assumed that only one enantiomer of the diester would react, a 2 : 1 molar ratio of diester to diol was used in the polycondensation. This remarkable reaction gives an optically active polyester, (−)-**16**, estimated to have incorporated ≥96% of one enantiomer of the racemic epoxide monomer. The second enantiomer, (+)-**14**, was left unchanged (estimated enantiomeric excess ≥95%). Moreover, the polymer retains the reactive epoxy functional groups which may be used for subsequent crosslinking. Although the butanediol had been consumed in only 6 h, the reaction was allowed to continue for 3.5 days to allow oligomeric intermediates to continue to higher polymers by reaction with each other, as is characteristic of a stepwise condensation polymerization (Saunders 1973; Sokolov 1968). An ether-soluble product comprising nearly 90% of the mass of unreacted diester **14** expected if only one of its enantiomers were involved in the polymerization, and an extremely viscous, ether-insoluble oil, comprising nealry 75% of the theoretical

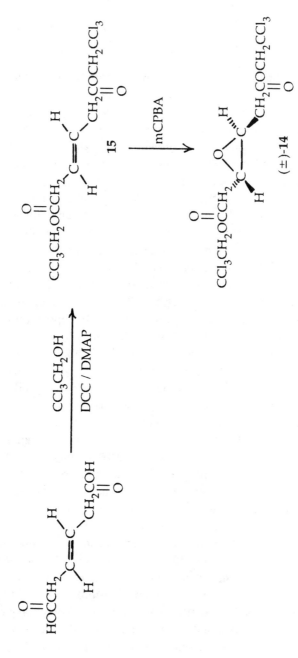

Figure 2.11. Synthesis of bis(2,2,2-trichloroethyl) *trans*-3,4-epoxyadipate. (Wallace and Morrow 1989a.)

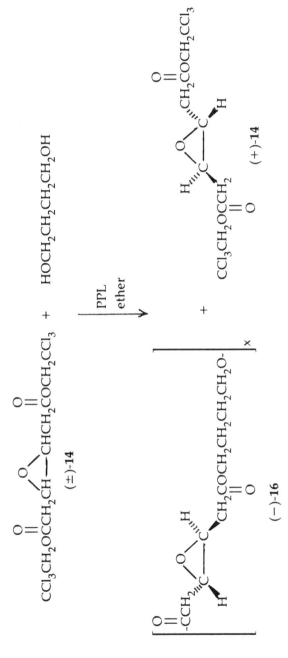

Figure 2.12. Enantioselective PPL-catalyzed polycondensation of bis(2,2,2-trichloroethyl) 3,4-epoxyadipate with 1,4-butanediol. (Wallace and Morrow 1989a.)

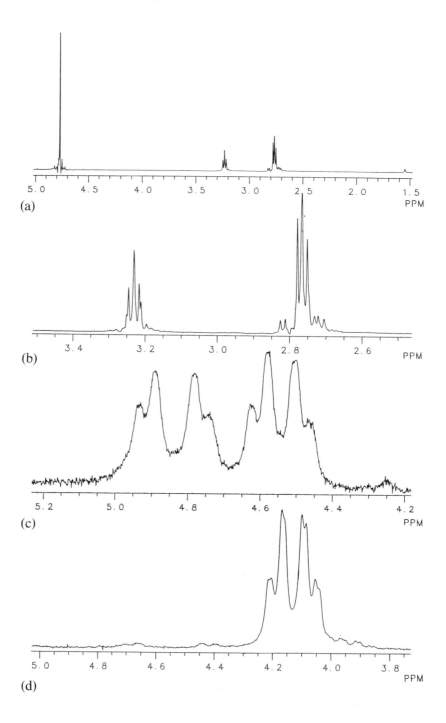

(a)

(b)

(c)

(d)

yield of polyester **16** were isolated from the reaction mixture. The unchanged diester, (+)-**14**, displayed $[\alpha]_D = +12.3°$ (methanol), as well as the expected chromatographic behavior and proton NMR spectrum. An end-group analysis, based on the proton NMR spectrum of the polymer, provided a M_N of 5,300 daltons. Gel permeation chromatographic analysis against polystyrene standards provided a M_W of 7,900 daltons. The polymer displayed $[\alpha]_D$ −13.4° ($CHCl_3$). It should be noted that the absolute configurations shown in Figure 2.12 for the optically active polymer, (−)-**16**, and for the unchanged (+)-**14** have not been proven independently, but are based on the stereoselectivity reported for PPL and other hydrolases when epoxy and other oxygenated functional groups are bonded at the chiral center of an ester undergoing transesterification or an acid undergoing esterification (Cambou and Klibanov 1984; Mohr, Rosslein, and Tamm 1987; Bianchi, Cabri et al. 1988).

Stereochemical Purity of the Optically Active Polymer and the Unchanged Monomer

To determine the enantiomeric purity of the products, and, hence, the degree of enantioselectivity displayed by the enzyme, both the unreacted diester, (+)-**14**, and the polyester product, (+)-**16**, were analyzed by proton NMR in the presence of the chiral shift reagent tris[3-(heptafluoropropylhydroxymethylene)-d-camphorato]europium (III), Eu(hfc)$_3$ (Whitesides and Lewis 1970; Sullivan 1978). For reasons we have detailed elsewhere (Wallace and Morrow 1989a), the analysis was performed on the complex absorption assigned to the methylene protons adjacent to the ester carbonyl. As seen in Figure 2.13(a), this

---

Figure 2.13. (a) 360 MHz ^{1}H NMR spectrum of racemic bis(2,2,2-trichloroethyl) 3,4-epoxyadipate, (±)-**14**, in CDCl$_3$; (b) expansion of the region from δ 2.2 to 2.3 ppm in (a), corresponding to the methylene protons adjacent to the carbonyl in **14**; (c) the new appearance and chemical shift of the absorption near δ 2.77 ppm in (b) on addition of 0.4 equivalent of the chiral lanthanide shift reagent Eu(hfc)$_3$ to the sample; (d) appearance of the absorption corresponding to the methylene protons adjacent to the carbonyl in the 360-MHz ^{1}H NMR spectrum of the optically active bis(2,2,2-trichloroethyl) 3,4-epoxyadipate, (+)-**14**, recovered as in Figure 2.12, in the presence of 0.4 equivalent of Eu(hfc)$_3$. (Wallace and Morrow 1989a.)

absorption appears at δ 2.77 ppm in the absence of the shift reagent. This region of the spectrum has been shown on an expanded scale in Figure 2.13(b). On analysis of racemic **14** in the presence of Eu(hfc)₃ at a substrate/shift reagent ratio of 2.5 : 1, this absorption had shifted to the region δ 4.4–5.0 ppm and was fully resolved into the pair of AB quartets centered at δ 4.55 and 4.85 ppm shown in Figure 2.13(c). When the experiment was repeated using the unchanged monomer recovered from the enzymatic polymerization, the downfield AB quartet had essentially disappeared as seen in Figure 2.13(d). Integrating the areas from δ 3.83 to 4.34 ppm and from δ 4.34 to 4.81 ppm in Figure 2.13(d) provides a ratio of 97.5 : 2.5 for the two regions. Using these relative areas as a measure of the amount of the major and minor enantiomer, respectively, in Eq. (2.1), an enantiomeric excess of 95% is established for the recovered (+)-**14**

$$\text{Enantiomeric excess (\%)} = \frac{A_{\text{major}} - A_{\text{minor}}}{A_{\text{major}} + A_{\text{minor}}} \times 100\% \qquad (2.1)$$

As we have detailed elsewhere (Wallace and Morrow 1989a), an estimate for the stereochemical purity of the polymer, **16**, may be obtained from theoretical consideration of the DP to be expected if varying amounts of the slower reacting enantiomer of **14** were to be incorporated into the polymer. Thus, if both enantiomers were incorporated equally, a DP of 1.5 would be expected, reflecting the fact that the ratio of racemic diester to diol is 2 : 1, and no hydroxyl end groups should be observed. Similarly, if the rate of incorporation of one enantiomer were 10 times that of the other, corresponding to a stereochemical purity of 91% in the polymer, a DP of 10.5 would be expected. Again, no free hydroxyl end groups should be seen. The observed DP of about 25, therefore, corresponds to a ratio of incorporation of 25 : 1 for the two enantiomers. Using Eq. (2.2), in which E_{fast} and E_{slow} refer to the amounts of the enantiomer that reacts faster or slower, respectively, during the polymerization, the stereochemical purity of the polymer can be estimated as 96%. However, this must be interpreted as a lower limit for the stereochemical purity of the polymer because the end-group analysis clearly shows the presence of hydroxyls that have not been end-capped. The term "stereochemical purity" is chosen to reflect that incorporation of the slower reacting enantiomer of diester **14** into a polymer molecule leads to a diastereomer, rather than the enantiomer, of polymer molecules containing only a single enantiomer of the monomer. In contrast with the case of enantiomeric excess, a fraction of the major enantiomer of the monomer need not be considered as cancelling the minor enantiomer so E_{slow} need not be subtracted in the numerator of the fraction as was done in Eq. (2.1).

$$\text{Stereochemical purity } (\%) = \frac{E_{fast}}{E_{fast} + E_{slow}} \times 100\% \qquad (2.2)$$

To illustrate the significance of incorporating 4% of the "wrong" enantiomer of the monomer on the probable structure of a molecule of **16**, the polymer molecules are conveniently modeled as being of a uniform size and consisting of 25 [AA–BB] repeat units (monodisperse, DP = 25). Using standard combinatorial algebra based on the binomial expansion of $(0.96 + 0.04)^{25}$ it can be shown that slightly more than one-third (36.0%) of the polymer molecules should contain no defects resulting from incorporation of the "wrong" enantiomer of the diester monomer. A second third (37.5%) of the polymer molecules would have a single defect. This defect will lead to formation of 25 different diastereomers of the defect-free molecules, and assuming that the defect is randomly located within the polymer chain, each diastereomer will contribute about 1.5% of the total number of polymer molecules. Two molecules of the "wrong" enantiomer of the monomer will be incorporated into about one-fifth (18.8%) of the polymer molecules. These will be distributed among some 600 different diastereomers, meaning that each will contribute <0.05% of the total. Three defects will be found in 6.0% of the molecules while four or more defects will occur in the remaining 1.7% of the polymer molecules. Because incorporation of a molecule of the "wrong" enantiomer should effectively terminate chain growth at one end of the polymer, the defects will tend to be located at the chain ends rather than randomly throughout the polymer chain.

It would be desirable to measure directly the amount of the "wrong" enantiomer incorporated into the polymer to confirm the value determined from the degree of polymerization. In the hope that the shift reagent might distinguish among individual diester monomer units within the polymer on the basis of their chirality rather than respond to the diastereomeric relationship among the optically active polymer molecules as a whole, a sample of (−)-**16** was analyzed by proton NMR in the presence of Eu(hfc)$_3$. Unfortunately, no absorption that could be assigned to incorporation of the minor enantiomer into the polymer appeared, possibly because the absorption of <4% of (+)-**14** would be too weak to be seen above the baseline noise, but probably because the shift reagent fails to separate lines solely on the basis of the configuration at each epoxide.

In an attempt to establish the characteristic behavior of a polymer containing equal amounts of the (R,R) and the (S,S) epoxides when in the presence of Eu(hfc)$_3$, a stereochemically random polymer, analogous to **16**, was synthesized by reversing the order of the two steps indicated by Figures 2.11 and 2.12. Thus, PPL-catalyzed polymerization of bis(2,2,2-trichloroethyl) *trans*-3-hexenedioate, **15**, with 1,4-butanediol was followed by epoxidation of the double bonds in the resulting polymer with mCPBA. Interestingly, the ability of the PPL to use **15** as a monomer is low, and the reaction was only partially complete, even after 14 days.

The stereochemical randomness of this new polymer and the relatively high diastereomeric purity of the optically active polymer, (−)-**16**, are easily seen in comparing their ^{1}H NMR spectra. Figure 2.14a shows the spectrum of the stereoregular polymer, (−)-**16**. Comparison of the absorption at δ 2.6 ppm in Figure 2.14(a) with that at δ 2.75 ppm in Figure 2.13(a) shows that one effect of polymerizing the epoxy diester is to exaggerate the difference in chemical shift between the two diastereotopic hydrogens comprising the methylene groups in the diester moiety of the polymer relative to those of the monomer. It is tempting to ascribe this effect to secondary structure in the polymer, but further experiments will be required to confirm this hypothesis. Figure 2.14(b) shows an expansion of the upfield region from Figure 2.14(a) while Figure 2.14(c) shows the comparible region of the spectrum displayed by the stereorandom polymer. The presence of a variety of diastereomeric arrangements for the epoxide in the latter polyester is seen in the absorption near 2.6 ppm, assigned to the methylene hydrogens of the diester moiety, no longer appearing as a clean pair of doubled doublets but as a complex multiplet. Although this absorption shifts significantly with a much lower concentration of Eu(hfc)$_3$ than is required for **16**, it does not resolve into two absorptions of equal area,

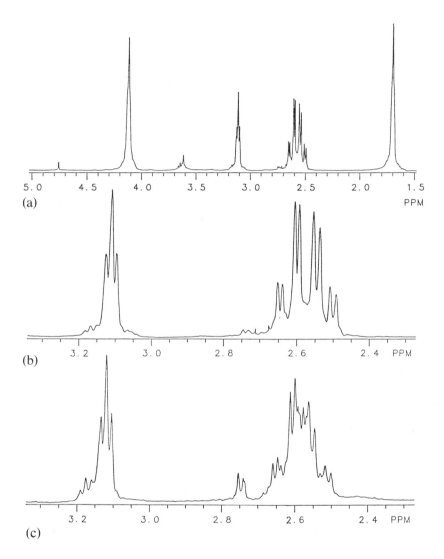

(a)

(b)

(c)

Figure 2.14. (a) 360 MHz ^{1}H NMR spectrum in CDCl$_3$ solution of the optically active polymer, (−)-**16**, prepared as in Figure 2.12, from PPL-catalyzed polymerization of (±)-**14** with 1,4-butanediol; (b) expansion of the region of (a) including the absorptions due to the methine protons adjacent to the epoxide (δ 3.11 ppm) and the methylene protons adjacent to the carbonyl (δ 2.57 ppm); (c) expansion of the same region as in (b) of the ^{1}H NMR spectrum of the analogous, stereochemically random polymer. (Wallace and Morrow 1989a.)

corresponding to the equal quantities of (R,R) and (S,S) epoxide that must be present in the polymer. Rather, it separates into two absorptions of unequal area (approximately $3:1$) indicating that certain sequences of (R,R) and (S,S) epoxides may be more important in determining how the polymer interacts with the shift reagent than the configurations of the individual epoxides.

It might appear, at first thought, that this problem in establishing the stereochemical purity of $(-)$-**16** directly could be resolved by preparing true $(\pm)$-**16** (an equimolar mixture of polymer molecules that are entirely (R,R) or entirely (S,S) in stereochemistry) and determining its behavior in the presence of Eu(hfc)$_3$. However, it must be recalled that incorporation of the "wrong" enantiomer of **14** into **16** probably results in formation of a collection of diastereomers, not the enantiomer, of the defect-free structure. Thus, the completely stereorandom polymer probably provides a better model for the expected behavior of a stereochemically impure polymer in the presence of a chiral shift reagent than would truly racemic **16**. Since the shift reagent was not able to distinguish between (R,R) and (S,S) epoxides within the stereorandom polymer, as had been hoped, no direct measure of the level of incorporation of the "wrong" enantiomer of the monomer into the polymer is currently available.

CONCLUSIONS

Polycondensation of diesters with diols at room temperature in an organic solvent using a hydrolytic enzyme as a transesterification catalyst represents a potentially valuable new method for preparing [AA–BB]$_x$ polyesters. The method appears to be general for the reaction of bis(2,2,2-trichloroethyl) alkanedioates with primary alkanediols in an ether solvent when porcine pancreatic lipase is the catalyst. The method also provides a means of preparing optically active [AA–BB]$_x$ polyesters having a highly reactive group, the epoxide, built into the polymer backbone. Though the molecular weights of the polymers are not yet as high as is desirable, the presence of both trichloroethyl ester and hydroxymethyl end groups in the products indicates that the source of the low molecular weight is not hydrolysis or some other process that destroys the reactive ends of the polymer, nor is the problem with the stoichiometry. Preliminary results using more active leaving groups in the diester to accelerate the rate of acylation of the enzyme suggest that the molecular weight is currently limited by this step. However, the effect of increasing the rate of reaction on the stereoselectivity of the polymerization remains to be determined. Other preliminary results suggest it may also be possible to improve molecular weights by suitable modification of the enzyme, the solvent, the reaction temperature, and/or the composition of the reaction mixture late in the polymerization.

Because of the rotational symmetry of the epoxide in **16**, it can be shown that cross-linking with a diamine leads always to the same stereochemical outcome, a β-amino alcohol in which the alcohol retains the stereochemistry of the epoxide while the carbon bearing the amine becomes inverted. The potential of such cross-linking reactions for preparing more rigid optically active polymers has just begun to be explored in our laboratory.

REFERENCES

Abramowicz, D. A., and C. R. Keese. 1989. Enzymatic transesterification of carbonates in water-restricted environments. *Biotechnol. Bioengineer.* **33**: 149–156.

Ajima, A., T. Yoshimoto, K. Takahashi, Y. Tamura, Y. Saito, and Y. Inada. 1985. Polymerization of 10-hydroxydecanoic acid in benzene with polyethylene glycol modified lipase. *Biotechnol. Lett.* **7**: 303–306.

Belan, A., J. Bolte, A. Fauve, J. G. Gourcy, and H. Veschambre. 1987. Use of biological systems for the preparation of chiral molecules. 3. An application in pheromone synthesis: preparation of sulcatol enantiomers. *J. Org. Chem.* **52**: 256–260.

Bianchi, D., W. Cabri, P. Cesti, F. Francalanci, and M. Ricci. 1988. Enzymatic hydrolysis of alkyl 3,4-epoxybutyrates. A new synthetic route to (*R*)-(−)-carnitine chloride. *J. Org. Chem.* **53**: 104–107.

Bianchi, D., P. Cesti, and E. Battistel. 1988. Anhydrides as acylating agents in lipase-catalyzed stereoselective esterification of racemic alcohols. *J. Org. Chem.* **53**: 5531–5534.

Bloembergen, S., D. A. Holden, T. L. Bluhm, G. K. Hamer, and R. H. Marchessault. 1987. Synthesis of crystalline hydroxybutyrate/β-hydroxyvalerate copolymers by coordination polymerization of β-lactones. *Macromolecules* **20**: 3086–3089.

Brandl, H., R. A. Gross, R. W. Lenz, and R. C. Fuller. 1988. *Pseudomonas oleovarans* as a source of poly(hydroxyalkanoates) for potential applications as biodegradable polyesters. *Appl. Environ. Microbiol.* **54:** 1977–1979.

Cambou, B. and A. M. Klibanov. 1984a. Preparative production of optically active esters and alcohols using esterase-catalyzed stereospecific transesterification in organic media. *J. Am. Chem. Soc.* **106:** 2687–2692.

Cambou, B., and A. M. Klibanov. 1984b. Comparison of different strategies for the lipase-catalyzed preparative resolution of racemic acids and alcohols: asymmetric hydrolysis, esterification, and transesterification. *Biotechnol. Bioengineer.* **26:** 1449–1454.

Chen, C.-S., S.-H. Wu, G. Girdaukas, and C. J. Sih. 1987. Quantitative analyses of biochemical kinetic resolution of enantiomers. 2. Enzyme-catalyzed esterifications in water-organic solvent biphasic systems. *J. Am. Chem. Soc.* **109:** 2812–2817.

Dahill, R. T., J. Dorsky, and W. Easter. 1970. A stereospecific synthesis of *trans*-3-(*exo*-5-*exo*-isocamphyl)cyclohexanol. *J. Org. Chem.* **35:** 251–253.

Degueil-Castaing, M., B. De Jeso, S. Drouillard, and B. Maillard. 1987. Enzymatic reactions in organic solvents. 2. Ester interchange of vinyl esters. *Tetrahedron Lett.* **28:** 953–954.

Francalanci, F., P. Cesti, W. Cabri, D. Bianchi, T. Martinengo, and M. Foà. 1987. Lipase-catalyzed resolution of chiral 2-amino 1-alcohols. *J. Org. Chem.* **52:** 5079–5082.

Gatfield, I. L. 1984. The enzymatic synthesis of esters in nonaqueous systems. *Ann. NY Acad. Sci.* 568–572.

Gil, G., E. Ferre, A. Meou, J. Le Petit, and C. Triantaphylides. 1987. Lipase-catalyzed ester formation in organic solvents. Partial resolution of primary allenic alcohols. *Tetrahedron Lett.* **28:** 1647–1648.

Gross, R. A., H. Brandl, H. W. Ulmer, M. A. Posada, R. C. Fuller, and R. W. Lenz. 1989. The biosynthesis and characterization of new poly(β-hydroxyalkanoates). *Polym. Prepr.* **30**(1): 492–493.

Gross, R. A., C. DeMello, R. W. Lenz, H. Brandl, and R. C. Fuller. 1989. The biosynthesis and characterization of poly(β-hydroxyalkanoates) produced by *Pseudomonas oleovarans*. *Macromolecules* **22:** 1106–1109.

Guo, Z.-W., and C. J. Sih. 1988. Enzymatic synthesis of macrocyclic lactones. *J. Am. Chem. Soc.* **110:** 1999–2001.

Gutman, A. L., K. Zuobi, and A. Boltansky. 1987. Enzymatic lactonisation of γ-hydroxyesters in organic solvents. Synthesis of optically pure γ-methylbutyrolactones and γ-phenylbutyrolactones. *Tetrahedron Lett.* **28:** 3861–3864.

Gutman, A. L., D. Oren, A. Boltanski, and T. Bravdo. 1987. Enzymatic oligomerization versus lactonization of ω-hydroxyesters. *Tetrahedron Lett.* **28:** 5367–5368.

Gutman, A. L., and T. Bravdo. 1989. Enzyme-catalyzed enantioconvergent polymerization of β-hydroxyglutarate in organic solvents. *J. Org. Chem.* **54:** 5645–5646.

Hassner, A. and V. Alexanian. 1978. Direct room temperature esterification of carboxylic acids. *Tetrahedron Lett.* 4475–4478.

Hemmerle, H. and H.-J. Gais. 1987. Asymmetric hydrolysis and esterification catalyzed by esterases from porcine pancreas in the synthesis of both enantiomers of cyclopentanoid building blocks. *Tetrahedron Lett.* **28:** 3471–3474.

Hennen, W. J., H. M. Sweers, Y.-F. Wang, and C.-H. Wong. 1988. Enzymes in carbohydrate synthesis: lipase-catalyzed selective acylation and deacylation of furanose and pyranose derivatives. *J. Org. Chem.* **53:** 4939–4945.

Kirchner, G., M. P. Scollar, and A. M. Klibanov. 1985. Resolution of racemic mixtures via lipase catalysis in organic solvents. *J. Am. Chem. Soc.* **107:** 7072–7076.

Kitaguchi, H., P. A. Fitzpatrick, J. E. Huber, and A. M. Klibanov. 1989. Enzymatic resolution of racemic amines: crucial role of the solvent. *J. Am. Chem. Soc.* **111:** 3094–3095.

Kitazume, T., T. Sato, and T. Kobayashi. 1988. An enzyme-assisted polymerization of chiral fluorinated materials in organic media. *Chem. Express* **3:** 135–138.

Klibanov, A. M. 1986. Enzymes that work in organic solvents. *CHEMTECH* 354–359.

Kruzinga, W. H., and R. M. Kellog. 1981. Preparation of macrocyclic lactones by ring closure of cesium carboxylates. *J. Am. Chem. Soc.* **103:** 5183–5189.

Ladner, W. E., and G. M. Whitesides. 1984. Lipase-catalyzed hydrolysis as a route to esters of chiral epoxy alcohols. *J. Am. Chem. Soc.* **106:** 7250–7251.

Langrand, G., M. Secchi, G. Buono, J. Baratti, and C. Triantaphylides. 1985. Lipase-catalyzed ester formation in organic solvents. An easy preparative resolution of α-substituted cyclohexanols. *Tetrahedron Lett.* **26:** 1857–1860.

Langrand, G., J. Barrati, G. Buono, and C. Triantaphylides. 1986. Lipase catalyzed reactions and strategy for alcohol resolution. *Tetrahedron Lett.* **27:** 29–32.

Lipinsky, E. S. 1985. Polymerizations that use biocatalysts. In *Biocatalysts in Organic Syntheses*, ed. (J. Tramper, H. C. van der Plas, and P. Linko, eds.), pp. 209–223. Amsterdam: Elsevier Science Publishers.

Makita, A., T. Nihira, and Y. Yamada. 1987. Lipase catalyzed synthesis of macrocyclic lactones in organic solvents. *Tetrahedron Lett.* **28:** 805–808.

Marchessault, R. H., T. L. Bluhm, Y. Deslandes, G. K. Hamer, W. J. Orts, P. R. Sundararajan, M. G. Taylor, S. Bloembergen, and D. A. Holden. 1988. Poly(β-hydroxyalkanoates); biorefinery polymers in search of applications. *Makromol. Chem., Macromol. Symp.* **19:** 235–254.

Marchessault, R. H. 1988. History of polyalkanoate research. *Polym. Prepr.* **29:** 584–585.

Margolin, A. L., J.-Y. Crenne, and A. M. Klibanov. 1987. Stereoselective oligomerizations catalyzed by lipases in organic solvents. *Tetrahedron Lett.* **28:** 1607–1610.

Margolin, A. L., and A. M. Klibanov. 1987. Peptide synthesis catalyzed by lipases in anhydrous organic solvents. *J. Am. Chem. Soc.* **109:** 3802–3804.

Matsumura, S., and J. Takahashi. 1986. Enzymic synthesis of functional oligomers. 1. Lipase-catalyzed polymerization of hydroxy acids. *Makromol. Chem. Rapid Commun.* **7:** 369–373.

Mohr, P., L. Rosslein, and C. Tamm. 1987. 3-Hydroxyglutarate and β,γ-epoxy esters as substrates for pig liver esterase (PLE) and α-chymotrypsin *Helv. Chim. Acta* **70:** 142–152.

Njar, V. C. O., and E. Caspi. 1987. Enzymatic transesterification of steroid esters in organic solvents. *Tetrahedron Lett.* **28:** 6549–6552.

Okumura, S., M. Iwai, and Y. Tominaga. 1984. Synthesis of ester oligomers by *Aspergillus niger* lipase. *Agric. Biol. Chem.* **48:** 2805–2808.

Peoples, O. P., S. Masamune, C. T. Walsh, and A. J. Sinskey. 1987. Biosynthetic thiolase from *Zoogloea ramigera*. III. Isolation and characterization of the structural gene. *J. Biol. Chem.* **262:** 97–102.

Ramos Tombo, G. M., H.-P. Schar, X. Fernandez i Busquets, and O. Ghisala. 1986. Synthesis of both enantiomeric forms of 2-substituted 1,3-propanediol monoacetates starting from a common prochiral precursor, using enzymatic transformations in aqueous and in organic media. *Tetrahedron Lett.* **27:** 5707–5710.

Riva, S., J. Chopineau, A. P. G. Kieboom, and A. M. Klibanov. 1988. Protease-catalyzed regioselective esterification of sugars and related compounds in anhydrous dimethylformamide. *J. Am. Chem. Soc.* **110:** 584–589.

Riva, S. and A. M. Klibanov. 1988. Enzymochemical regioselective oxidation of steroids without oxidoreductases. *J. Am. Chem. Soc.* **110:** 3291–3295.

Saunders, K. J. 1973. *Organic Polymer Chemistry*, pp. 22–24. London: Chapman and Hall.

Selegny, E., Ed. 1979. *Optically Active Polymers*. Dordrecht, Holland: D. Reidel Publishing Co.

Sokolov, L. B. 1968. *Synthesis of Polymers by Polycondensation*, pp. 2–7. Jerusalem, Israel: Israel Program for Scientific Translation.

Sonnet, P. E. 1987. Kinetic resolutions of aliphatic alcohols with a fungal lipase from *Mucor miehei*. *J. Org. Chem.* **52:** 3477–3479.

Stokes, T. M., and A. C. Oehlschlager. 1987. Enzyme reactions in apolar solvents: the resolution of ($\pm$)-sulcatol with porcine pancreatic lipase. *Tetrahedron Lett.* **28:** 2091–2094.

Sullivan, G. R. 1978. Chiral lanthanide shift reagents. *Top. Stereochem.* **10:** 287–329.

Theisen, P. D., and C. H. Heathcock. 1988. Improved procedure for preparation of optically active 3-hydroxyglutarate monoesters and 3-hydroxy-5-oxoalkanoic acids. *J. Org. Chem.* **53:** 2374–2378.

Therisod, M., and A. M. Klibanov. 1986. Facile enzymatic preparation of monoacylated sugars in pyridine. *J. Am. Chem. Soc.* **108:** 5638–5640.

Therisod, M., and A. M. Klibanov. 1987. Regioselective acylation of secondary hydroxyl groups in sugars catalyzed by lipases in organic solvents. *J. Am. Chem. Soc.* **109:** 3977–3981.

Vergara, J., and R. V. Figini. 1977. Synthesis of Poly(DL-β-hydroxybutyric acid). *Makromol. Chem.* **178:** 267–270.

Vogl, O. and G. D. Jaycox. 1986. Molecular asymmetry can produce optical activity. *CHEMTECH* 698–703.

Wallace, J. S., F. E. Arnold, and L. S. Tan. 1989. *In situ* generation of rigid-rod polyimides from DMAC soluble polyisoimides: a new model for the study of molecular composites fabrication. *Polymer* (in press).

Wallace, J. S., and C. J. Morrow. 1989a. Biocatalytic synthesis of polymers. Synthesis of an optically active, epoxy substituted polyester by lipase catalyzed polymerization. *J. Polym. Sci. Polym. Chem. Ed.* **27:** 2553–2567.

Wallace, J. S., and C. J. Morrow. 1989b. Biocatalytic synthesis of polymers. 2. Preparation of [AA–BB]$_x$ polyesters by porcine pancreatic lipase catalyzed transesterification in anhydrous, low polarity organic solvents. *J. Polym. Sci. Polym. Chem. Ed.* **27:** 3271–3284.

Walsh, C. T. 1987. Enzymes in the biosynthesis of polyesters. In *Proceedings of the Robert A. Welch Foundation on Chemical Research. XXXI. Design of Enzymes and Enzyme Models*, pp. 185–203.

Wang, Y.-F., and C.-H. Wong. 1988. Lipase-catalyzed irreversible transesterification for preparative synthesis of chiral glycerol derivatives. *J. Org. Chem.* **53:** 3127–3129.

Wang, Y.-F., J. J. Lalonde, M. Momongan, D. E. Bergbreiter, and C.-H. Wong. 1988. Lipase-catalyzed irreversible transesterification using enol esters as acylating reagents: preparative enantio- and regioselective syntheses of alcohols, glycerol derivatives, sugars, and organometallics. *J. Am. Chem. Soc.* **110:** 7200–7205.

West, J. B. and C.-H. Wong. 1987. Use of nonproteases in peptide synthesis. *Tetrahedron Lett.* **28:** 1629–1632.

Whitesides, G. M., and D. W. Lewis. 1970. Tris[3-(*tert*-butylhydroxymethylene)-*d*-camphorato]europium (III). A reagent for determining enantiomeric purity. *J. Am. Chem. Soc.* **92:** 6979–6980.

Wilson, W. K., S. B. Baca, Y. J. Barber, T. J. Scallen, and C. J. Morrow.

1983. Enantioselective hydrolysis of 3-hydroxy-3-methylalkanoic acid esters with pig liver esterase. *J. Org. Chem.* **48:** 3960–3966.

Yamamoto, K., T. Nishioka, J. Oda, and Y. Yamamoto. 1988. Asymmetric ring opening of cyclic acid anhydrides with lipase in organic solvents. *Tetrahedron Lett.* **29:** 1717–1720.

Zaks, A. and A. M. Klibanov. 1984. Enzymatic catalysis in organic media at 100°C. *Science* **224:** 1249–1251.

Zaks, A., and A. M. Klibanov. 1985. Enzyme-catalyzed processes in organic solvents. *Proc. Natl. Acad. Sci. USA* **382:** 3192–3196.

3
Regiospecific Hydroxylation of Biphenyl and Analogs by *Aspergillus parasiticus*

Daniel A. Abramowicz, Charles R. Keese, and
Suzanne H. Lockwood

The regiospecific conversion of biphenyl to 4,4′-dihydroxybiphenyl (biphenol) by various fungi has been studied for several years. These selective enzymatic hydroxylation syntheses of biphenols are of interest as model xenobiotics for the mammalian metabolism of aromatics (Smith and Rosazza 1974) and as screening systems for carcinogens utilizing 4,4′-dihydroxybiphenyl as the probe (McPherson et al. 1976). Our interest in these systems concerns the production of 4,4′-dihydroxybiphenyl as an intermediate in the plastics and dye industries (Schwartz 1979; Fewson 1981; Cain 1980). For example, the biphenol diol is a monomer in Union Carbide's polymer Radel℠ and GE's engineering polymer Ultem II℠. The organism *Aspergillus parasiticus* catalyzes the specific conversion of biphenyl to 4,4′-dihydroxybiphenyl in a one-step process.

Early work utilizing fungal systems for the production of 4,4′-dihydroxybiphenyl surveyed several different biological systems and focused on the filamentous fungus *A. parasiticus* as the leading candidate (Schwartz et al. 1980; Smith et al. 1980). This organism was shown to catalyze the specific conversion of biphenyl to almost exclusively the desired 4,4′-dihydroxybiphenyl product (Smith et al. 1980). More recently, workers at Martin Marietta Laboratories have concentrated on optimizing the production of the biphenol from biphenyl with this organism (Golbeck et al. 1983; Golbeck and Cox 1984; Cox and Golbeck 1985). These workers found that the biphenol itself at low concentrations would stimulate higher activity (Golbeck et al. 1983). Optimal yields of >80% were achieved, but at rates considered too low to support an economical process (average of 2.5 mg/g dry weight/day) (Cox and Golbeck 1985).

In this chapter, we describe methods utilized to increase the rate of hydroxylation and product concentration for the biconversion of biphenyl to 4,4'-dihydroxybiphenyl. In our systems we have achieved rates of >1,000 mg/g dry weight/day at >2 g/L and yields of >95%. This rate is nearly three orders of magnitude greater than that reported by others (Cox and Golbeck 1985) and this activity has been successfully scaled to 220-L systems. In addition, we have identified compounds other than 4,4'-dihydroxybiphenyl that can induce the hydroxylating system. We have also extended the substrate range of this system to include substituted biphenyls and biphenyl analogs. Indeed, the system displayed a broad substrate specificity that was relatively independent of electronic effects. These improvements provide motivation to reevaluate this biocatalytic synthesis of 4,4'-dihydroxybiphenyl and other analogs.

MATERIALS AND METHODS

Culture Conditions

ATCC strain #15517 of the fungus *A. parasiticus*, formerly *A. toxicarius*, was obtained from the American Type Culture Collection. Cultures were incubated at 30°C into sterilized complex Sabouraud–dextrose liquid media (20 g/L of glucose, 10 g/L of neopeptone) at pH 5.6 in submerged cultures or on solid media (1.5% Bacto agar). Liquid cultures were inoculated with spores (45×10^6 spores/L of culture) and incubated in a rotary shaker (350 rpm, 30°C, New Brunswick model G25). Substrate was added as an aqueous emulsion after full growth was achieved. Spores were grown on thin agar plates (1 mm) to ensure high spore yield (see Fig. 3.1). In submerged culture, pellets were formed in preference to a more disperse morphology as shown in Figure 3.2.

Figure 3.1. Production of spores of *A. parasiticus* on thin agar plates. The sequence of photographs depicts early, intermediate, and late stages of growth. The spores formed are harvested by suspension in liquid medium.

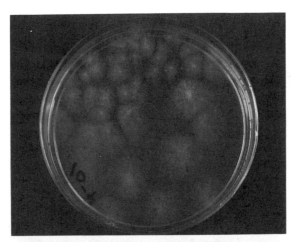

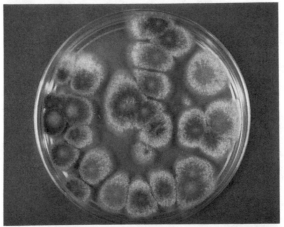

65

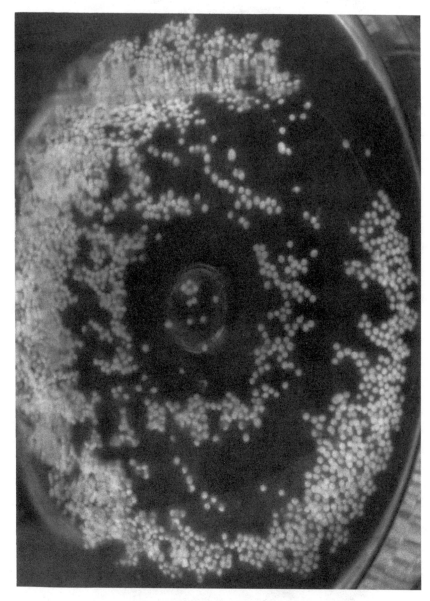

Figure 3.2. Production of pellet morphology in submerged culture. Pellet size varied from 3 to 8 mm.

Substrate Emulsions

Molten substrate in water (200 mg/mL) was dispersed by ultrasonic energy (Branson Cell Disrupter) while stirring into an emulsion stabilized by 10 mg/mL of high-quality gelatin in the aqueous phase (Grayslake Gelatin Co., Grayslake, IL, U.S.A.). The purity of the gelatin proved critical for obtaining fine stable emulsions in our hands. The emulsion was allowed to cool while stirring to yield a fine dispersion of particles (approximately 5 μm diameter).

Analytical Techniques

Aliquots were removed at various times to analyze the extent of conversion. Methanol was added to the cell suspension to 80%, then the material was acidified and extracted at 60°C for 1 h. The insoluble cell debris was removed by filtration or centrifugation, and the supernatant fraction analyzed by HPLC [Waters 840 liquid chromatograph, Millipore Partisphere C18 reverse-phase column, 30–100% (vol/vol) gradient of acetonitrile in water]. This technique gave quantitative recovery of substrate and product from the cellular material.

Product Isolation

Products were purified to allow characterization and use in polymerization reactions. The general procedure involved solvent extraction from the fungus, silica gel chromatography, and recrystallization to obtain pure material. Characterization included mass spectrometry and nuclear magnetic resonance.

Specific details for the purification of one product (4-bromo-4'-hydroxybiphenyl) from a scale up run follow. After the reaction is complete, the culture broth (220 L) was filtered through cheesecloth or a continuous centrifuge (Tolhurst center slung centrifuge with a 12-inch solid bowl) to recover the cellular material. This biomass was extracted three times with pure methanol to recover all of the product. The methanol fractions were pooled and evaporated to dryness. The residual solid was dissolved in 60:40 hexane/toluene and eluted from a 5-kg silica gel column with the same solvent mixture. Under these conditions, unreacted 4-bromobiphenyl is eluted and can be recovered. Elution with 80:20 toluene/ethyl acetate selectively removes the product,

4-bromo-4'-hydroxybiphenyl. The silica gel retains nearly all of the colored polar contaminants and in one step yields relatively pure product. Product fractions were pooled and heated to drive off some of the ethyl acetate. Recrystallization in this manner provided a high yield of pure product (>99%).

Protein Isolation

Many attempts were made to obtain cell free enzymatic hydroxylation activity, without success. Variables investigated included: pH (6, 7.4, 8.3); growth stage of the organism before isolation (30, 50, 80 h); various cell disruption techniques (french press, bead beater, liquid nitrogen, sonication); isolation temperature (4, 23°C); cofactors (NADH, NADPH, FAD); detergents (Triton X-100, β-octylglucoside, 3-[(3-cholamidopropyl)dimethylammonio]-2-hydroxy-1-propanesulfonate (CHAPS); protease inhibitors in various combinations [phenyl-methylsulfonyl fluoride (PMSF), pA-PMSF, bestatin, chymostatin, EDTA, leupeptin, α-macroglobulin, and pepstatin, all obtained from Boehringer Mannheim].

Inducers

4-Hydroxy- and 4,4'-dihydroxybiphenyl (Aldrich Chemical Co.; 5–250 ppm) were added along with the substrate biphenyl emulsion 40 h after inoculation. Cultures grown on soybean flour (Sigma Chemical Co.) included 5 g/L of the flour in the medium. To investigate the inductive effect of genistein (K & K Laboratories, Plainview, NY, U.S.A.), emulsions were prepared as described above containing 200 mg/mL biphenyl plus 10 mg/mL of 4,4'-dihydroxybiphenyl or 15 mg/mL of genistein. 0.25 mL emulsion was added to 50 mL of fully grown culture (1,000 ppm biphenyl, 50 ppm 4,4'-dihydroxybiphenyl, 75 ppm genistein) and products monitored by HPLC as described above.

Scale-Up

Initial laboratory studies typically involved 50–100 mL of culture in a 250–500 mL Erlenmeyer shake flask. The first scale-up increased the culture volume to 5 L. The apparatus involved simply a 5-gallon plastic bucket covered with aluminum foil to minimize contamination. Air saturated with water was bubbled into the reactor, and the system was stirred with a mechanical stirrer at 500 rpm. A 100-mL sterile culture at 24 h growth was used to inoculate the 5-L volume. The medium was sterilized separately: the neopeptone was autoclaved at 50× and the glucose filter sterilized at 25×. This was necessary as no sterilizable fermentor of adequate size was available and concentrated solutions of glucose cannot be autoclaved.

The 220-L scale-up was performed in a 100-gallon Nalgene container. Holes were made in the lid for the stirrer and temperature control. Air was bubbled into the bottom of the vessel at 20–60 lpm, with sufficient flow to maintain nearly saturating oxygen levels during growth. Steam-filled temperature coils were utilized to maintain a temperature of 30°C. A Lightnin 1/3 HP stirrer spinning at 1,700 rpm provided mixing (single marine-type impellor, 3.5 inch). The addition of antifoam was required to maintain a constant foam level (Sigma Antifoam B emulsion). A 4-L sterile culture at 24 h growth was used to inoculate the 220-L volume. The antibiotic tetracycline (Sigma Chemical Co.) was added at 10 μg/mL to minimize bacterial contamination in the unsterilized vessel.

RESULTS AND DISCUSSION

Hydroxylation Activity

The filamentous fungi *A. parasiticus* develops through a complex life cycle, a portion of which is displayed in Figure 3.3. Germination from spores occurs after approximately 8 h, with final growth achieved 40 h after inoculation. Typically substrate is added at this point to minimize any effect the substrate may have on growth. Although biphenyl is a known fungistat (Hakkinen et al. 1973), most of the substituted biphenyls tested did not detectably alter sporulation or growth. For example, 4-bromobiphenyl can be added at the time of inoculation without significantly affecting growth or hydroxylation activity. The activity profile obtained from this reaction is shown in Figure 3.4. Note that the hydroxylation activity initiates as growth ceases, suggesting that *para*-hydroxylation is a secondary metabolic activity of the fungus. At this point in the growth cycle of the fungus, many new genes are expressed leading to spore formation. For example, estimates from RNA/DNA hybridization studies indicate that approximately 1,300

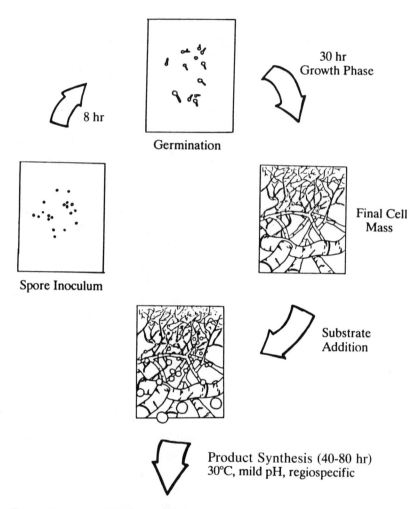

Figure 3.3. Partial life cycle of the filamentous fungus *A. parasiticus*. After full growth is achieved, the addition of a fine dispersion of substrate allows hydroxylation.

genes are activated only during sporulatoin in *A. nidulans* (Timberlake 1980). It is speculated that the monooxygenase responsible for this hydroxylation activity is one of these gene products.

The reaction with biphenyl involves two separate hydroxylations, as shown in Figure 3.5. Using 4-hydroxybiphenyl as the substrate, we have shown that the same product is isolated, confirming that the monohydroxylated species is an intermediate in the reaction sequence. The sulfate esters are detected after prolonged incubation as previously reported (Golbeck et al. 1983). These conjugates are only found in very old cultures (>4 days) and can easily be hydrolyzed back to the phenols. Therefore these side reactions are not a problem in the system.

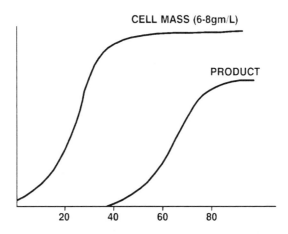

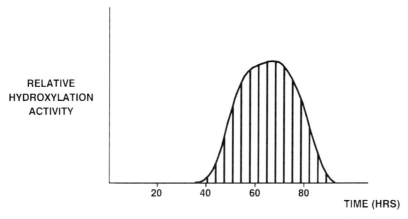

Figure 3.4. Hydroxylation activity profile of the fungus. Note that the activity begins after growth has ceased.

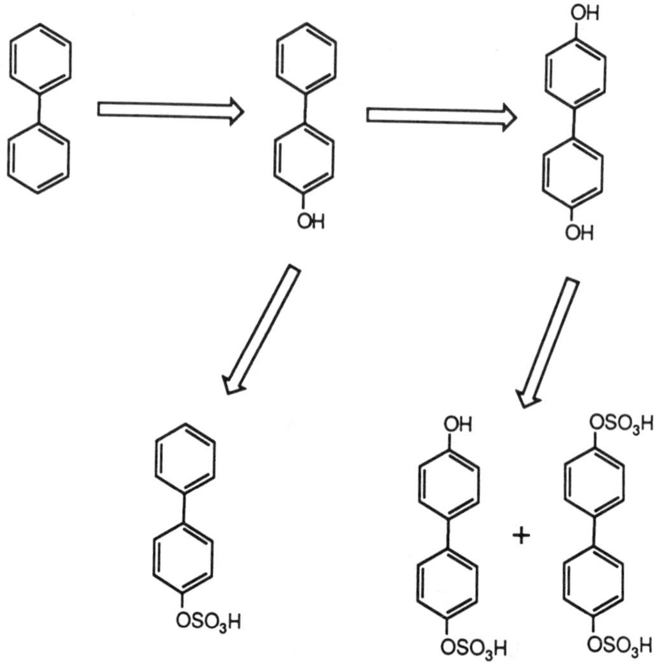

Figure 3.5. Hydroxylation of biphenyl catalyzed by the fungus *A. parasiticus*. Both the mono- and dihydroxy products are converted to the sulfate esters by old cultures. These side products, if formed, are easily acid-hydrolyzed back to the phenols.

Induction

Golbeck et al. (1983) also reported that low concentrations of the intermediate or the product, 4-hydroxybiphenyl and 4,4'-dihydroxybiphenyl respectively, "prime" the system resulting in greater rates and product levels. We have confirmed this effect as shown in Figure 3.6. Concentrations >50 ppm display no increased benefit. We discovered that the addition of soybean flour (5 g/L) can cause a similar beneficial effect (see Figure 3.7). Note that without "inducer," product levels are nearly undetectable. The addition of either 4,4'-dihydroxybiphenyl or soybean meal results in a dramatic increase in activity.

It is speculated that the enzyme responsible for this activity is an inducible P450 monooxygenase, although classic P450 inducers such as

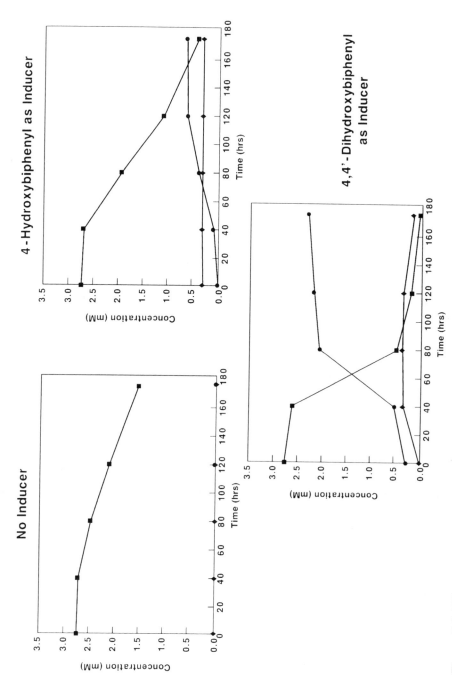

Figure 3.6. Effect of "inducer" on the hydroxylation of the substrate biphenyl. Note the dramatic increase in the rate and final level of the product 4,4'-dihydroxybiphenyl with the addition of the mono- and dihydroxybiphenyl. (■), Biphenyl; (♦), 4-OH-BP; (●), 4,4'-OH-BP.

73

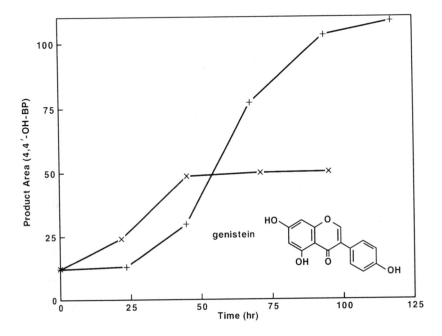

Figure 3.7. Beneficial effect of soybean flour (5 g/L) on the hydroxylation of biphenyl. Both samples contain 50 ppm of the 4,4'-dihydroxybiphenyl as an inducer. Genistein is a component of the soybean flour. +, soybean (5 g/L); ×, control.

phenobarbitol and 3-methylcholanthrene do not induce the system (Golbeck et al. 1983). It is known that soybean flour or a component isoflavoniod genistein will induce cytochrome P450 activity in *Streptomyces griseus* (Sariaslani and Kunz 1986). Genistein, when added alone as a pure compound, displayed a capacity to "induce" the hydroxylase activity similar to that of the biphenol, as shown in Figure 3.8. We have also shown that 4-bromo-4'-hydroxybiphenyl acts in a similarly beneficial manner (data not shown). Therefore at least three compounds have now been shown to induce greater para hydroxylation activity for biphenyl. It is not known whether the benefit observed with these compounds is indeed induction at the nucleic acid level, an allosteric effect on the enzyme itself, or some other unknown mechanism.

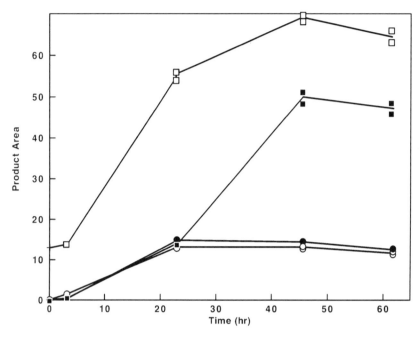

Figure 3.8. ''Induction'' of biphenyl hydroxylation activity with 50 ppm of 4,4'-dihydroxybiphenyl or genistein. Note that the production of the mono- and disubstituted products is similar. (□), 4,4'-OH-BP (+4,4'-OH-BP); (■), 4,4'-OH-BP (+ genistein); (○), 4-OH-BP (+4,4'-OH-BP); (●), 4-OH-BP (+ genistein).

A classic lag time is observed before activity is detected, as shown in Figure 3.9 (arrows indicate the time of substrate addition). Note the reproducible 8-hour delay before product is detected. If a low concentration of one of the ''inducers'' (50 ppm of 4-bromo-4'-hydroxy-biphenyl) is added prior to substrate addition, this characteristic lag disappears as shown in Figure 3.10. This result supports the concept of classic induction or increased gene expression as the cause of the increased activity. Genistein may induce a P450 monooxygenase enzyme in this organism as well, or its structural likeness to 4,4'-dihydroxybiphenyl may explain the similarity of its effect on the hydroxylation activity.

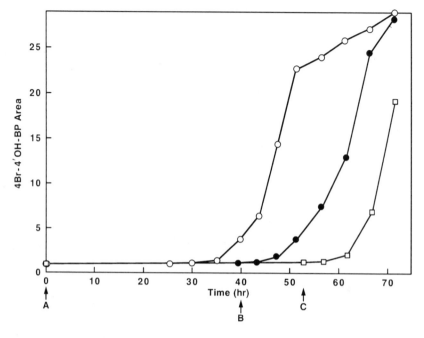

Figure 3.9. Observed lag time for the conversion of 4-bromobiphenyl to 4-bromo-4'-hydroxybiphenyl. Arrows indicate the time of substrate addition. Full growth is achieved in 35–40 h; Substrate addition at (A) 0, (B) 40, and (C) 53 h to separate flasks. (○), A activity; (●), B activity; (□), C activity.

Protein Isolation

Attempts to isolate the active monooxygenase from growing cells have not been successful. Initially a range of parameters were utilized as described in the methods section. Cell-free preparations and membrane fragments were incubated with substrate for several hours and analyzed for the formation of hydroxylated products. No combination of cell rupture techniques, protease inhibitors, detergents, cofactors, pH, or substrate attempted yielded a reproducibly significant activity with cell free extracts from cultures of various ages. Two possible explanations for these results are presented below. First, the detection of hydroxylated products required enough material to analyze by HPLC chromatography. Ultraviolet spectrophotometric detection is relatively insensitive and several hours of activity were required to "see"

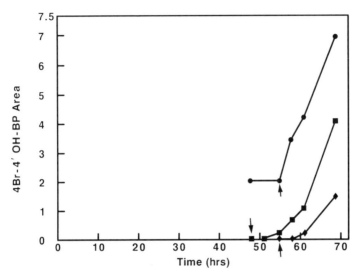

Figure 3.10. Disappearance of lag time with the addition of the "inducer" 4-bromo-4'-hydroxybiphenyl. Arrows indicate time of substrate (4-bromo-biphenyl) addition. All samples had attained full growth by 40 h. Sample that received 50 ppm of the "inducer" at 48 h displayed no lag in hydroxylation activity. (■), Substrate at 48 h; (●), substrate at 55 h (4Br-4'OH-BP at 48 hr); (◆), substrate at 55 hr.

the products. The development of more sensitive detection schemes could overcome this limitation. Second, none of the commercially available protease inhibitors completely prevent protease activity. Figure 3.11 schematically displays the protease activity found in the presence of a complex inhibitor mixture. This assay involves clearing a spot on a fibrinogen plate with a solution containing protease activity. Note that approximately 40% of the original protease activity in cell extracts (B and C) remained viable in the presence of protease inhibitors (D). Therefore there exist proteases that are resistant to commercial protease inhibitors, possibly allowing degradation of the enzyme after cell rupture.

Rate Enhancement

Oxygen availability was investigated by varying the culture volume in the flask. Various volumes of fully grown *Aspergillus* culture (300, 200,

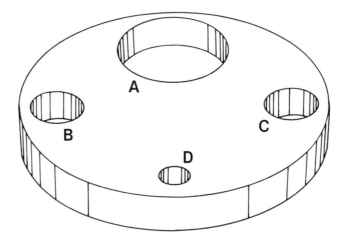

Figure 3.11. Measure of protease activity on a Fibrinogen plate (Boehringer-Mannheim). Solution containing protease was added dropwise and incubated at 37°C for 1.5 h. Cell extracts produced by grinding fully grown cells in a Ten Brock homogenizer and centrifuging at 15,000 × *g* for 1 min (A) control, 0.06 U papain; (B) and (C) 10 μL cell extract supernatant; (D) 10 μL cell extract in presence of protease inhibitors (130 μ*M* bestatin, 100 μg/mL chymostatin, 0.2 mg/mL EDTA, 1 μ*M* leupeptin, 1 μ*M* pepstatin, and 250 μ*M* PMSF; all obtained from Boehringer Mannheim).

100, 50 mL) were added to 500-mL Erlenmeyer flasks. At this point the substrate biphenyl was added with inducer and hydroxylation monitored. The results of this experiment are shown in Figure 3.12. Note that as the volume of the culture decreases, the rate of hydroxylation increases significantly. This is most probably due to the increased surface area to volume ratio in the flasks with a smaller culture volume. Therefore, with biphenyl as the substrate, oxygen availability can become a rate-limiting factor. Methods which improve the mass transport or solubility of oxygen should improve hydroxylation rates. For example, increasing the concentration of oxygen in the headspace of the flask increases the rate of biphenol production (data not shown). Substrates other than biphenyl, including 4-bromobiphenyl, azobenzene, and diphenyl acetylene, do not display the same sensitivity to oxygen availability.

It was determined that cell mass, and therefore enzyme concentration, are not limiting the rate of hydroxylation, as shown in Figure 3.13.

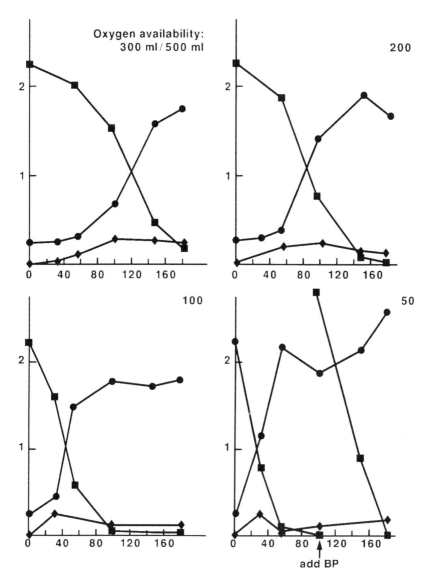

Figure 3.12. Effect of oxygen availability on the rate of biphenyl hydroxylation in *A. parasiticus*. All samples contained 50 ppm of 4,4'-dihydroxybiphenyl as "inducer" and 1000 ppm of biphenyl in the culture medium. 300, 200, 100, or 50 mL of fully grown *Aspergillus* culture was added to a 500-mL Erlenmeyer flask. (■), Biphenyl; (♦), 4-OH-BP; (●), 4,4'-OH-BP.

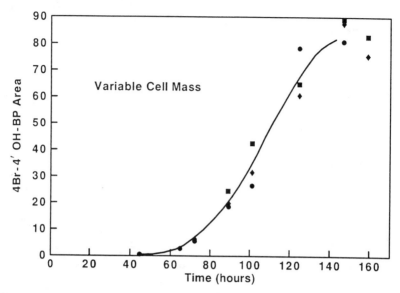

Figure 3.13. Effect of cell population density on the rate of 4-bromobiphenyl hydroxylation. Fully grown cells were washed and resuspended at 0.5, 1.0, and 2.0× cell density (1× = 8–10 g fungus/L). (■), Control; (◆), 0.5×; (●), 2×.

After full growth, the cells were recovered by filtration and resuspended at various cell densities to equal volumes in growth buffer. Note that no significant change in hydroxylation rate is observed for cell densities of 0.5×, 1.0×, and 2.0× in the hydroxylation of 4-bromobiphenyl. A cell density of 1.0× indicates full growth at 8–10 g fungus/L. The experiment was also performed by resuspension in a media containing no nitrogen source to eliminate any additional growth. Similar results were obtained and this data is presented in Table 3.1. At cell densities down to 0.125×, the rate of hydroxylation was relatively unchanged, but at 0.063×, the rate did indeed decrease, and under these conditions the specific activity exceeded 1,400 mg/g dry weight/day with retention of high yields. The greatest previous hydroxylation rate reported in the literature was 25 mg/g dry weight/day (Golbeck et al. 1983). Other factors known to be rate limiting in the hydroxylation of biphenyl include the solubility of the substrate and oxygen availability, as previously described.

Table 3.1. Effect of cell population density on the specific activity of 4-bromobiphenyl hydroxylation.

CELL MASS	SPECIFIC ACTIVITY
2×	105 mg/g/day
1×	170
0.5×	290
0.25×	460
0.125×	1,020
0.063×	1,440

Fully grown cells were washed and resuspended in nitrogen-free media to prevent further growth. (1× = 8–10 g/L.)

One method to increase substrate concentration could involve the addition of miscible organic solvents. Unfortunately we have determined that this organism is very sensitive to such solvents, as shown in Figure 3.14. Note that with only 0.25% dimethyl sulfoxide (DMSO) [Fig. 3.14(A)] or 0.5% DMF [Fig. 3.14(B)], the hydroxylation activity is significantly affected. With the addition of up to 1% DMSO, growth was unaffected while the hydroxylation activity was reduced nearly to zero [Fig. 14(B)]. As substrates had previously been added as concentrated solutions in such solvents, our use of fine dispersions in place of these solvents eliminates this limitation. The preparation of such dispersions from stabilized emulsions is described in Materials and Methods.

The combination of the factors previously described, including the elimination of solvents, increased substrate availability, and increased oxygen availability result in consistently high activity as shown in Figure 3.15. Here, 2 g/L 4-bromobiphenyl is rapidly converted in high yield to the specific 4-bromo-4-hydroxybiphenyl. In this system, 1× cell density displays a specific activity of greater than 250 mg/g dry weight/day (see Table 3.1). The hydroxylation is compleed in less than 24 h.

Alternate Substrates

It had previously been shown (Schwartz et al. 1980; Smith et al. 1980; Golbeck et al. 1983; Golbeck and Cox 1984; Cox and Golbeck 1985) that the fungus *A. parasiticus* will specifically convert biphenyl to 4,4'-dihydroxybiphenyl at low rates. In addition to developing methods to enhance reaction rates and concentrations, we have extended the substrate range for this hydroxylation. The system displayed considerable flexibility and many analogs of biphenyl were shown to be substrates for this specific para hydroxylation. Potential substrates were added after full growth to avoid potential problems due to growth inhibition by the substrate. These compounds were added with and without inducers such as 4,4'-dihydroxybiphenyl and 4-bromo-4'-hydroxy-biphenyl. The results of these experiments are shown in Figure 3.16.

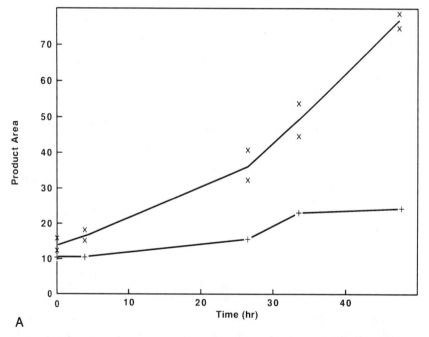

A

Figure 3.14. Effect of water-miscible organic solvents on the rate of 4-bromo-biphenyl hydroxylation. (A) Addition of dimethylformamide (DMF). + = 0.5% DMF; × = no DMF. (B) Addition of dimethylsulfoxide (DMSO). (Top) (●), dry weight. Bottom: (●), 0% DMSO; (◆), 0.25%; (■), 1.0%; (▲), 4.0%.

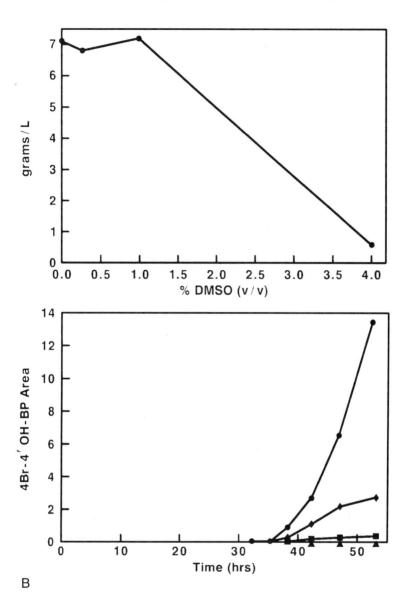

B

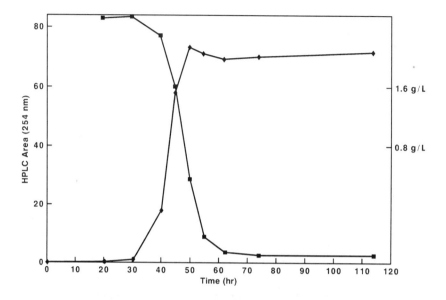

Figure 3.15. Hydroxylation activity profile for the conversion of 4-Bromo-biphenyl to 4-bromo-4'-hydroxybiphenyl. (■), 4Br-BP; (◆), 4Br-4' OH-BP.

The compounds are subdivided into three classes. Class 1 includes only biphenyl at the present. This class requires an inducer for high activity (see Fig. 3.6). Inducers include 4,4'-dihydroxybiphenyl, genistein, and 4-bromo-4'-hydroxybiphenyl. Class 2 includes compounds that are not substrates. The common feature among this class involves a kink or bend between the two phenyl rings. This steric constraint may prevent these compounds from binding properly in the active site. Class 3 includes a range of compounds that do not require an inducer for high activity. This class contains a large number of substituted biphenyls and biphenyl analogs with extended linear linkers such as the acetylene bridge or trans-ethylene linker.

The range of structures in class 3 implies that the electronic effect of the ring substitutent has little effect on activity. For example, both electron releasing (—OH, —OCH$_3$, —CH$_3$) and electron withdrawing groups (—NO$_2$, —Cl, —Br) on the biphenyl moiety do not significantly affect the hydroxylation rate. In addition, substituents can occupy the

2-, 3-, or 4-positions on the ring. All of the substrates in class 3 do not require an inducer for high activity, although inducers may still display positive effects. These compounds can be added during inoculation, since typically they display no fungistatic effects. There does appear to be a correlation of activity with solubility. Substrates with very poor water solubility such as *p*-terphenyl and *trans*-stilbene are hydroxylated at slower rates.

Scale-Up

To prepare enough of the hydroxylated products for potential applications, the laboratory scale was increased to 5 L and then 220 L. The 5-L scale was carried out in a 5 gallon plastic bucket as described in Materials and Methods. The substrate 4-bromobiphenyl (2 g/L) was added during inoculation of the 5-L volume. The results of this experiment are shown in Figure 3.17. Note that the activity begins after growth plateaus (full growth at 60 h) and continues with a rate similar to the laboratory scale experiment. The yield was only 60%, compared to >95% for the laboratory experiments. This may have been the result of the considerable bacterial contamination which occurred in this unsterilized system. This contaminant represented approximately 50% of the biomass. Therefore antibiotics were tested for their effect on the growth and hydroxylation activity of the fungus [Fig. 3.18(A), 10 μg/mL tetracycline, and 10 μg/mL of carbicillin]. As no significant change in rate or final level was observed, tetracycline was included in subsequent large scale runs in unsterilized vessels. It was also noted that considerable foaming occurred during the experiment. The addition of antifoam slightly inhibited growth and delayed the onset of hydroxylation activity, but the rate and product levels were unchanged as shown in Figure 3.18(B).

The apparatus used for the 220-L fermentation is shown in Figure 3.19 and described in more detail in Materials and Methods. Tetracycline was added at the time of inoculation along with 2 g/L of the substrate 4-bromobiphenyl as a fine dispersion. The results of this experiment are shown in Figure 3.20. Note that again the activity begins after growth plateaus and continues with a rate similar to the laboratory scale experiment. The recovery of the product 4-bromo-4'-hydroxybiphenyl was straightforward and as described in the Materials

Substrate | Class 1 | Product

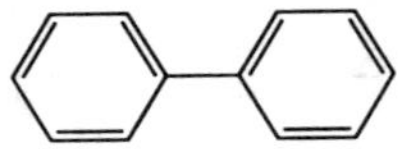

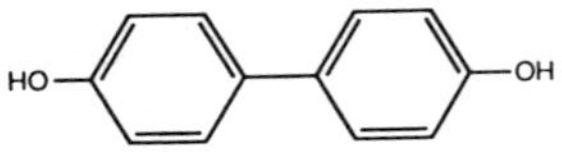

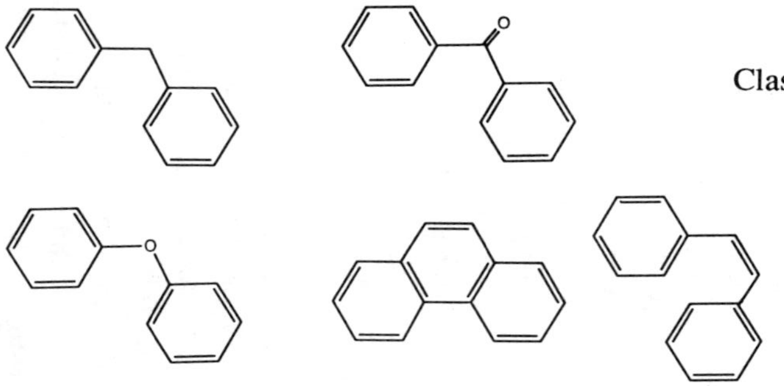

Class 2

Class 3

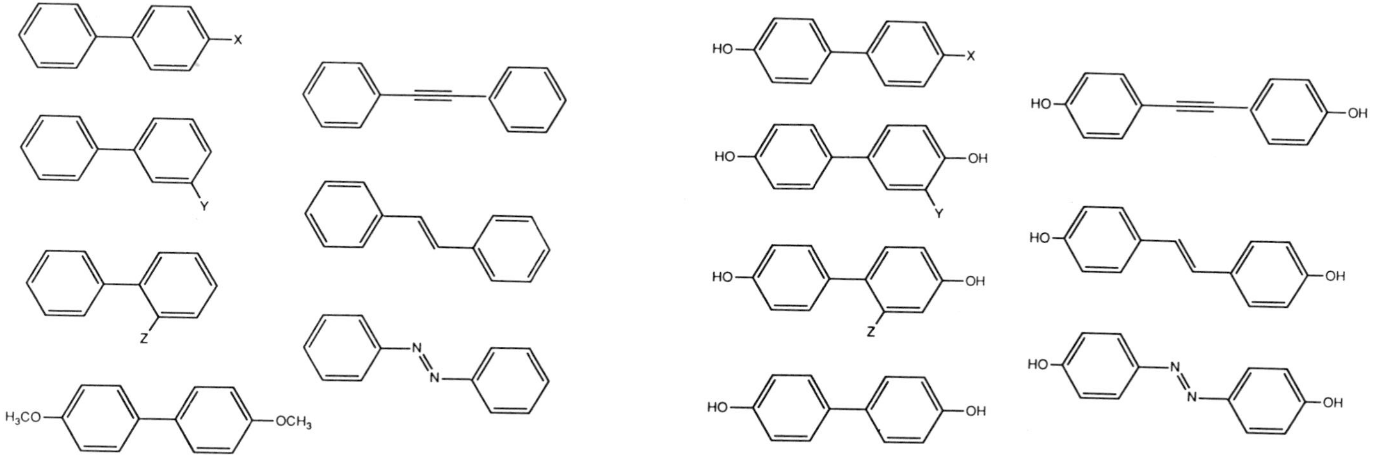

Figure 3.16. Chemical structures of potential substrates for regiospecific hydroxylation in *A. parasiticus*. X = OH, NO$_2$, CH$_3$, OCH$_3$, COOH, COCH$_3$, Br, Cl, F, H, CH(CH$_3$)$_2$, COH(CH$_3$)$_2$, NH$_2$? Y = Br, CH$_3$, OH. Z = Cl, CH$_3$, OCH$_3$.

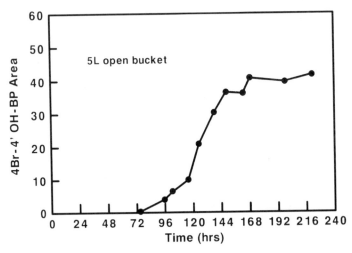

Figure 3.17. Results of fermentation scale-up to 5 L. (●), 4Br-4′ OH-BP.

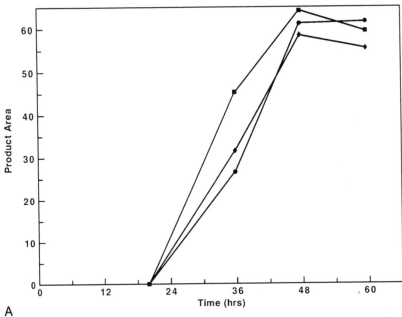

A

Figure 3.18. The effect of antibiotics and antifoaming agents on the *para*-hydroxylation activity of *A. parasiticus*. (A) The growth medium included 10 μg/mL of an antibiotic, tetracycline (tet) or carbicillin (Cb). (■), Control; (♦), + tet; (●), + Cb. (B) The growth medium included 0.1% of Sigma antifoam B emulsion. (○), Control; (●), + antifoam (0.1%).

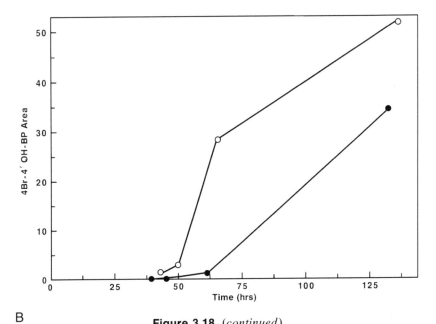

Figure 3.18. (*continued*)

B

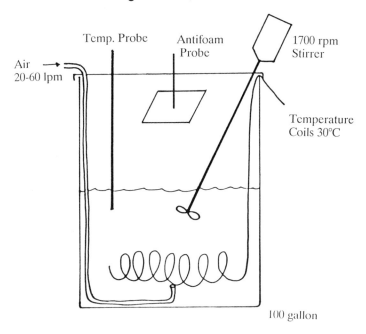

V_T = 220 liters

Figure 3.19. Apparatus utilized for the 220-L scale-up.

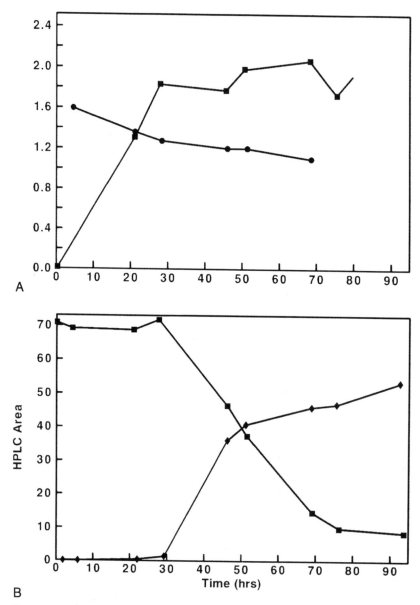

Figure 3.20. Results of the conversion of 4-bromobiphenyl to 4-bromo-4'-hydroxybiphenyl on the 220-L scale. (A) Growth and glucose utilization. (■), dry wt (g/L); (●), glucose (g/0.1 L). (B) Product formation. (◆), 4Br-BP; (■), 4Br-4'OH-BP.

and Methods section. The yield has improved to >85%, as contamination was reduced by the addition of the antibiotic. The success of the 4,400× scale-up in simple homemade vessels implies that results analogous to the laboratory scale are attainable in large sterile fermentors with good mass transport and oxygen availability.

CONCLUSIONS

The fungus *A. parasiticus* catalyzes the conversion of biphenyl and biphenyl analogs to the corresponding *para*-hydroxylated diols. With biphenyl as the substrate, low concentrations of "inducers" dramatically increase the reaction rate. Other methods to improve oxygen and substrate availability have resulted in significant additional rate enhancement. The reaction has been demonstrated at rates >1400 mg/g dry weight/day, concentrations >2 g/L, and yields of >95%.

The monooxygenase responsible for this bioconversion accepts a wide range of substrates. This enzyme has not been isolated in its active form and currently only whole cell systems are being employed. Such fermentations are suggested by the instability of the enzyme observed in protein isolations and the potential requirement for a reducing cofactor such as NAD(P)H. Scale-up of the fermentation to 220 L has been demonstrated successfully.

ACKNOWLEDGMENTS

The authors would like to especially thank Dave Mobley and Dave Dietrich for their invaluable assistance in the design and operation of the 220-L reactor. We would also like to thank Bruce Johnson for his advice on the purification of the hydroxylated products.

REFERENCES

Cain, R. B. 1980. In *Hydrocarbons in Biotechnology* (D. E. F. Harrison, I. J. Higgins, and R. Watkinson, Eds.) London: Heydon and Son.

Cox, J. C., and J. H. Golbeck 1985. Hydroxylation of biphenyl by *Aspergillus parasiticus:* approaches to yield improvement in fermenter cultures. *Biotechnol. Bioeng.* 27: 1395–1402.

Fewson, D. A. 1981. In *Microbial Degradation of Xenobiotics and Recalcitrant Compounds* (T. Leisinger, R. Hutter, A. M. Cook, and J. Nuesch, Eds.), pp. 141–179. New York: Academic Press.

Golbeck, J. H., S. A. Albaugh, and R. Radmer 1983. Metabolism of biphenyl by *Aspergillus toxicarius:* induction of hydroxylating activity and accumulation of water-soluble conjugates. *J. Bacteriol.* **156:** 49–57.

Golbeck, J. H., and J. C. Cox 1984. Hydroxylation of biphenyl by *Aspergillus toxicarius:* conditions for a bench scale fermentation process. *Biotechnol. Bioeng.* **26:** 434–441.

Hakkinen, I., S. Hernberg, P. Karli, and E. Vikkula 1973. Diphenyl poisoning in fruit paper production: new health hazard. *Arch. Environ. Health* **26:** 70–74.

McPherson, F. J., J. W. Bridges, and D. V. Parke 1976. The effects of benzopyrene and safrole on biphenyl 2-hydroxylase and other drug-metabolizing enzymes. *Biochem. J.* **154:** 773–780.

Sariaslani, F. S., and D. A. Kunz 1986. Induction of cytochrome P-450 in *Streptomyces griseus* by soybean flour. *Biochem. Biophys. Res. Commun.* **141:** 405–410.

Schwartz, R. D. 1979. Microbial production of hydroxylated biphenyl compounds. U.S. Patent no. 4,153,509.

Schwartz, R. D., A. L. Williams, and D. B. Hutchinson 1980. Microbial production of 4,4'-dihydroxybiphenyl: biphenyl hydroxylation by fungi. *Appl. Environ. Microbiol.* **39:** 702–708.

Smith, R. V., and J. P. Rosazza 1974. Microbial models of mammalian metabolism: aromatic hydroxylation. *Arch. Biochem. Biophys.* **161:** 551–558.

Smith, R. V., P. J. Davis, A. M. Clark, and S. Glover-Milton 1980. Hydroxylations of biphenyl by fungi. *J. Appl. Bacteriol.* **49:** 65–73.

Timberlake, W. E. 1980. Developmental gene regulation in *Aspergillus nidulans. Dev. Biol.* **78:** 497.

4
Simple Carbohydrates as Starting Materials in Organic Synthesis
A Comparison of Strategies

J. W. Frost, L. M. Reimer, K. M. Draths, D. L. Pompliano, and D. L. Conley

Chemists have had a long-standing interest in using the carbon chains and stereocenters of simple carbohydrates as starting materials for organic synthesis. Nature often exploits such a strategy as illustrated by the biosynthesis of aromatic amino acids and related secondary metabolites from carbohydrate precursors (Haslam 1974; Weiss and Edwards 1980). A variety of molecules are now being synthesized that inhibit specific aromatic amino acid biosynthetic enzymes. This is partially a consequence of the discovery of plant death associated with chemical inhibition of enzymes in plant aromatic amino acid biosynthesis (Grossbard and Atkinson 1985; Pompliano et al. 1989). Analogs of substrates of the aromatic amino acid biosynthetic enzymes are logical inhibitors. The question then becomes whether these inhibitors, like the naturally occurring substrates, can be derived from simple carbohydrates.

An aromatic amino acid biosynthetic enzyme whose inhibition has received considerable attention (Le Maréchal et al. 1980; Widlanski, Bender, and Knowles 1989; Myrvold et al. 1989) is dehydroquinate (DHQ) synthase, which catalyzes the conversion (Fig. 4.1) of 3-deoxy-D-*arabino*-heptulosonic acid 7-phosphate (DAHP) to dehydroquinate (DHQ). Inorganic phosphate is formed as a byproduct. Substrate DAHP consists of a 3-deoxy-2-ulosonic acid "head," a 7-carbon "backbone," D-*arabino* "ribs," and a phosphate monoester "tail." One analog, originally synthesized by Le Maréchal et al. (1981), which has been investigated extensively is 3-deoxy-D-*arabino*-heptulosonic

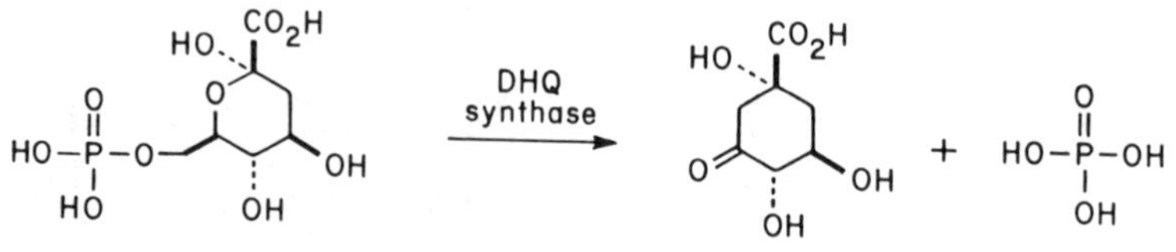

Figure 4.1. Conversion of DAHP to DHQ and inorganic phosphate catalyzed by the enzyme DHQ synthase.

acid 7-phosphonic acid **11** (Fig. 4.2). The synthesis of **11**, the phosphonic acid analog of DAHP, illustrates a number of the strategy options that are available when simple carbohydrates are used as starting materials for synthesis of enzyme inhibitors.

The first approach taken (Fig. 4.2; see pages 96 and 97) involved synthetic conversion of D-glucose **1** to D-glucose 6-phosphonic acid **8**, oxidative degradation of the D-glucose 6-phosphonic acid to D-erythrose 4-phosphonic acid **9**, and final coupling of the D-erythrose 4-phosphonic acid with phosphoenolpyruvate (PEP) **10** in a reaction catalyzed by the enzyme DAHP synthase (Reimer 1988). Synthesis of D-glucose 6-phosphonic acid began with sequential treatment of D-glucose with benzyl alcohol and benzaldehyde to form benzyl 4,6-benzylidene **2**. The low yields of this reaction were not problematic because the starting materials were inexpensive and the reaction could easily be run on a kilogram scale. Benzylation of the remaining alcohols provided **3** which was reacted with lithium aluminum hydride–aluminum trichloride to cleave selectively the benzylidene acetal and yield the primary alcohol **4**. Bromination and subsequent reaction with an excess of triethyl phosphite gave fully protected D-glucose 6-phosphonate **6**. Transesterification of the phosphonate diester with bromotrimethylsilane, aqueous hydrolysis of the resulting silyl esters, and catalytic hydrogenation over palladium on carbon yielded D-glucose 6-phosphonic acid **8**. Lead (IV) tetraacetate oxidation of **8** gave the needed D-erythrose 4-phosphonic acid **9**. Unfortunately, D-erythrose 4-phosphonate was a poor substrate for DAHP synthase. Only trace levels of the DAHP phosphonic acid analog **11** could be produced.

The failure of the first approach to provide sufficient levels of phosphonate DAHP analog **11** highlights one of the problems in attempting to use unnatural substrates in biocatalytic synthetic steps. Often, the ability of an unnatural substrate to be accepted by the active site of an enzyme can be appraised only after synthesis of the molecule. For DAHP synthase the synthetic efforts in making D-erythrose 4-phosphonic acid were largely wasted as the aldose was a poor substrate for DAHP synthase. This suggests either complete avoidance of biocatalysis or a more prudent strategy where enzyme-catalyzed transformations of unnatural substrates are avoided during biocatalytic assistance in DAHP analog synthesis.

These solutions are incorporated into synthesis of phosphonic acid DAHP analog **11** from methyl (methyl 3-deoxy-D-*arabino*-heptulopyranosid)onate **12** as shown in Figure 4.3 (see page 98) (Reimer et al. 1986). Selective conversion of the primary alcohol to bromide **13**, condensation with an excess of triethyl phosphite to give fully protected phosphonate **14**, and stepwise deprotection affords phosphonate **11**. Choice of methyl (methyl 3-deoxy-D-*arabino*-heptulopyranosid)onate follows from expectations that this molecule could be made via a variety of different strategies utilizing simple carbohydrates as starting materials. Construction of methyl (methyl 3-deoxy-D-*arabino*-heptulopyranosid)onate by chemical synthesis using D-arabinose as starting material completely avoiding biocatalysis will be examined first. Next, immobilized enzyme synthesis and microbial synthesis using D-fructose and D-glucose, respectively, as starting carbohydrates will be examined. These biocatalytic options avoid enzyme-catalyzed transformations of unnatural substrates. Development and comparison of (1) chemical synthesis, (2) immobilized enzyme synthesis, and (3) microbial whole cell synthesis provides an opportunity to evaluate critically various options for practical utilization of simple carbohydrates in enzyme inhibitor construction.

CHEMICAL SYNTHESIS

Most chemical syntheses of 3-deoxy-D-*arabino*-heptulosonates require expensive 2-deoxy-D-glucose as starting material. In contrast, a new chemical synthesis (Fig. 4.4; see page 99) of intermediate **12** utilizes D-arabinose **17** as a comparatively inexpensive, simple carbohydrate starting material. Reaction of the D-arabinose with 1,3-propanedithiol to obtain dithioacetal **18** is followed by benzylation of the hydroxyl

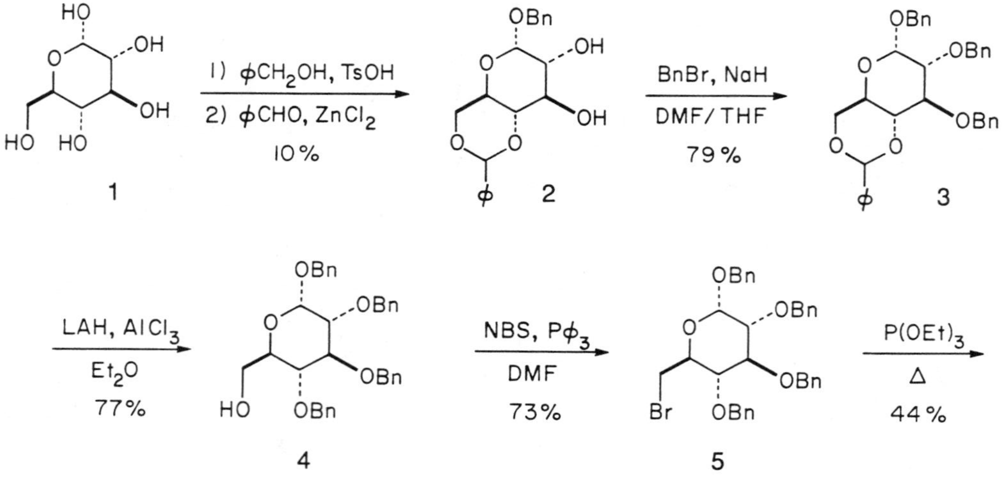

HO
OH
OH
HO
HO
1
1) φCH2OH, TsOH
2) φCHO, ZnCl2
10%
OBn
OH
OH
φ
2
BnBr, NaH
DMF/THF
79%
OBn
OBn
OBn
φ
3
LAH, AlCl3
Et2O
77%
OBn
OBn
OBn
HO
OBn
4
NBS, Pφ3
DMF
73%
OBn
OBn
OBn
Br
OBn
5
P(OEt)3
Δ
44%

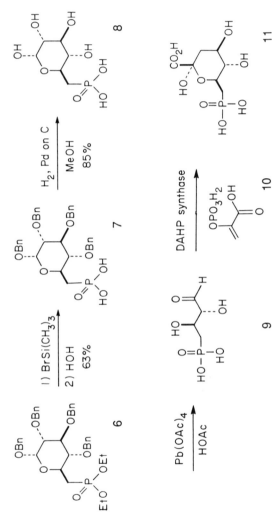

Figure 4.2. Construction of 3-deoxy-D-*arabino*-heptulosonate 7-phosphonic acid **11** where chemical synthesis precedes biocatalysis. D-Glucose **1** is chemically converted to D-erythrose 4-phosphonic acid **9** which is then coupled with phosphoenolpyruvate **10** in a reaction catalyzed by the enzyme DAHP synthase.

97

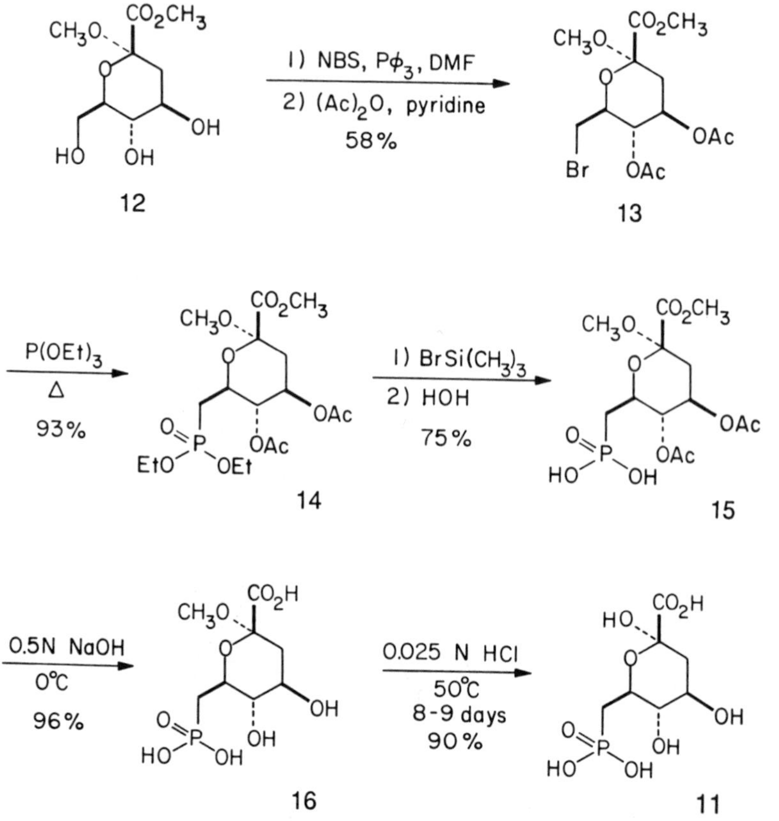

Figure 4.3. Chemical conversion of methyl (methyl 3-deoxy-D-*arabino*-hep-tulopyranosid)onate **12** to 3-deoxy-D-*arabino*-heptulosonate 7-phosphonic acid **11**. Starting material **12** can be derived from chemical synthesis, enzymatic synthesis, or microbial whole cell synthesis.

groups to yield fully protected D-arabinose **19**. Oxidative removal of the dithioacetal gives acyclic, protected D-arabinose **20** which will provide five of the backbone carbons and all but one of the stereocenters of intermediate **12**. The remainder of the backbone carbons as well as the 3-deoxy-2-ulosonic acid follows from condensation of acyclic, protected **20** with the anion of protected phosphonoglycolate triesters **24** (Table 4.1). Product 3-deoxy-D-*arabino*-heptulosonic acid (DAH) is obtained as acyclic, enol ether **21**. Deprotection requires hydrolysis of

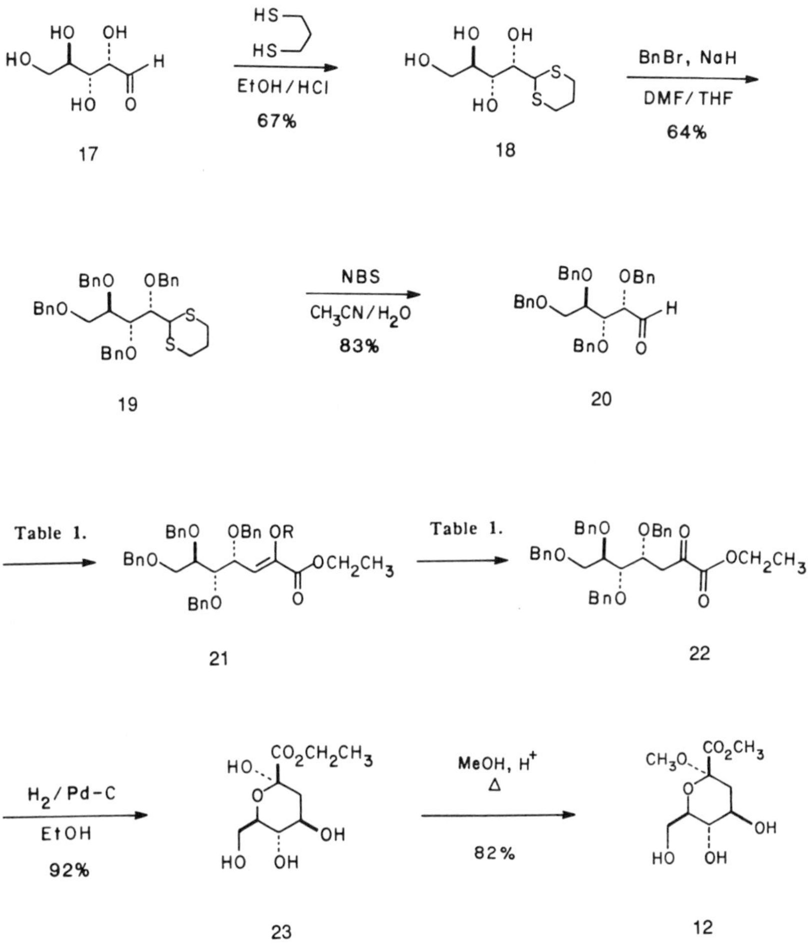

Figure 4.4. Chemical synthesis of methyl (methyl 3-deoxy-D-*arabino*-heptu-lopyranosid)onate **12** from D-arabinose **17**. With this route, biocatalysis is avoided during construction of phosphonic acid analog **11**.

the enol ether and removal of the benzyl ethers of **22** to yield DAH ethyl ester **23**. Reaction of **23** in refluxing acidic methanol yields methyl (methyl 3-deoxy-D-*arabino*-heptulopyranosid)onate **12**.

Phosphonoglycolate triesters used in the chain extension reaction are derived from tartaric acid (Nakamura 1981; Horne, Gaudino, and

Table 4.1 Chain extension of protected D-arabinose.

$(CH_3O)_2P$—C(=O)—CH(OR)—OCH_2CH_3 (24)	CONDENSATION[a] $20 + 24 \rightarrow 21$ % YIELD	DEPROTECTION[b] $21 \rightarrow 22$ % YIELD
a —Si+	77	88
b —C(=O)CH₃	89	81
c —C(=O)—O—C(CH₃)₂—CCl₃	74	64
d —C(CH₃)₂—OCH₃ (with CH₃)	—	59

[a] General procedure for condensation: Phosphonate **24** (1.20 mmol) in 0.5 ml of tetrahydrofuran (THF) was added dropwise to a stirred solution of LiN(TMS)₂ (1.30 mmol) in 2 ml of THF at −45°C and then for 10 min at −78°C. Aldehyde **20** (0.510 g, 1.00 mmol) in 1.0 ml of THF was added and the solution stirred for 30 min at −78°C, 30 min at 0°C, and a further 30 min at room temperature. The mixture was poured into ether and sequentially extracted with 10% HCl, saturated sodium bicarbonate, and brine. The organic layer was dried (MgSO₄) and the solvent removed under reduced pressure. In all cases purification by flash chromatography (hexane/ethyl acetate, 9:1, yielded pure enol ether **21a–d**.

[b] Deprotection of **21a**: Enol ether **21a** (0.483 g, 0.680 mmol) in 1 ml of chloroform and 4 ml of pyridine was cooled in an ice bath under a nitrogen atmosphere. After the addition of 0.40 ml of HF in pyridine, the solution stirred for 1 h, and was then poured into water. The aqueous layer was extracted twice with chloroform and the combined organic layers dried (MgSO₄). The product 2-keto ester **22** was obtained after concentration and flash chromatography (hexane/ethyl acetate, 9:1, vol/vol) as a clear oil. Deprotection of **21b**: Enol ether **21b** (0.320 g, 0.501 mmol) was treated with 15 ml of freshly prepared 50 mM sodium ethoxide at 4°C for 10 min. The solution was neutralized with Dowex 50 (H⁺ form) and decanted into ethyl acetate. The organic layer was extracted three times with brine, dried (MgSO₄), and then purified by flash chromatography (hexane/ethyl acetate, 9:1, vol/vol) to obtain pure **22**. Deprotection of **21c**: See Horne, Gaudino, and Thompson (1984). Deprotection of **21d**: See text.

Thompson 1984). Diethyl tartrate is oxidized with periodic acid to yield glyoxylate ester which is condensed with dimethyl phosphite to obtain the unprotected phosphonomethyl triester. Choice of the protecting group for the hydroxyl group is a critical consideration. As α-hydroxyl organophosphonates, phosphonoglycolate triesters are prone to acid- and base-catalyzed carbon to phosphorus bond cleavage. Such conditions have to be avoided while attaching the hydroxyl protecting groups during phosphonoglycolate triester formation. The same protecting groups must be removed from the enol ether product of chain extension. Unmasking of the 3-deoxy-2-ulose portion of fully protected DAH can be complicated by elimination of the protected C-4 hydroxyl group to yield a 3,4-unsaturated-2-ulosonate. These attachment and removal constraints are adequately addressed by protection of the phosphonoglycolate triester alcohol as dimethyl-t-butylsilyl ether **24a**, acetate **24b**, carbonate **24c**, or ketal **24d** (Table 4.1). 2-Methoxyisopropyl ether protection of phosphonoglycolate triester (**24d**) is unique in that a separate deprotection step is not required after chain extension.

Hydroxyl group removal or substitution along with alteration of stereocenters can be accomplished with the synthesis shown in Figure 4.4 by introducing these changes into the five-carbon aldose prior to chain extension. Introducing similar changes into methyl (methyl 3-deoxy-D-*arabino*-heptulopyranosid)onate **12** would require convoluted protection/deprotection schemes. Therefore, as a route to convert a simple carbohydrate into DAHP analogs, the synthesis shown in Figure 4.4 is likely to be very useful. On the other hand, Figure 4.4 synthesis of methyl (methyl 3-deoxy-D-*arabino*-heptulopyranosid)onate **12** is a multistep chemical synthesis requiring purifications at each step of the synthesis. This interferes with the researcher's ability to pursue other synthetic tasks concurrent with accumulating the key intermediate **12**. Such labor intensity constitutes a significant shortcoming for construction of DAHP analogs such as phosphonate **11** which are most conveniently derived from methyl (methyl 3-deoxy-D-*arabino*-heptulopyranosid)onate **12**.

IMMOBILIZED ENZYME SYNTHESIS

An alternative to chemical synthesis of 3-deoxy-D-*arabino*-heptulosonate is to assemble a multienzyme system that mimics nature's ability to convert D-fructose into DAHP. The enzymes hexokinase, transketolase, and DAHP synthase are essential to DAHP production by plants, bacteria, and fungi. Transketolase and hexokinase are available commercially. DAHP synthase can be easily purified from an engineered strain of *Escherichia coli* which overproduces the tyrosine sensitive isozyme of DAHP synthase (Ogino et al. 1982). Using these enzymes and the natural biosynthesis of DAHP as a model, the enzyme synthesis shown in Figure 4.5 was assembled as a single reaction vessel process (Reimer et al. 1986). Each of the enzymes shown in Figure 4.5 is immobilized on a polymeric support.

The first step in Figure 4.5 synthesis of DAHP is hexokinase-catalyzed conversion of D-fructose to its C-6 phosphate monoester utilizing catalytic quantities of ATP. ADP, as it is formed, is rephosphorylated by PEP in a reaction catalyzed by immobilized pyruvate kinase. Transketolase-catalyzed transfer of the C-1, C-2 ketol portion of D-fructose 6-phosphate to D-ribose 5-phosphate results in formation of D-sedoheptulose 7-phosphate and D-erythrose 4-phosphate. Thiamine pyrophosphate as a cofactor is essential to the ketol transfer catalyzed by transketolase. Subsequent condensation of the D-erythrose 4-phosphate with PEP is catalyzed by DAHP synthase to yield the desired DAHP. When the DAHP production is complete, centrifugation separates the immobilized enzymes from the DAHP-containing supernatant. Recovered immobilized enzymes can be stored at refrigerator temperature and then reused on resuspension in water with addition of cofactors and cosubstrates.

Conversion (Fig. 4.6) of the immobilized enzyme-produced DAHP to a synthetically useful form begins with addition of immobilized alkaline phosphatase to the crude supernatant. After the DAHP has been completely converted to DAH, the immobilized phosphatase is recovered by centrifugation and the DAH purified from the crude supernatant with anion-exchange chromatography. This DAH is converted to methyl (methyl 3-deoxy-D-*arabino*-heptulopyranosid)onate **12** in refluxing acidic methanol followed by crystallization of the key intermediate. Complete anion-exchange chromatographic purification of DAH is not essential due to the significant purification achieved during crystallization of key intermediate **12**.

Two molecules, D-erythrose 4-phosphate and PEP, occupy pivotal positions in the immobilized enzyme synthesis. Aldoses and aldose phosphates such as D-erythrose 4-phosphate are very difficult to ma-

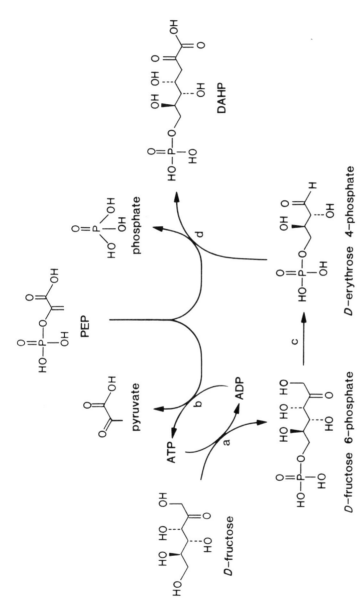

Figure 4.5. Immobilized enzyme synthesis of DAHP from D-fructose. The reactions catalyzed by pyruvate kinase and DAHP synthase, both of which utilize phosphoenolpyruvate as a substrate, are responsible for the high yielding conversion of starting material into product. Enzymes: (a) hexokinase; (b) pyruvate kinase; (c) transketolase; (d) DAHP synthase.

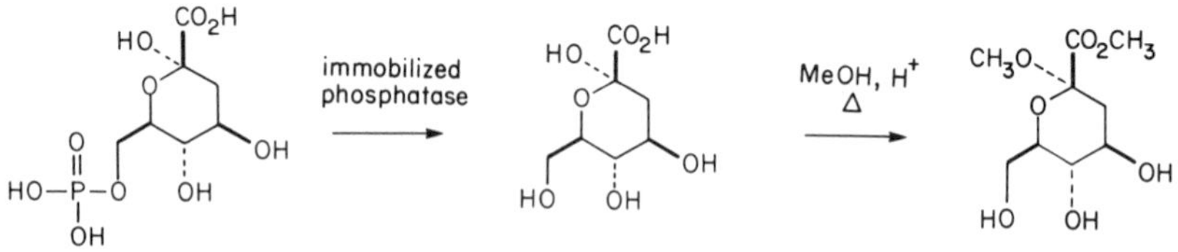

Figure 4.6. Conversion of immobilized enzyme-produced DAHP into a form suitable for subsequent chemical synthesis of 3-deoxy-D-*arabino*-heptulosonate 7-phosphonic acid **11**. Immobilized phosphatase removes the phosphate monoester of DAHP. The resulting DAH can then be converted into crystalline methyl (methyl 3-deoxy-D-*arabino*-heptulopyranosid)onate **12**.

nipulate (Duke, MacLeod, and Williams 1981). Dimerization and polymerization set strict limits on the pH, temperature, and concentration at which D-erythrose 4-phosphate exists in monomeric form. Within these constraints the immobilized enzyme synthesis shown in Figure 4.5 is ideal because D-erythrose 4-phosphate is enzymatically generated under conditions of high dilution. While the immobilized enzyme synthesis is proceeding, D-erythrose 4-phosphate cannot even be detected. The role of PEP in the enzyme synthesis shown in Figure 4.5 is twofold. Transfer of the phosphate group from PEP to ATP and then on to D-fructose is energetically a very favorable reaction. PEP reaction with D-erythrose 4-phosphate catalyzed by DAHP synthase is also favorable energetically. These two essentially irreversible reactions of PEP largely account for the 85% conversion of starting D-fructose into product DAHP. This is the highest yield of any reported method for DAHP synthesis (Sprinson, Rothschild, and Sprecher 1963; Trigalo, Level, and Szabo 1975; Frost and Knowles 1984; Turner and Whitesides 1989).

However, the immobilized enzyme system cannot be indefinitely reused. Freshly immobilized DAHP synthase usually must be added after the third recycling of the immobilized enzyme system. The assembled immobilized enzyme synthesis is much shorter than the chemical synthesis (Fig. 4.4) of methyl (methyl 3-deoxy-D-*arabino*-heptulo-pyranosid)onate **12**, but setting up this system requires enzyme purification, enzyme immobilization, and cosubstrate preparation. In addition, several steps are required to convert immobilized enzyme synthetic DAHP into intermediate **12**. This results in many of the purely chemical steps of Figure 4.4 being replaced with biochemical operations. The immobilized enzyme synthesis shown in Figures 4.5 and 4.6 is thus a useful complement but is not clearly superior to chemical synthesis in terms of the labor intensity associated with amassing methyl (methyl 3-deoxy-D-*arabino*-heptulopyranosid)onate.

MICROBIAL WHOLE CELL SYNTHESIS

Microbes possess a series of different enzymes that catalyze successive, intermediate steps in the biosynthesis of aromatic amino acids. Introduction of a mutation into the gene that encodes for one of these intermediate enzymes may lead to the host microbe's loss of the encoded, catalytically functional enzyme. For example, *E. coli aroB* is a strain carrying a mutation in the *aroB* gene which results in loss of DHQ synthase activity. *E. coli aroB* cells accumulate the substrate of the missing DHQ synthase and thus constitute an important alternative to immobilized enzyme and chemical synthesis of 3-deoxy-D-*arabino*-heptulosonates (Frost and Knowles 1984; Reimer et al. 1986). Typically, substrates of missing aromatic amino acid biosynthetic enzymes are exported out of the microbial cell into the growth supernatant. Exported substrate will often accumulate to a concentration of 1–2 m*M*. *E. coli aroB* export 3-deoxy-D-*arabino*-heptulosonates (DAH and DAHP) into the growth supernatant as the organism moves into the early stationary portion of its growth curve. DAH is usually present in considerable excess relative to DAHP.

Purification of the DAH begins with removal of the *E. coli* cells via centrifugation. The decanted growth supernatant is passed through a large cation-exchange column and then neutralized with lithium hydroxide. Different solubilities of lithium salts lead to selective precipitation of the inorganic salts on addition of methanol to the neutralized growth supernatant. Anion-exchange chromatography separates the DAH from the remaining components of the growth supernatant. Refluxing this DAH in acidic methanol and crystallization affords pure methyl (methyl 3-deoxy-D-*arabino*-heptulopyranosid)onate **12**.

Growth of *E. coli aroB*, DAH/DAHP accumulation, and DAH purification do take time. However, these operations only require periodic attention. The chemist can be synthesizing analogs of DAHP *while* bringing up more intermediate 12. This stands in marked contrast to chemically synthesizing methyl (methyl 3-deoxy-D-*arabino*-heptulopyranosid)onate 12. Unlike immobilized enzyme synthesis, microbial synthesis does not require enzyme purification, enzyme immobilization, or preparation of cofactors and cosubstrates. The dominance of DAH production by *E. coli arob* precludes the need for phosphatase digestion of DAHP as required at the end of the immobilized enzyme synthesis.

One significant disadvantage of microbial DAH synthesis is that lower quantities of DAH are produced on a per volume basis relative to immobilized enzyme synthesis. Liters of microbial mutant growth supernatant are required to obtain the same amounts of 3-deoxy-D-*arabino*-heptulosonate produced in hundreds of milliliters of immobilized enzyme synthetic supernatant. Microbial growth and DAH accumulation are fortuitously ammenable to large-scale operation although increasing the concentration of DAH produced in *E. coli arob* growth supernatants would be a useful development. Consideration of how aromatic amino acid biosynthesis is regulated in *E. coli* leads to several approaches appropriate for attaining such a goal (Herrmann and Somerville 1983).

The flow of carbohydrates into aromatic amino acid biosynthesis is substantially determined by the activity and in vivo concentration of DAHP synthase, the first enzyme of aromatic amino acid biosynthesis (Herrmann 1983; Camakaris and Pittard 1983). Three isozymes of DAHP synthase are present in *E. coli* and are encoded by the *aroF*, *aroG*, and *aroH* genes. AroF encoded DAHP synthase contributes 20% of the total DAHP synthase in *E. coli* (Herrmann 1983) and is regulated at the transcriptional and posttranslational level by intracellular tyrosine levels. Posttranslational regulation determines the activity of this isozyme via tyrosine's inhibition of the enzyme (Dusha and Dénes 1976; Schoner and Herrmann 1976). The concentration of *aroF* encoded DAHP synthase is transcriptionally regulated by the *tyrR* gene product (Fig. 4.7). Tyrosine binding activates the aporepressor encoded by *tyrR*. Activated repressor, in turn, binds to the *aroF*$_o$ operator portion of the *aroF* gene, preventing RNA polymerase reading of the gene (Garner and Herrmann 1985; Cobbett and Delbridge 1987).

For *E. coli arob* to grow, all three aromatic amino acids must be added to the culture medium. This presents a complication because

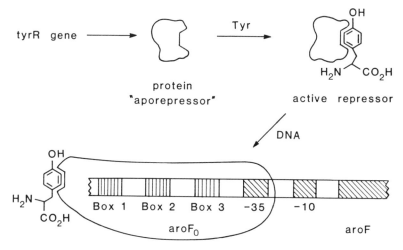

Figure 4.7. Transcriptional control of the tyrosine-sensitive isozyme of DAHP synthase. Boxes 1, 2, and 3 are regions of DNA essential to binding of active repressor (encoded by *tyrR*) to the operator region *aroF*$_0$. The hashed-in areas of DNA designated -35 and -10 are essential to RNA polymerase binding and subsequent transcription of the *aroF* gene. Because of the overlap between these two regions of DNA, active repressor binding interferes with RNA polymerase binding.

tyrosine will also reduce the concentration and activity of the DAHP synthase isozyme, thus restricting the flow of carbohydrates into aromatic amino acid biosynthesis. One approach to increase DAH/DAHP production by *E. coli aroB* is to disarm tyrosine's transcriptional control of the levels of the *aroF* encoded DAHP synthase isozyme. This can be accomplished by introducing multiple copies of the *aroF* gene to ensure that substantial copies of the gene escape binding by the tyrosine-activated *tyrR* gene product. Introduction of pKB45, a multicopy plasmid (Zurawski et al. 1978; Shultz et al. 1984) containing the *aroF* gene, into *E. coli aroB* does result in a fivefold overproduction of DAH/DAHP (Fig. 4.8). Alternatively, the *tryR* gene product could be mutated such that repressor protein is no longer capable of binding to the *aroF*$_0$ region of *aroF*. Levels of DAH/DAHP overproduction similar to those shown in Figure 4.8 are achieved in *E. coli aroB*, *tyrR* (Reimer et al. 1986). Unfortunately, combination of these strategies in *E. coli* (*aroB*, *tyrR*, pKB45) has not yet led to additional levels of DAH/

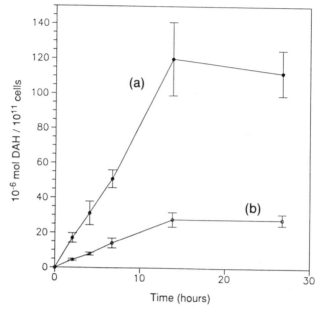

Figure 4.8. Accumulation of DAH/DAHP by [curve (a)] *E. coli aroB* pKB45 and [curve (b)] *E. coli aroB* as a function of time. Bars denote the observed variation (one standard deviation) in DAH/DAHP accumulation by *E. coli* grown from different single colonies. Growth, DAH/DAHP accumulation, and quantitation according to Frost and Knowles (1984) and Reimer et al. (1986).

DAHP overproduction. This is likely due to plasmid instability (Rood, Sneddon, and Morrison 1980).

Another possible limitation on the levels of DAH/DAHP produced by *E. coli aroB* is the intracellular availability of D-erythrose 4-phosphate. Curiously, the concentration of D-erythrose 4-phosphate is so low in many biological systems that it evades detection (Paoletti, Williams, and Horecker 1979). Transketolase apparently has a significant impact on intracellular concentrations of D-erythrose 4-phosphate. This is reflected by the need to add aromatic amino acid supplement to the culture medium of *E. coli* BJ502, a mutant possessing reduced concentrations of transketolase (Josephson and Fraenkel 1969, 1974). Presumably, reduced levels of transketolase in *E. coli* BJ502 lower the intracellular concentration of D-erythrose 4-phosphate to a point that insufficient DAHP is generated to fulfill the aromatic nutritional requirement of the microbe. Introduction into BJ502 of low copy cosmid pKD44b removes the requirement for aromatic amino acid supplementation (Pompliano 1987). Cosmid pKD44b carries the gene that encodes *E. coli* transketolase.

To gauge the impact of transketolase levels on DAH/DAHP production, a mutation was first introduced into the *aroB* gene of *E. coli* BJ502. This construct, as expected, accumulated 3-deoxy-D-*arabino*-heptulosonates in its growth medium (Fig. 4.9). The micromoles of DAH/DAHP produced per cell reflect the reduced level of D-erythrose 4-phosphate that is available in *E. coli* BJ502 due to the attenuated concentration of transketolase. Cosmid pKD44b was then introduced into *E. coli* BJ502 *aroB*. As can be seen in Figure 4.9, a substantial increase in DAH/DAHP concentration was observed in the growth medium of *E. coli* BJ502 *aroB*, pKD44b. This is the first evidence at the

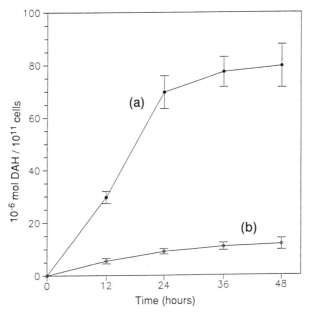

Figure 4.9. Accumulation of DAH/DAHP by [curve (a)] *E. coli* BJ502 *aroB* pKD44b and [curve (b)] *E. coli* BJ502 *aroB* as a function of time. Bars denote the observed variation (one standard deviation) in DAH/DAHP accumulation by *E. coli* grown from different single colonies. Growth, DAH/DAHP accumulation, and quantitation according to Frost and Knowles (1984) and Reimer et al. (1986).

molecular level that the carbon flow into aromatic amino acid biosynthesis is affected by the concentration of transketolase in the intact *E. coli* cell. Work is currently in progress to evaluate the impact on DAH/DAHP production by *E. coli aroB* on simultaneous amplification of the in vivo concentrations of transketolase and DAHP synthase.

FINAL RESULTS

Due to the limited range of substrates tolerated by DAHP synthase, inhibitors of DHQ synthase such as the phosphonic acid analog of DAHP are best constructed via routes where chemical synthesis follows biocatalysis (Fig. 4.3) as opposed to the reversed order of events (Fig. 4.2). Three different approaches have been developed to obtain intermediate methyl (methyl 3-deoxy-D-*arabino*-heptulopyranosid)-onate from simple carbohydrates. Chemical synthesis derives the intermediate from D-arabinose. Immobilized enzyme synthesis can mimic the systems employed by plants, bacteria, and fungi to make DAHP. This route utilizes D-fructose as starting material. Finally, an *E. coli* mutant when grown on D-glucose exports DAH and DAHP into its growth supernatant where these 3-deoxy-D-*arabino*-heptulosonates accumulate. Of the various approaches, whole cell synthesis with *E. coli aroB* is the least labor intensive method for deriving methyl (methyl 3-deoxy-D-*arabino*-heptulopyranosid)onate from simple carbohydrates.

Use of simple carbohydrates as starting materials in chemical synthesis has been crippled by the typically long syntheses that are required to convert carbohydrate to target organic. Use of microbial whole cell synthesis can provide a solution to this problem. The synthesis of methyl (methyl 3-deoxy-D-*arabino*-heptulopyranosid)onate from D-arabinose is an efficient synthesis. Yet this 11-step synthesis with a product purification at the end of each step pales in comparison to the two step *E. coli aroB* synthesis of intermediate which requires only two purifications. The key to the microbial assisted synthesis is that the microbe is performing many of the assembly tasks which are typically executed by the chemist during de novo synthesis. Obviously if microbial assisted synthesis is to be relevant, a molecule targeted for synthesis must be constructed via an intermediate readily derived from biosynthetic enzyme substrate. For synthesis of molecules that inhibit specific biosynthetic enzymes, utilization of the carbohydrate-derived enzyme substrate is an appealing option.

Substrates of biosynthetic enzymes have been largely ignored by synthetic chemists because these molecules are often commercially unavailable or available only in small quantities at exorbitant prices. Delineation of the factors that control carbohydrate-derived carbon flow into various biosynthetic pathways will certainly have an impact on this state of affairs. With *E. coli aroB* production of DAH, various genetic constructs are pointing to what might be engineered to achieve increased accumulation of DAH/DAHP by *E. coli aroB*. Factors identified to date include the concentrations of DAHP synthase and transketolase in *E. coli* cells. Although the increases in DAH/DAHP production achieved in *E. coli aroB* are modest, each advance is a step toward achieving massive overproduction of DAH/DAHP (and likely the rest of the common pathway enzyme substrates) by a microbe grown on simple carbohydrate feedstocks. It seems reasonable that this type of genetically engineered biocatalysis will move the use of simple carbohydrates as starting materials in chemical synthesis into a new era.

ACKNOWLEDGMENT

The research detailed in this chapter was funded by the Searle Scholars Program, the Alfred P. Sloan Foundation, and a Camille and Henry Dreyfus Teacher–Scholar Grant.

REFERENCES

Camakaris, H., and J. Pittard. 1983. Tyrosine biosynthesis. In *Amino Acids: Biosynthesis and Genetic Regulation*, pp. 339–378. (K. M. Herrmann and R. L. Somerville, eds.). Reading, MA.: Addison-Wesley.

Cobbett, C. S., and M. L. Delbridge. 1987. Regulatory mutants of the *aroF-tyrA* operon of *Escherichia coli* K-12. *J. Bacteriol.* **169:** 2500–2506.

Duke, C. C., J. K. MacLeod, and J. F. Williams. 1981. Nuclear magnetic resonance studies of D-erythrose 4-phosphate in aqueous solution. Structures of the major contributing monomeric and dimeric forms. *Carbohydr. Res.* **95:** 1–26.

Dusha, I., and G. Dénes. 1976. Purification and properties of tyrosine-sensitive 3-deoxy-D-*arabino*-heptulosonate 7-phosphate synthetase of *Escherichia coli* K-12. *Biochim. Biophys. Acta* **438:** 563–573.

Frost, J. W., and J. R. Knowles. 1984. 3-Deoxy-D-*arabino*-heptulosonic acid

7-phosphate: chemical synthesis and isolation from *Escherichia coli* auxotrophs. *Biochemistry* **23**: 4465–4469.

Garner, C. C., and K. M. Herrmann. 1985. Operator mutations of the *Escherichia coli aroF* gene. *J. Biol. Chem.* **260**: 3820–3825.

Grossbard, E., and D. Atkinson, 1985. *The Herbicide Glyphosate*. Boston: Butterworths.

Haslam, E. 1974. *The Shikimate Pathway*. New York: John Wiley & Sons.

Herrmann, K. M. 1983. The common aromatic biosynthetic pathway. In *Amino Acids: Biosynthesis and Genetic Regulation*, pp. 301–322. (K. M. Herrmann and R. L. Somerville, eds.). Reading, MA.: Addison-Wesley.

Herrmann, K. M., and R. L. Somerville. 1983. *Amino Acids: Biosynthesis and Genetic Regulation*. Reading, MA.: Addison-Wesley.

Horne, D., J. Gaudino, and W. J. Thompson. 1984. A convenient method for the synthesis of α-ketoesters from aldehydes. *Tetrahedron Lett.* **25**: 3529–3532.

Josephson, B. L., and D. G. Fraenkel. 1969. Transketolase mutants of *Escherichia coli*. *J. Bacteriol.* **100**: 1289–1295.

Josephson, B. L., and D. G. Fraenkel. 1974. Sugar metabolism in transketolase mutants of *Escherichia coli*. *J. Bacteriol.* **118**: 1082–1089.

Le Maréchal, P., C. Froussios, M. Level, and R. Azerad. 1980. The interaction of phosphonate and homophosphonate analogues of 3-deoxy-D-*arabino*-heptulosonate 7-phosphate with 3-dehydroquinate synthetase from *Escherichia coli*. *Biochem. Biophys. Res. Commun.* **92**: 1104–1109.

Le Maréchal, P., C. Froussios, M. Level, and R. Azerad. 1981. Synthesis of phosphono analogues of 3-deoxy-D-*arabino*-hept-2-ulosonic Acid 7-phosphate. *Carbohydr. Res.* **94**: 1–10.

Myrvold, S., L. M. Reimer, D. L. Pompliano, and J. W. Frost. 1989. Chemical inhibition of dehydroquinate synthase. *J. Am. Chem. Soc.* **111**: 1861–1866.

Nakamura, E. 1981. New acyl anion equivalent. A short route to the enol lactam intermediate in cytochalasin synthesis. *Tetrahedron Lett.* **22**: 663–666.

Ogino, T., C. Garner, J. L. Markley, and K. M. Herrmann. 1982. Biosynthesis of aromatic compounds: ^{13}C NMR spectroscopy of whole *Escherichia coli* cells. *Proc. Nat. Acad. Sci. USA* **79**: 5828–5832.

Paoletti, F., J. F. Williams, and B. L. Horecker. 1979. An enzymic method for the analysis of D-erythrose 4-phosphate. *Anal. Biochem.* **95**: 250–253.

Pompliano, D. L. 1987. *Enzyme-Targeted Disruption of Aromatic Amino Acid Biosynthesis in Plants*. Ph.D. Dissertation. Stanford University.

Pompliano, D. L., L. M. Reimer, S. Myrvold, and J. W. Frost. 1989. Probing lethal metabolic perturbations in plants with chemical inhibition of dehydroquinate synthase. *J. Am. Chem. Soc.* **111**: 1866–1871.

Reimer, L. M. 1988. *Inhibitors of Dehydroquinate Synthase: Synthesis and Evaluation of Herbicidal Activity.* Ph.D. Dissertation. Stanford University.

Reimer, L. M., D. L. Conley, D.L. Pompliano, and J. W. Frost. 1986. Construction of an enzyme-targeted organophosphonate using immobilized enzyme and whole cell synthesis. *J. Am. Chem. Soc.* **108:** 8010–8015.

Rood, J. I., M. K. Sneddon, and J. F. Morrison. 1980. Instability in *tyrR* strains of plasmids carrying the tyrosine operon: isolation and characterization of plasmid derivatives with insertions or deletions. *J. Bacteriol.* **144:** 552–559.

Schoner, R., and K. M. Herrmann. 1976. 3-Deoxy-D-*arabino*-heptulosonate 7-phosphate synthase: purification, properties, and kinetics of the tyrosine-sensitive isozyme from *Escherichia coli. J. Biol. Chem.* **251:** 5440–5447.

Shultz, J., M. A. Hermodson, C. C. Garner, and K. M. Herrmann. 1984. The nucleotide sequence of the *aroF* gene of *Escherichia coli* and the amino acid sequence of the encoded protein, the tyrosine sensitive 3-deoxy-D-*arabino*-heptulosonate 7-phosphate synthase. *J. Biol. Chem.* **259:** 9655–9661.

Sprinson, D. B., J. Rothschild, and M. Sprecher. 1963. The synthesis of 3-deoxy-D-*arabino*-heptulosonic acid 7-phosphate. *J. Biol. Chem.* **238:** 3170–3175.

Trigalo, F., M. Level, and L. Szabo. 1975. Phosphorylated sugars. Part XVII. Synthesis of 3-deoxy-D-*arabino*-[1-^{14}C]heptulosonic acid 7-(dihydrogen phosphate). *J. Chem. Soc. Perkin I* 600–602.

Turner, N. J., and G. M. Whitesides. 1989. A combined chemical-enzymatic synthesis of 3-deoxy-D-*arabino*-heptulosonic acid 7-phosphate. *J. Am. Chem. Soc.* **111:** 624–627.

Weiss, U., and J. M. Edwards. 1980. *The Biosynthesis of Aromatic Compounds.* New York: John Wiley & Sons.

Widlanski, T., S. L. Bender, and J. R. Knowles. 1989. Dehydroquinate synthase: a sheep in wolf's clothing? *J. Am. Chem. Soc.* **111:** 2299–2300.

Zurawski, G., K. Brown, D. Killingly, and C. Yanofsky. 1978. Nucleotide sequence of the leader region of the phenylalanine operon of *Escherichia coli. Proc. Natl. Acad. Sci. USA* **75:** 4271–4275.

5
The Production of Amino Acids by Transamination

STEVEN P. CRUMP, JEFFERY S. HEIER, AND J. DAVID ROZZELL

Amino acids have numerous industrial applications, including use as feed additives, ingredients in infusion solutions and other pharmaceutical products, components in flavor compositions, and chiral intermediates. Although amino acids are produced by a range of methods including chemical and fermentation techniques, biocatalytic routes are being proposed for the production of L-amino acids with increasing frequency. For example, Chibata and co-workers as well as several other groups have developed processes for the production of L-aspartic acid from ammonium fumarate catalyzed by whole cells containing the enzyme aspartase (Tosa et al. 1973; Wood and Calton 1984; Fusee et al. 1981). Genex Corporation commercialized a process for the production of L-phenylalanine based on the stereoselective addition of ammonia to trans-cinnamic acid catalyzed by the enzyme phenylalanine ammonia lyase (Hamilton et al. 1985). The enzyme serine hydroxymethyltransferase has been shown to be useful in the production of L-serine from glycine and formaldehyde (Hamilton et al. 1985). Porcine aminoacylase has been used to resolve D,L mixtures of certain amino acids such as alanine, tryptophan, valine, and methionine (Leuchtenberger et al. 1984). This chapter will focus on the application of a class of enzymes known as aminotransferases (transaminases) to the production of amino acids; depending on the availability of precursor, transamination has potential broad applicability to the production of a number of different amino acids, both naturally occurring and nonnatural.

MECHANISM OF TRANSAMINATION

Transaminases catalyze the transfer of an amino group from an L-amino acid donor to a 2-ketoacid acceptor. This amino group transfer is mediated by the cofactor pyridoxal phosphate, which is reversibly bound to the enzyme through a Schiff-base linkage to the ε-amino group of an active-site lysine. The mechanism of the reaction is well understood as a result of the detailed studies of Meister (Meister 1955, 1956; Christen and Meister 1985).

Transaminases characteristically exhibit ping-pong kinetics (Christen and Meister 1985). Mechanistically, the transamination reaction can be thought of as the result of two discrete steps: the first step is the transfer of an amino group from the amino group donor to pyridoxal phosphate, generating a 2-ketoacid byproduct which dissociates from the enzyme and an enzyme-bound pyridoxamine phosphate intermediate; the second step involves the transfer of the amino group from the pyridoxamine phosphate to the 2-ketoacid acceptor, regenerating the pyridoxal phosphate cofactor and producing the corresponding amino acid product. The net result is that a desired L-amino acid can be produced from a given 2-ketoacid precursor using an inexpensive L-amino acid as the amino group donor (Fig. 5.1). As a co-product of this reaction, a second 2-ketoacid is produced in equimolar amounts along with the desired L-amino acid.

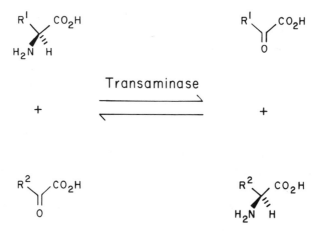

Figure 5.1. General reaction catalyzed by transaminases.

To gain further understanding of the catalytic mechanism of trans-aminases, the crystal structure of the aspartic transaminase from *E. coli* has recently been determined (Smith et al. 1986), and studies are underway using recombinant DNA techniques to investigate the effect of changes in the primary structure of this enzyme on its properties (Cronin et al. 1987).

ADVANTAGES AND DISADVANTAGES OF TRANSAMINASES FOR BIOCATALYTIC REACTIONS

Transaminases have been known since the work of Needham (Needham 1930) and Szent-Gyorgi (Annau et al. 1936) in the 1930s. These enzymes have been found widely in nature and are known to catalyze key steps in the biosynthesis and metabolism of amino acids. In principle, almost any amino acid can be produced by transamination from the appropriate 2-ketoacid.

Among the advantages of using transaminases as biocatalysts are the following:

1. Optically active amino acids are produced stereoselectively from achiral 2-ketoacids; no resolution of the product is required.
2. In most cases the catalytic rates of the enzyme-catalyzed reactions are relatively rapid.
3. A large number of the required 2-ketoacid precursors can be produced inexpensively by conventional chemical synthesis.
4. Transaminases are potentially applicable to the production of a wide range of D- and L-amino acids, both naturally occurring and nonnatural, because enzymes are known with different and wide-ranging specificities, e.g., for D-amino acids; L-amino acids; and amino acids with aromatic side chains, acidic side chains, branched alkyl side chains, etc.
5. The 2-ketoacid byproducts may also have significant value in ad-dition to the L-amino acids themselves.

As an example of a simple transamination process, we have studied the production of L-alanine and 2-ketoglutarate from the inexpensive precursors L-glutamate and pyruvic acid (Fig. 5.2). This reaction is catalyzed with a high degree of specificity by the enzyme glutamic–

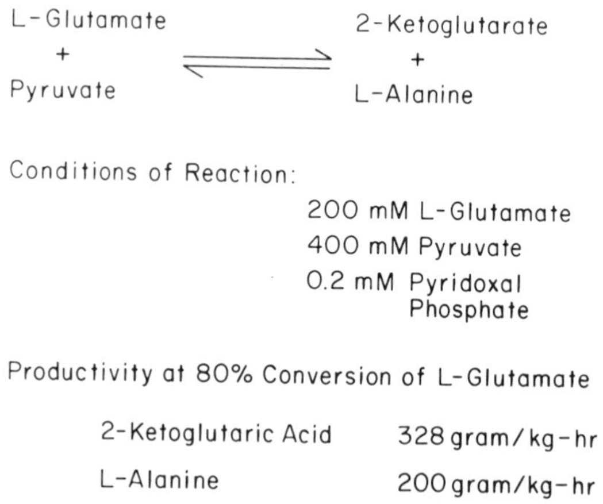

L-Glutamate　　　　　　　2-Ketoglutarate
　　+　　⇌　　　　+
Pyruvate　　　　　　　　L-Alanine

Conditions of Reaction:

　　200 mM L-Glutamate
　　400 mM Pyruvate
　　0.2 mM Pyridoxal
　　　　Phosphate

Productivity at 80% Conversion of L-Glutamate

　　2-Ketoglutaric Acid　　328 gram/kg-hr

　　L-Alanine　　　　　　200 gram/kg-hr

Figure 5.2. Scheme for the production of L-alanine and 2-ketoglutarate by transamination.

pyruvic transaminase (GPT). The porcine enzyme is available commercially.

The reaction to produce L-alanine and 2-ketoglutarate from pyruvate and L-glutamate slightly favors the desired products thermodynamically. The equilibrium constant for the reaction was measured by incubating starting mixtures of L-glutamate and pyruvate or L-alanine and 2-ketoglutarate with GPT in the presence of 0.1 mM pyridoxal phosphate at pH 7.5 and allowing the reaction to proceed until no further change in concentration of the components could be detected. The results shown in Table 5.1 give rise to a calculated K_{eq} of 1.9.

Table 5.1. Determination of K_{eq} for glutamate-pyruvate transamination.

	T_O	T_F	T_O	T_F	T_O	T_F
[L-Glutamate]	200	49	100	40	0	40
[Pyruvate]	400	147	100	42	0	40
[L-Alanine]	0	152	0	58	100	59
[2-Ketoglutarate]	0	151	0	60	100	60

Concentrations in mM.

The enzyme was immobilized by covalent attachment to a porous silica support material (Weetall and Filbert 1974). In a typical experiment, aminopropyl silica (Corning) was suspended in an aqueous solution, pH 7.0, containing ethyl dimethylaminopropyl carbodiimide (10 mg/ml) and pyridoxal phosphate (0.5 mM). Glutamic–pyruvate transaminase was added, and the mixture was mixed gently for 1 h at room temperature. At the end of this time, the silica particles were washed with phosphate buffer (pH 7.0) followed by 0.2 M NaCl solution to remove any enzyme that had been adsorbed but not covalently bound. The results for a typical experiment are summarized in Table 5.2.

When the immobilized GPT was used for the continuous production of L-alanine and 2-ketoglutarate, the porous silica particles were loaded into a column and substrate was pumped upward through the column with a peristaltic pump. The reaction was monitored by quantitating in parallel the amount of pyruvate consumed using lactate dehydrogenase in the presence of NADH and the amount of 2-ketoglutarate produced using the enzyme saccharopine dehydrogenase in the presence of NADH and L-lysine (Rozzell 1987). L-Alanine was determined by amino acid analysis of the effluent of the immobilized enzyme column.

A column containing 500 mg of immobilized GPT was operated continuously for 6 months at 25°C. The substrate solution contained 200 mM L-glutamate, 400 mM sodium pyruvate, 0.1 mM pyridoxal phosphate, pH 7.5. Under these conditions, approximately 75% of the limiting reactant (L-glutamate) was converted to 2-ketoglutarate along with the production of stoichiometric amounts of L-alanine. The immobilized GPT biocatalyst produced 100 mg of L-alanine and 164 mg of 2-ketoglutarate per hour. After over 200 days of continuous operation, the biocatalyst retained approximately 50% of its activity (see Fig. 5.3).

Table 5.2.   Immobilization of glutamic–pyruvic transaminase.

Aminopropyl silica weight	500 mg
Enzyme offered	30 mg
Activity offered	1,530 U
Enzyme bound	10 mg
Activity bound	200 U

Activity retained after immobilization: 40%

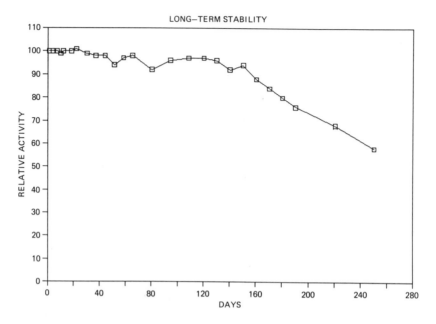

Figure 5.3. Stability of glutamic–pyruvic transaminase coupled to porous glass.

DRIVING THE REACTION TO COMPLETION

There is one disadvantage to the transamination reaction as described to this point: because the transamination involves an amino acid reacting with a 2-ketoacid to generate products consisting of a 2-ketoacid and an amino acid, the equilibrium constant for the general reaction would be expected to be near unity. Indeed, the K_{eq} measured for the glutamic–pyruvic transamination was 1.9. As a result, the net conversion of substrates to products for the general transamination process is thermodynamically limited. A key to the development of an efficient transamination system lies in overcoming the problem of incomplete conversion of the 2-ketoacid precursor to the desired L-amino acid product.

Our approach to this problem is to couple the transamination reaction to a second reaction that consumes the byproduct ketoacid in an essentially irreversible step; this has the effect of driving the transamination reaction to completion. We accomplished this objective by making one important observation: for certain transaminases, L-aspartic acid can function in place of L-glutamic acid as the amino group donor. Using the glutamic–aspartic transaminase from *E. coli*, the rates of transamination of several different 2-ketoacids were compared, including that of phenylpyruvate, *p*-hydroxyphenylpyruvate, and indoyl-3-pyruvate. Both L-glutamate and L-aspartate were tested as the amino donors. In all cases, L-aspartic acid was utilized as a substrate at approximately 70–80% the rate of utilization of L-glutamate.

The important feature of the process is that if L-aspartic acid is used as the amino donor for transamination rather than L-glutamic acid, then the corresponding 2-ketoacid co-product is oxaloacetate (rather than 2-ketoglutarate); oxaloacetate is a β-ketoacid and may be easily decarboxylated to pyruvate. This decarboxylation can be accomplished thermally, chemically, or as shown in Figure 5.4, enzymatically using the enzyme oxaloacetate decarboxylase. The essentially irreversible decarboxylation of oxaloacetate to pyruvate drives the entire process to completion, allowing the transamination of 2-ketoacids to L-amino acids in yields approaching 100% of theoretical (Rozzell 1985, 1987; Wood and Calton 1984).

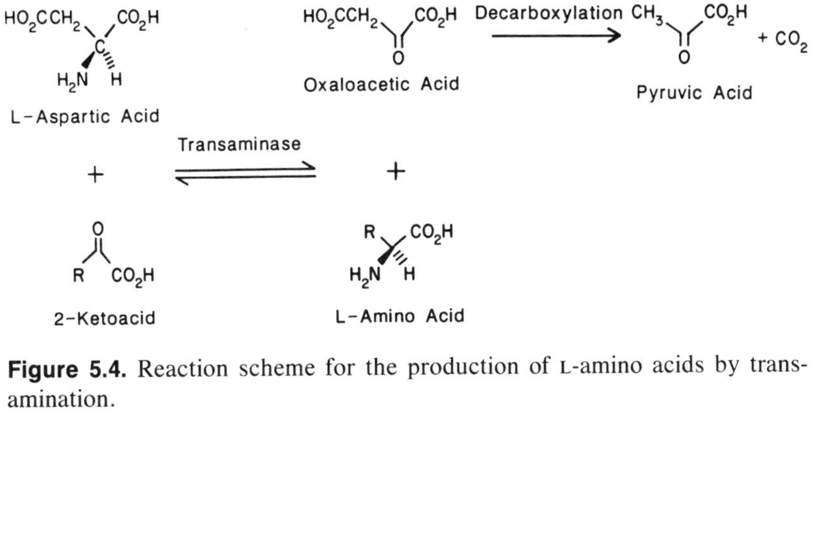

Figure 5.4. Reaction scheme for the production of L-amino acids by transamination.

PRODUCTION OF AMINO ACIDS USING THE ASPARTIC TRANSAMINASE FROM E. COLI

The aspartic transaminase from *E. coli* is the product of the *aspC* gene. The usefulness of this enzyme for amino acid production results from its relatively broad range of substrate acceptability. As shown in Table 5.3, this enzyme catalyzes the transamination of a number of different 2-ketoacids to produce the corresponding L-amino acids in the presence of L-aspartic acid as the amino group donor.

Interestingly, for the production of aromatic amino acids, the aspartic transaminase (EC 2.6.1.1) from *E. coli* is preferable to the enzyme that carries out this function metabolically, the so-called aromatic amino acid transaminase (EC 2.6.1.5). The aspartic transaminase is more stable than the aromatic transaminase, and in addition, its catalytic rate constants for transamination of aromatic 2-ketoacids are similar. Although the aspartic transaminase has approximately a 10-fold higher K_m for the corresponding aromatic 2-ketoacids, the high concentrations of substrates used in a biocatalytic process ensure that the enzyme is functioning at its maximal catalytic rate.

L-Phenylalanine, a component in infusion solutions and an intermediate used in the manufacture of the high-intensity sweetener aspartame, can be produced from the 2-ketoacid phenylpyruvate by the transaminase-catalyzed reaction process just described. Figure 5.5 shows the beneficial effect of decarboxylation in driving the reaction to completion. In this experiment, phenylpyruvate sodium salt and

Table 5.3. Substrate specificity of aspC.

L-AMINO ACID	RELATIVE RATE
L-Phenylalanine	100
L-Tyrosine	130
L-Tryptophan	150–190
L-2-Aminoadipic acid	22
L-4-Phenyl-2-aminobutanoic acid	18
L-*m*-Tyrosine	~40
L-Methionine	20–40
L-Serine	3
L-Cysteine	2–3

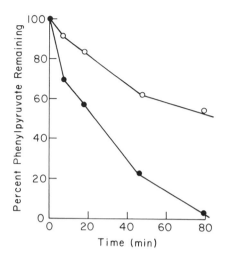

Figure 5.5. Effect of oxaloacetate decarboxylase on the conversion of phenyl-pyruvate to L-phenylalanine by transamination. (Upper): Transaminase alone. (Lower): Transaminase plus oxaloacetate decarboxylase. $[Mg^{2+}]$ = 6 mM; [Tris] = 50 mM; pH = 7.5.

L-aspartate were incubated with *E. coli* aspartate transaminase (EC 2.6.1.1) at room temperature at pH 7.5 in both the presence and absence of oxaloacetate decarboxylase isolated from *Pseudomonas putida*. Magnesium ion, which is a cofactor for the decarboxylase, was also present in both reaction mixtures. The transamination reaction was monitored by following the disappearance of phenylpyruvate spectrophotometrically. The reaction in which oxaloacetate decarboxylase was included to accelerate the decarboxylation of oxaloacetate to pyruvate proceeded to completion much more rapidly than the reaction without the decarboxylase.

IMMOBILIZATION OF THE TRANSAMINASE

Immobilization of the aspartic transaminase from *E. coli* has been accomplished by several different methods. A typical procedure involves the covalent binding of the enzyme to a glutaraldehyde-activated silica-ceous support. Sixteen grams of a polyvinylchloride-silica composite that had been pretreated with polyethyleneimine was washed with 250 ml of 50 mM potassium phosphate buffer, pH 7.0, and then contacted for 1 h with 250 ml of 5% glutaraldehyde (wt/vol) dissolved in the same buffer. The support was next washed with potassium phosphate buffer until no glutaraldehyde could be detected in the effluent by color reaction with 2,4-dinitrophenylhydrazine. A solution of the transaminase in pH 7.0 potassium phosphate buffer containing 0.5 mM pyridoxal phosphate and 10 mM 2-ketoglutarate (150 ml, 10 mg/ml of protein) was recirculated through the support for 90 min at room temperature, and the immobilized enzyme was then washed with 500 ml of potassium phosphate buffer containing 0.3 M NaCl to remove any enzyme that had been adsorbed but not covalently attached. The results of the immobilization of *aspC* transaminase are shown in Table 5.4.

The immobilized transaminase was used to study the production of L-phenylalanine. A substrate solution containing the following composition was used for the experiments:

Sodium phenylpyruvate	200 mM
L-Aspartate	300 mM
Pyridoxal phosphate	0.1 mM
MgCl$_2$	100 mM
pH	7.0–7.5
Temperature	37°C

Under these conditions, phenylpyruvate was converted to L-phenylalanine in 90% yield using an apparatus in which the substrate solution was recycled through the immobilized enzyme at a flow rate of approximately 15 ml/min. The reaction was monitored by removing aliquots at various time intervals and assaying for phenylpyruvate remaining and L-phenylalanine produced (Rozzell 1987). The productivity of the biocatalytic system was 0.51 g of L-phenylalanine produced per gram total biocatalyst weight per hour (510 g/kg biocatalyst-hour).

Table 5.4. Immobilization of *aspC* transaminase.

Protein offered	1.5 g
Activity offered	120,000 U
Protein bound	1.22 g
Activity of immobilized enzyme	78,100 U

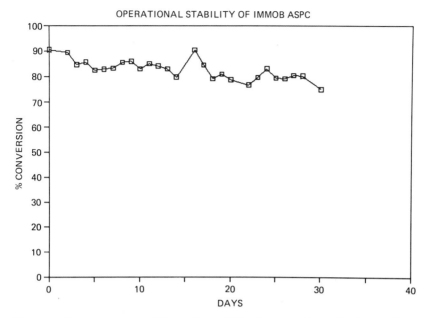

Figure 5.6. Long-term stability of immobilized transaminase for the production of L-phenylalanine.

The stability of the biocatalyst was measured under the operating conditions described above. For these experiments, fresh batches of the substrate solution shown above were prepared every 48 h. The temperature of the reaction was maintained at 37°C. The substrate solution was pumped continuously through the immobilized enzyme matrix, and the effluent was assayed periodically for L-phenylalanine content to determine the residual transaminase activity. The results are shown in Figure 5.6. The half-life of the immobilized transaminase under these conditions was found to be approximately 1,500–2,000 h.

PRODUCTION OF BRANCHED-CHAIN L-AMINO ACIDS

One commercially interesting group of L-amino acids on which the aspartic transaminase has little or no activity is the group of branched-chain amino acids consisting of L-leucine, L-isoleucine, and L-valine. For the production of these compounds it was necessary to identify another transaminase; we isolated a potentially useful enzyme from *E.*

coli, the so-called branched-chain amino acid transaminase, encoded by the *ilvE* gene (Lee-Peng et al. 1979).

After isolation of the *E. coli* branched-chain transaminase from wild-type *E. coli* K-12 grown on a rich medium, we found that the ketoacids 2-ketoisocaproic acid, 2-keto-3-methylbutanoic acid, and 2-ketoisovaleric acid could all be transaminated by the amino donor L-glutamic acid to L-leucine, L-isoleucine, and L-valine, respectively. Unfortunately, L-aspartic acid proved to be completely inert as the amino donor for this particular transaminase; the enzyme had a strict requirement for L-glutamate as substrate. The drawback to the use of L-glutamate was that the reaction was equilibrium limited in the production of the desired branched-chain L-amino acids.

Despite the narrowness in substrate specificity of the enzyme for the donor amino acid, a scheme was devised to overcome this apparent problem; it is outlined in Figure 5.7 using L-valine as an example. The conversion makes use of a coupled three-step reaction system. In the first reaction, the *ilvE* branched-chain transaminase catalyzes the reaction of L-glutamate wiht 2-ketoisovalerate to produce 2-ketoglutarate and L-valine. In the same reaction mixture, the *aspC* transaminase is used to catalyze the transamination between L-aspartate and 2-ketoglutarate to produce oxaloacetate and reform the L-glutamate to participate in another transamination step. This allows the L-glutamate and 2-ketoglutarate to be continuously recycled as the reaction proceeds, and renders L-glutamate catalytic rather than stoichiometric in the conversion of branched-chain ketoacids to the corresponding L-amino acids. Finally, oxaloacetate, produced as a byproduct of the transamination of L-aspartate, is decarboxylated to pyruvate; as before, the essentially irreversible decarboxylation drives the entire sequence of reactions to completion. The net reaction is the transamination of 2-ketoisovalerate to L-valine with L-aspartate, using 2-ketoglutarate as an intermediary amino transfer agent.

Using this type of reaction scheme branched-chain amino acids can be produced efficiently. There is no detectable inhibition of the branched-chain transaminase by L-aspartate; in addition, there is only a small amount of cross-reactivity of the *ilvE* transaminase with pyruvate, giving rise to L-alanine. The enzyme was immobilized as described earlier for the aspartic transaminase, and the immobilized biocatalyst was found to be stable. The half-life of the immobilized ilvE

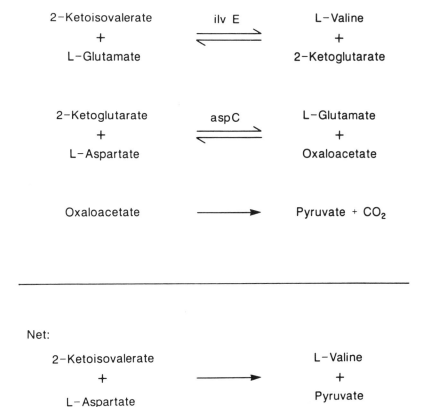

Figure 5.7. Coupled enzyme system for the production of L-alanine.

transaminase was approximately 3 months at 37°C under operational conditions. The enzyme was immobilized with approximately a 40% retention of activity. Solutions containing 200–600 mM 2-ketoisovalerate were transaminated to L-valine, using an equimolar concentration of L-aspartic acid and a concentration of L-glutamate of 50–100 mM. The conversion of 2-ketoisovalerate to L-valine was approximately 90% (Rozzell 1989).

A special case of this coupled transaminase system is illustrated by the production of L-alanine as shown in Figure 5.8. In this sequence of

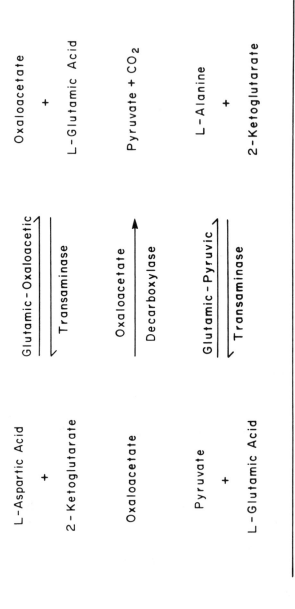

Figure 5.8. Coupled enzyme system for the production of L-alanine.

reactions, the reaction mixture contains L-aspartate in high concentrations, and the reaction is initiated with a small amount of pyruvate and L-glutamate. Glutamic-pyruvic transaminase catalyzes the formation of 2-ketoglutarate and L-alanine. The 2-ketoglutarate is recycled to L-glutamate by transamination with L-aspartate using a glutamate transaminase such as the *aspC* gene product. Again as before, the reaction is driven to completion by decarboxylation of oxaloacetate, which in this case also continuously supplies the starting 2-ketoacid pyruvate to be transaminated to L-alanine. The net reaction is the decarboxylation of L-aspartate to L-alanine and CO_2. Although other superior methods exist for the production of L-alanine in practice, this sequence of reactions illustrates nonetheless the potential for coupled enzymatic reaction systems in the production of amino acids.

ENZYME PRODUCTION

Transaminases are indeed widely distributed in nature, but the normal levels of enzymatic activity are too low to allow the economical production of large quantities of enzyme for the preparation of biocatalysts. To facilitate the production of both the *aspC* and *ilvE* transaminases, the genes encoding both transaminases were cloned from *E. coli* K-12. The *aspC* transaminase subunit contains 396 amino acids and exists as a homodimer (Kuramitsu et al. 1985a); the *ilvE* transaminase exists as a hexamer of identical subunits containing 309 amino acids (Kuramitsu et al. 1985b). There is little sequence homology between the two enzymes.

For production of the desired enzymes, the wild-type regulatory and promoter sequences were removed from the 5'-portion of the genes, and these were replaced with sequences for a desired promoter system. One promoter that may be used is the P_L promoter, a strong, temperature-inducible promoter system. The *aspC* and *ilvE* genes from *E. coli* K-12 were placed in a direct reading frame with the P_L promoter using standard techniques (Maniatis et al. 1982). *E. coli* host strains harboring plasmids containing ampicillin resistance and the desired transaminase genes under the control of the P_L promoter were grown to the desired cell density, and inductions were effected by a temperature shift from 30 to 40°C. Synthesis of the desired enzyme was rapid, reaching maximum levels in approximately 2 h. Figures 5.9 and 5.10

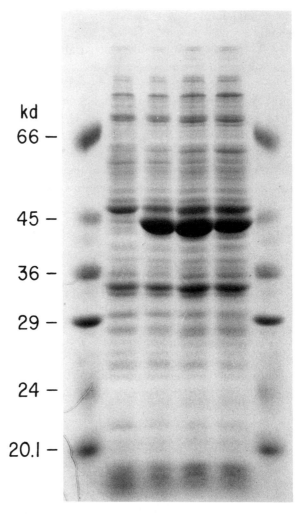

Figure 5.9. Gel electrophoresis of the whole-cell lysate of *E. coli* harboring aspC gene expressed under the P_L promoter.

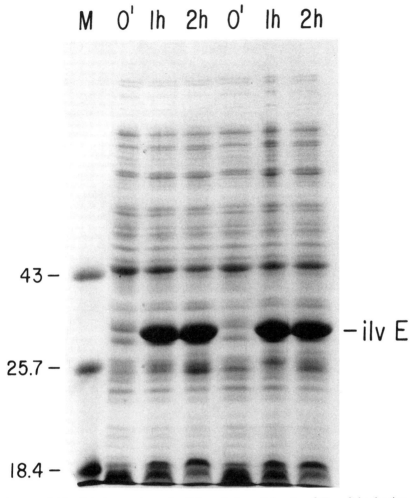

Figure 5.10. Gel electrophoresis of the whole-cell lysate of *E. coli* harboring *ilvE* gene expressed under the P$_L$ promoter.

show the production of the recombinant transaminases. Activity assays confirmed that all of the enzyme produced was soluble and catalytically active. The desired transaminase corresponded to approximately 40% of the total protein produced in the recombinant *E. coli* cell.

CONCLUSIONS

Transaminase enzymes were investigated for the biocatalytic production of amino acids. The aspartic and branched chain transaminases from *E. coli* were immobilized on silica-based support materials and found to be useful for the production of a range of L-amino acids, both naturally occurring and unnatural. The immobilized enzyme systems could catalyze the conversion of 2-ketoacids to the corresponding optically active L-amino acids in yields of greater than 90%. The immobilized enzymes displayed half-lives of greater than 1000 h. The genes encoding the enzymes were cloned to facilitate the production of highly active enzyme preparations. The enzymes could be produced at levels of approximately 40% of the total cell protein in recombinant *E. coli* under the control of the P_L promoter.

ACKNOWLEDGMENTS

We would like to thank our co-workers Jeffery Deetz, Gwynneth Edwards, Jasbir Seehra, Paul Schendel, Tony Schuber, David Hill, and Melinda King-Morris for their experimental work, technical assistance, and valuable suggestions during the course of this work, and Genetics Institute, Inc. for sponsoring this research.

REFERENCES

Annau, E., I. Banga, A. Blazo, V. Bruckner, K. Laki, F. B. Staub, and A. Szent-Gyorgi. 1936. *Z. Physiol. Chem.* **224**: 105.

Christen, P., and D. E. Metzler. 1985. *Transaminases*. New York: John Wiley & Sons.

Cronin, C. N., B. A. Malcolm, and J. F. Kirsch. 1987. *J. Am. Chem. Soc.* **109**: 2222–2223.

Fusee, M. C., W. E. Swann, and G. J. Calton. 1981. *Appl. Environ. Microbiol.* **42**: 672–676.

Hamilton, B. K., H. Y. Hsiao, W. E. Swann, D. M. Anderson, and J. J. Delente. 1985. *Trends Biotechnology* **3**: 64–68.

Kuramitsu, S., S. Okuno, T. Ogawa, H. Ogawa, and H. Kagamiyama. 1985a. *J. Biochem.* **97**: 1259–1262.

Kuramitsu S., T. Ogawa, H. Ogawa, and H. Kagamiyama. 1985b. *J. Biochem.* **97**: 993–999.

Lee-Peng, F.-C., M. A. Hermodson, and G. B. Kohlhaw. 1979. *J. Bacteriol.* **139:** 339–345.

Leuchtenberger, W., M. Karrenbauer, and U. Plocken. 1984. *Ann. NY Acad. Sci.* **434:** 78–86.

Maniatis, T., E. F. Fritsch, and J. Sambrook Molecular Cloning: A Laboratory Manual. Cold Spring Harbor (1982).

Meister, A. 1955. *Adv. Enzymol.* **16:** 185–246.

Meister, A. 1956. *Annu. Rev. Biochem.* **25:** 29–56.

Needham, D. M. 1930. *Biochem. J.* **24:** 208.

Rozzell, J. D. 1985. U.S. Patent 4,518,692

Rozzell, J. D. 1987. *Methods Enzymol.* **136:** 479–497.

Rozzell, J. D. 1989. U.S. Patent 4,826,766.

Smith, D. L., D Ringe, W. L. Finlayson, and J. F. Kirsch. 1986. *J. Mol. Biol.* **191:** 301–302.

Tosa, T., T. Sato, T. Mori, Y. Matuo, and I. Chibata. 1973. *Biotech. Bioengin.* **15:** 69–84.

Weetall, H. H. and A. M. Filbert. 1974. *Methods Enzymol.* **34:** 59–72.

Wood, L. L., and G. J. Calton. 1984. *Biotechnology* **2:** 1081–1084.

Wood, L. L., and G. J. Calton. 1988. U.S. Patent 4,728,611.

6
A Biocatalytic Approach to Vitamin C Production
Metabolic Pathway Engineering of *Erwinia herbicola*

ROBERT A. LAZARUS, JANA L. SEYMOUR, R. KEVIN STAFFORD,
MARK S. DENNIS, MARGERY G. LAZARUS, CARA B. MARKS, AND
STEPHEN ANDERSON

Although the primary focus of the biotechnology industry has been on the overproduction of new proteins, primarily for pharmaceutical purposes, there has been growing interest in other fields such as agriculture, diagnostics, and the biocatalytic production of organic chemicals. We have recently been working to develop a novel biosynthetic process for the production of vitamin C (L-ascorbic acid, ASA). This approach has involved the study of the enzymes, coenzymes, and metabolic pathways of different bacteria with the goal of creating new metabolic routes to make a new product (Fig. 6.1). This metabolic pathway engineering approach, which required the identification and characterization of several new enzymes and the cloning and expression of the gene coding for one of these enzymes, has led to a successful one-step bioconversion of D-glucose (G) into 2-keto-L-gulonic acid (2-KLG), a key intermediate in the synthesis of ascorbic acid.

The strategy for the approach we have taken was constrained by the structure of ASA (Fig. 6.2). First, the synthesis must be chiral because only the L-enantiomer is biologically active and second, the final step should be nonoxidative because ascorbate can be readily oxidized. Finally, the process must be efficient and simple to operate to be economically attractive. Although ASA is a natural product that is amenable to a bioconversion process, it did not seem to be a good final target because it could be oxidatively unstable during the course of a micro-

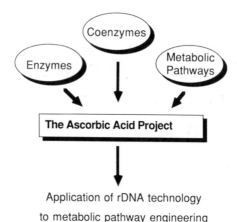

Application of rDNA technology
to metabolic pathway engineering

Figure 6.1. Ascorbic acid project overview.

bial fermentation. We therefore chose the more stable 2-KLG as our target and sought to develop a strategy that could utilize our technology and be commercially viable.

HISTORICAL PERSPECTIVE

The synthesis of ASA has received considerable attention over many years due to its relatively large market volume and high value as a specialty chemical (Crawford and Crawford 1980; Gaffe 1984). At present, most vitamin C is produced by a modification of the Reichstein–Grussner synthesis, first developed in 1934 (Reichstein and Grussner 1934). This process is a somewhat lengthy and capital-inten-

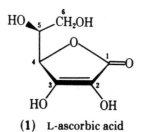

(1) L-ascorbic acid

Figure 6.2. Synthetic considerations for ASA. (1) Must be chiral. (2) Final step should be nonoxidative. (3) Must be economical for commercialization.

sive route that involves the chemical reduction of D-glucose to D-sorbitol, followed by a microbial oxidation to L-sorbose, which is then protected by acetonization, chemically oxidized at C1, and deprotected to give 2-KLG. 2-KLG, a stable compound, can then be readily and efficiently converted into L-ascorbate via either an acid- or base-catalyzed cyclization (Fig. 6.3). Many modifications and improvements to the process over the years have resulted in an overall yield of ASA from glucose of >50%.

While there continues to be improvement in the chemical routes, many of the more recent advances have taken advantage of the ability of microorganisms to carry out the individual steps that are required by the unique stereochemical demands of this molecule (Crawford and Crawford 1980; Kulhanek 1970). An important breakthrough was the discovery of organisms belonging to the coryneform group of bacteria that were capable of carrying out the conversion of 2,5-diketo-D-gluco-

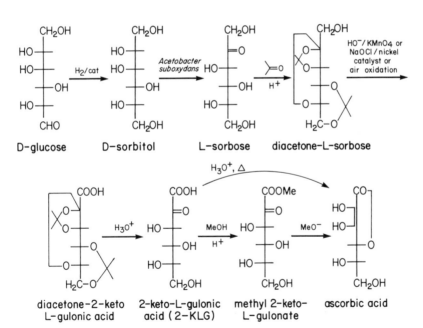

Figure 6.3. Reichstein–Grussner synthesis of ASA (Reichstein and Grussner, 1934; Crawford and Crawford 1980; Jaffe 1984). The curved line from 2-keto-L-gulonic acid to ascorbic acid represents a nonaqueous acid-catalyzed cyclization.

nate (2,5-DKG) to 2-KLG (Sonoyama et al. 1982); recently, this bioconversion has been observed in a number of different bacteria (Sonoyama et al. 1987). In addition, several species of *Acetobacter*, *Gluconobacter*, and *Erwinia* are known to efficiently oxidize D-glucose to 2,5-DKG (Sonoyama et al. 1988; Weenk et al. 1984). These developments have led to a tandem fermentation process that utilizes mutant strains of *Erwinia* and *Corynebacterium*. This process, which has been shown to carry out the glucose to 2-KLG bioconversion (Fig. 6.4) very effectively (Sonoyama et al. 1982), involves an initial fermentation of glucose to 2,5-DKG by *Erwinia*; the 2,5-DKG produced is then fed into a second fermenter containing the *Corynebacterium* sp. and converted into the desired 2-KLG.

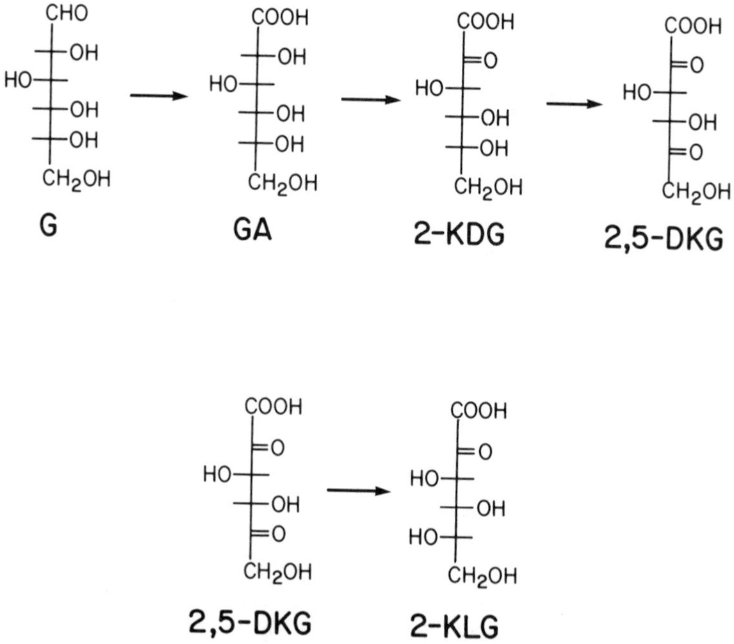

Figure 6.4. The chemical intermediates produced in the tandem fermentation process from (top) D-glucose to 2,5-DKG (*E. herbicola*) followed by (bottom) the conversion to 2-KLG (*Corynebacterium* sp.).

RECOMBINANT STRATEGY

Although the tandem fermentation coupled with one chemical step is a considerable simplification over the multistep Reichstein route, we felt that we could further improve on this process (Anderson et al. 1985). We reasoned that by combining the relevant traits of both the *Erwinia* sp. and *Corynebacterium* sp. into a single organism, we could create a novel two-step ASA process: a fermentation of glucose to 2-KLG catalyzed by a recombinant "metabolically engineered" organism, followed by the chemical conversion of 2-KLG to ASA. Although the conversion of 2,5-DKG to 2-KLG by whole cells of a mutant strain of *Corynebacterium* sp. ATCC 31090 had been demonstrated (Sonoyama et al. 1982), essentially nothing was known about the biochemical pathway(s) responsible for this bioconversion. We speculated that this reaction was catalyzed by a single nicotinamide adenine nucleotide-dependent reductase (2,5-DKGR). To accomplish the construction of the "metabolically engineered" strain, we first purified the 2,5-DKGR that catalyzes the conversion of 2,5-DKG into 2-KLG from *Corynebacterium* sp. and then cloned the gene encoding this enzyme and expressed it in *Erwinia herbicola*, an efficient 2,5-DKG producer. The resulting organism is able to convert glucose to 2-KLG in a single fermentation (Fig. 6.5). Our approach utilizes the same inexpensive starting material as the Reichstein process and a similar 2-KLG to ASA cyclization step; however, it should be considerably less capital intensive and more efficient to operate.

DEVELOPMENT OF THE RECOMBINANT BACTERIAL STRAIN

Selection and Screening Strategies

The first methodology used in attempting to clone the 2,5-DKG reductase in *Corynebacterium* sp. was based on a selection strategy (Fig. 6.6). This relies upon the fact that the *E. herbicola* ATCC21998 host grows on a minimal 2,5-DKG media but does not grow on a minimal 2-KLG media. Therefore expression of the 2,5-DKGR gene in *Erwinia* might allow growth on 2-KLG via oxidation to 2,5-DKG. Recombinant DNA plasmids containing randomly cut *Corynebacterium* DNA were

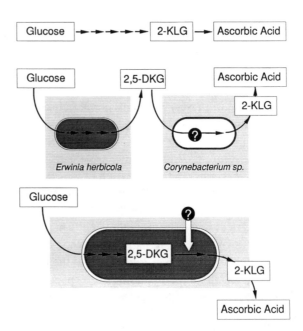

Figure 6.5. Comparison of (top) the Reichstein–Grussner chemical process, (middle) the tandem fermentation process, and (bottom) the recombinant DNA-based process.

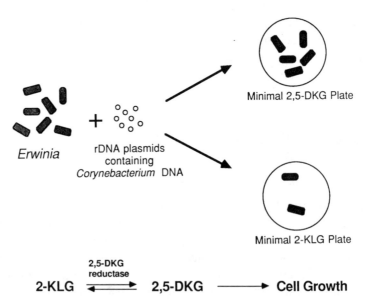

Figure 6.6. Selection strategy for cloning the 2,5-DKG reductase gene from *Corynebacterium* sp.

introduced into the *Erwinia* strain, which was then plated out on minimal 2,5-DKG and 2-KLG media plates; however, no growth on minimal 2-KLG media was observed.

An alternative approach to cloning the 2,5-DKGR gene was based on a screening strategy (Fig. 6.7). The same recombinant DNA plasmids containing randomly cut *Corynebacterium* DNA were introduced into *E. coli*. Transformants were plated out, permeabilized by treatment with $CHCl_3$ vapor, and assayed for activity with a filter paper overlay containing 2-KLG, NADP (known to be the 2,5-DKGR cofactor; see below), phenazine methosulfate (PMS), and the tetrazolium dye, 3-(4,5-dimethylthiazol-2-yl)-2,5-diphenyltetrazolium bromide (MTT). This approach also relies on the oxidation of 2-KLG to 2,5-DKG with the concomitant reduction of NADP to NADPH. The reduced cofactor is then coupled via PMS to the colorless MTT_{ox} to give the colored MTT_{red}. This approach was also unsuccessful in isolating the 2,5-DKGR gene. The reason for the failure of these methods to clone the 2,5-DKGR gene may be related to the thermodynamics of the 2,5-DKG/2-KLG redox couple. The equilibrium constant for this reaction

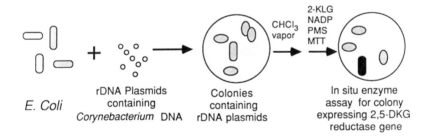

E. Coli   rDNA Plasmids   Colonies   In situ enzyme
                        containing   containing   assay for colony
                        Corynebacterium DNA   rDNA plasmids   expressing 2,5-DKG
                                                                         reductase gene

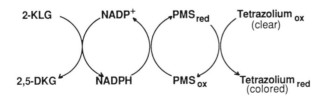

Figure 6.7. Screening strategy for cloning the 2,5-DKG reductase gene from *Corynebacterium* sp.

indicates that oxidation of 2-KLG at neutral pH is thermodynamically highly unfavorable (see below). It is also a kinetically slow reaction, so that very little 2,5-DKG may be formed in vivo, i.e., the reaction is essentially irreversible.

Identification and Characterization of 2,5-DKG Reductase

In conjunction with the cloning approaches described above, the isolation and characterization of the enzyme was undertaken. The cytosolic fraction of a crude lysate from *Corynebacterium* sp. ATCC31090 was assayed for reduction of 2,5-DKG using either NADH or NADPH as a cofactor; activity was detected in both cases. During the course of our efforts to purify and characterize this 2,5-DKGR (Miller et al. 1987), we observed several peaks of activity eluting from an anion-exchange column. It is important to note that the term 2,5-DKG reductase is somewhat ambiguous because reduction can occur at C5 to give either 2-keto-D-gluconate (2-KDG) or the desired 2-KLG, at C2 to give either 5-keto-D-gluconate (5-KDG) or 5-keto-D-mannonate (5-KDM), or at C1 to give the aldehyde; the enzyme as defined in this chapter catalyzes the reduction of 2,5-DKG stereospecifically to 2-KLG (Fig. 6.8). We therefore needed to develop the analytical methodology that would unambiguously identify which activity peak produced 2-KLG as the reduction product. The methods developed to identify the different

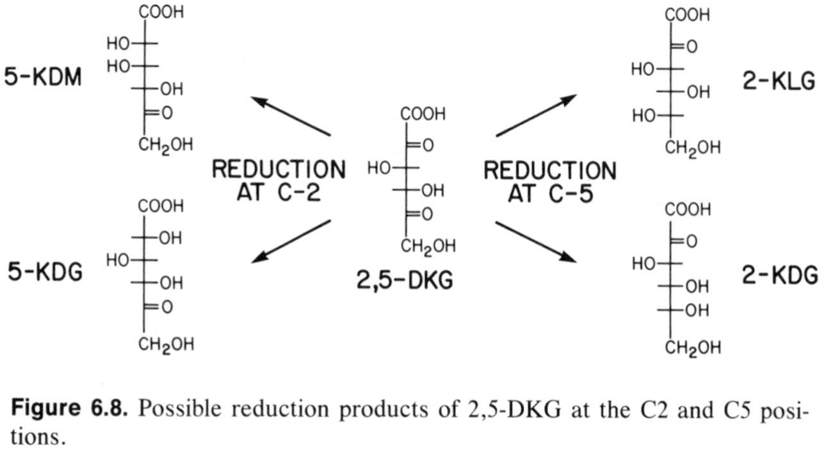

Figure 6.8. Possible reduction products of 2,5-DKG at the C2 and C5 positions.

ketoaldonic and aldonic acids that can be derived from reduction of 2,5-DKG (Lazarus and Seymour 1986) included both anion-exchange HPLC (Aminex A-27 resin/0.2 M ammonium formate pH 3.2 mobile phase) and gas chromatography–mass spectrometry (GC–MS) of the pertrimethylsilylated derivatives (5% cross-linked phenylmethylsilicone-fused silicon-bonded capillary column). These methods were used to identify and quantitate 2-KLG and other related metabolites produced in the fermentation process.

Using this analytical methodology, we were able to identify an NADPH-dependent 2,5-DKGR that produced 2-KLG. This enzyme was purified to apparent homogeneity from cytosolic extracts of lysed cells by consecutive chromatography using DEAE-cellulose ion-exchange, Cibacron blue F3GA agarose, and HPLC TSK gel permeation columns. To our knowledge, this was the first example of an enzyme that catalyzes this reaction; there have since been reports on other 2,5-DKG reductases (Sonoyama and Kobayashi 1987). The purified reductase was active as a monomer with an estimated native molecular weight of 35,000 daltons based on gel permeation chromatography; the molecular weight was 34,000 daltons based on sodium dodecyl sulfate (SDS)-polyacrylamide gels and 29,992 based on the translated DNA sequence of the gene (Anderson et al. 1985). As expected, a stoichiometric conversion of 2,5-DKG and NADPH to 2-KLG and NADP by this enzyme was observed based on HPLC and UV quantitative methods. The apparent K_m values for 2,5-DKG and NADPH are 26 mM and 10 μM, respectively; NADPH is favored as a cofactor over NADH by 170-fold. The enzyme is active over a broad pH range, with maximal activity observed from pH 6.0 to 7.0 for the reduction of 2,5-DKG. The oxidation of 2-KLG has a much narrower pH range with optimal activity observed at pH 9.2. The relative V_{max} values, at the respective pH optima for each direction, favor 2,5-DKG reduction by ca. 400-fold. The measured equilibrium constant for the reduction of 2,5-DKG by NADPH was 5.6×10^{-13}. This may account for the failure of the selection and screening approaches taken previously to clone the 2,5-DKGR gene, because the 2-KLG oxidation reaction would be quite unfavorable at physiological pH values. Because *Corynebacterium* sp. does not produce 2,5-DKG as a metabolite, the true physiological substrate for 2,5-DKGR remains unknown; both dihydroxyacetone and 5-keto-D-fructose are also substrates for this enzyme.

Cloning the 2,5-DKGR Gene

The 2,5-DKGR gene was isolated from *Corynebacterium* sp. DNA using probes based on the amino terminal sequence of the purified protein. Because no regions of minimal codon degeneracy existed for the construction of mixed oligonucleotide probes, a cloning strategy was adopted in which long synthetic DNA probes were used. Based on thermal melting curves obtained for *Corynebacterium* DNA, an extremely high GC content (71%) was predicted. Therefore, two 43-mer DNA probes were designed using codons known to be prevalent in bacterial DNA of high GC content. The 43-mer probes hybridized to DNA fragments from a *Corynebacterium* genomic library, enabling us to clone the 2,5-DKGR gene into *E. coli*. The entire DNA sequence of the gene was obtained and was consistent with amino acid sequences obtained from the N-terminus, C-terminus, and internal fragments generated by tryptic digests.

Expression of the 2,5-DKGR Gene in *E. herbicola*

Expression of the 2,5-DKGR gene was accomplished by the insertion of transcriptional and translational control sequences that work efficiently in *E. coli*, and by the deletion of the *Corynebacterium* sp. DNA upstream of the ATG start codon. The 2,5-DKGR gene was inserted into a tetracycline resistant (Tc^r) pBR322 based plasmid immediately downstream from the *E. coli trp* promoter and a synthetic ribosome binding site. The resulting ptrp1-35 plasmid was used to transform *E. herbicola* ATCC 21998 and tetracycline-resistant colonies were isolated. There was a high level of active 2,5-DKGR expressed (Fig. 6.9) as assayed by SDS-polyacrylamide gel electrophoresis, Western blot analysis, and spectrophotometric enzyme activity measurements of cytosolic extracts of the lysed *E. herbicola* transformants (Anderson et al. 1985).

We also wanted an inducible expression system, such that both the level and timing of 2,5-DKGR expression could be regulated. Therefore, we constructed a new expression plasmid (p269, Fig. 6.10) that placed the 2,5-DKGR gene under the control of the *tac*II promoter (DeBoer et al. 1983). With the assumption that the regulatory elements operative in *E. coli* would also function in *Erwinia*, the *lac*I repressor gene, under the control of the *B. licheniformis* penicillinase promoter

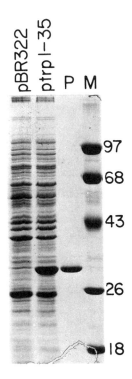

Strain	Relative 2,5-DKGR Activity
21998pBR322	1.0
21998ptrp1-35	46.1

Figure 6.9. Expression of 2,5-DKG reductase in *E. herbicola* 21998pBR322 and ptrp1-35 cultures analyzed by 10% SDS gel electrophoresis and enzymatic activity assays.

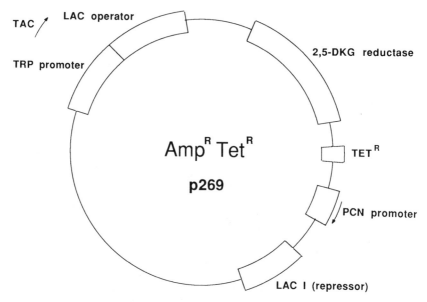

Figure 6.10. Regulatable expression plasmid for 2,5-DKG reductase in *E. herbicola*.

was also inserted into the plasmid. This allowed expression of 2,5-DKGR to be controlled by lactose or its analogs (e.g., isopropyl-β-D-thiogalactopyranoside, IPTG). Expression was tightly regulated in the *E. herbicola* p269 strain; no 2,5-DKGR activity could be detected in the absence of IPTG. A direct correlation was observed between the amount of IPTG added, the level of 2,5-DKGR activity in the cytosolic fraction, and the rate of 2,5-DKG reduction by whole cells. The fully induced p269 strain expressed 2,5-DKGR at approximately 20% of the level of the constitutively expressing ptrp1-35 strain.

CARBOHYDRATE METABOLISM IN THE RECOMBINANT STRAINS

Bioconversion of Glucose to 2-KLG

The bioconversion of D-glucose to 2-KLG by *E. herbicola* (ATCC 21998) harboring the ptrp1-35 plasmid was carried out in 10-L fermentors. After the culture had grown to ca. 20 OD_{550}, the addition of

D-glucose (30 g/L) resulted in a rapid and efficient oxidation to D-gluconate (GA), 2-KDG, and finally 2,5-DKG. This was followed by a slower and somewhat less efficient reduction of 2,5-DKG to 2-KLG, catalyzed by the cloned 2,5-DKGR (Fig. 6.11). No 2-KLG was observed in control cultures that lacked the 2,5-DKGR gene. The identity of 2-KLG was confirmed by HPLC, GC–MS, ^{1}H and ^{13}C nuclear magnetic resonance (NMR), optical rotation, and melting point analysis.

Although the conversion of glucose to 2-KLG had been demonstrated, a number of interesting observations needed to be explained. In particular, the conversion of 2,5-DKG to 2-KLG was not quantitative and we were unable to close the carbon balance by accounting for the CO_2. In addition to the expected metabolites, an unknown metabolite, which eluted from an ion-exchange HPLC column with a retention time in the aldonic acid region (Fig. 6.12), was observed. The fact that this metabolite disappeared as 2-KLG appeared suggested that the two compounds were chemically related (Fig. 6.11). The unknown metabolite was identified as L-idonate (IA) based on the following observations: (1) identical retention times on both the ion-exchange and organic acids HPLC analytical systems and (2) identical GC retention times and mass spectra of the pertrimethylsilated derivative of the

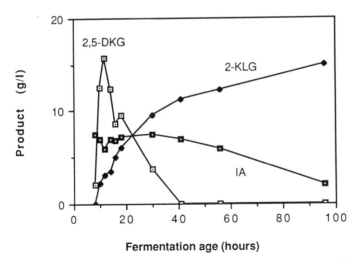

Figure 6.11. Bioconversion of D-glucose to 2-KLG by *E. herbicola* ptrp1-35 in a 10-L fermenter.

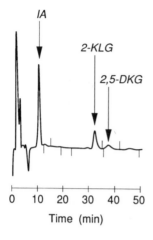

Figure 6.12. HPLC (Aminex-A-27) trace of *E. herbicola* ptrp1-35 fermentation broth.

unknown peak as compared with authentic standards of IA (Lazarus and Seymour 1986). The identification of IA and its subsequent quantitation allowed a complete closure of the carbon balance for the fermentation.

Metabolism of IA

The observation of IA as a significant metabolite in *E. herbicola* ptrp1-35 fermentations prompted an investigation of the metabolic pathways involved in both its synthesis and degradation. Although a large number of potential metabolic routes from glucose exist, it seemed reasonable that IA was produced by reduction of 5-KDG or 2-KLG. Evidence for both the 5-KDG route in *Fusarium* species (Takagi 1962) and the 2-KLG route in a variety of microorganisms has been reported (Ameyama and Adachi 1982; Makover et al. 1975). In shake flasks of whole cells of *E. herbicola* plus 5-KDG, all the 5-KDG was consumed and no other metabolites were observed. Shake flask incubations of *E. herbicola* plus 2-KLG produced IA; the sum of the IA and 2-KLG concentrations remained constant throughout the bioconversion. Furthermore, *E. herbicola* is able to grow using either 5-KDG, G, GA, 2-KDG, or 2,5-DKG as the sole carbon source in minimal media; no growth is observed with either 2-KLG or IA as the sole carbon source. We therefore concluded that the IA produced in these fermentations is derived from 2-KLG catabolism (Fig. 6.13).

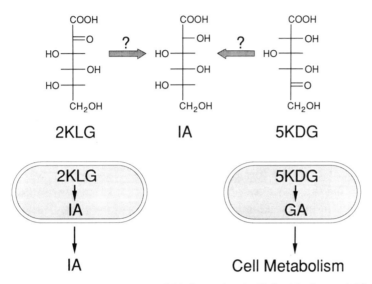

Figure 6.13. Metabolic route of IA formation in *E. herbicola* ptrp1-35.

Carbohydrate Reductive and Oxidative Enzymatic Pathways

To control the metabolic flux from D-glucose to 2-KLG in the recombinant *E. herbicola* strains, an understanding of the key metabolic pathways involved was critical. In addition, the metabolic diversion of 2-KLG to IA in fermentations was surprising because neither of these carbohydrates was known to be a natural metabolite of *Erwinia*. Also, it appeared that 2-KLG was produced from IA oxidation (Fig. 6.11), in the absence of any other carbon source, raising the question as to the specific metabolic pathways involved for both 2-KLG and IA metabolism.

The metabolism of carbohydrates by ketogenic bacteria has been investigated in a wide variety of strains (Asai 1968; Lessie and Phibbs 1984; Ameyama et al. 1987). Oxidation of glucose to ketogluconates has been shown to proceed via membrane-bound dehydrogenases that are linked to the electron transport chain. Subsequent reduction of the ketogluconates or their phosphorylated forms is catalyzed by cytosolic NAD(P)H-requiring reductases; the products can then enter into central metabolism. We therefore looked for cytosolic NAD(P)H-linked reductases in *E. herbicola* utilizing 2-KLG, 2-KDG, 2,5-DKG, and 5-KDG as substrates. Enzymatic activity was observed in crude lysates; the enzymes were separated by elution from a DEAE-cellulose column using a linear salt gradient (Fig. 6.14). The first peak, a 5-KDG reductase [5KR(G)], required NADPH and stereospecifically reduced 5-KDG to GA as identified by HPLC and GC–MS (Lazarus and Seymour 1986). The second peak of activity, a 2-ketoaldonate reductase (2-KR), catalyzed the stereospecific reduction of 2-KDG to GA, 2-KLG to IA, and 2,5-DKG to 5-KDG, using either NADPH or NADH as the cofactor, the former being preferred. No evidence was found supporting the existence of any 2-ketoaldonate kinases, such as those observed in the pseudomonads (Lessie and Phibbs 1984).

Purification and Characterization of 2-KR

2-Ketoaldonate reductase was purified by successive chromatography on DE52 ion exchange, Reactive Red 120 Agarose, and Mono Q ion-

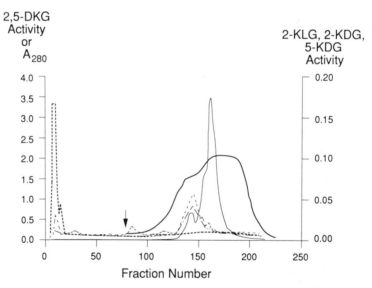

Figure 6.14. Identification of ketoreductase activities in *E. herbicola* ptrp1-35 by ion-exchange chromatography on DEAE-52 cellulose. (——), A_{280}; (···), 5-KDG activity; (----), 2-KLG activity; (——), 2,5-DKG activity; (—·—), 2-KDG activity. A linear NaCl gradient was started at the arrow.

exchange resins, and finally by preparative native gel electrophoresis (Seymour and Lazarus 1989) to apparent homogeneity as assessed by the presence of a single band on a silver-stained SDS polyacrylamide gel. The molecular weight of the enzyme was 36,000 ± 2,000 daltons based on SDS gels and 70,000 ± 3,000 daltons based on its elution as a single, symmetrical peak of active enzyme by gel filtration chromatography on an HPLC TSK column. These results suggest that the enzyme is active as a dimer of identical subunits. The enzyme is active from pH 5–10, with maximal activity for 2-KLG reduction at pH 6.5–7.5 and for IA oxidation at pH 9.8. The measured equilibrium constant for the reduction of 2-KLG by NADPH was 2.6×10^{-11}.

2-Ketoaldonate reductase is a relatively nonspecific enzyme in its requirements for both substrate and cofactor. This was first observed from activity profiles during the purification procedure (Fig. 6.14) in which the 2-KLG, 2-KDG, and 2,5-DKG dependent reductase activities from *E. herbicola* coeluted, using either NADH or NADPH as the cofactor. The only other substrate found for the enzyme, hydroxypyruvate, had the highest turnover number of all the compounds tested; pyruvate, D-fructose, or L-sorbose were not substrates for the enzyme.

The products of the enzymatic reduction of 2-KLG, 2-KDG, and 2,5-DKG by NAD(P)H are IA, GA, and 5-KDG, respectively (Fig. 6.15). This is based on data obtained from both HPLC and GC–MS, where retention times and masses matched those of authentic standards. Reduction of 2,5-DKG is specific for the 2-keto position because no 2-KLG or 2-KDG, the potential products from reduction at the 5-position, were observed. In addition, reduction is stereospecific because only 5-KDG, and no 5-keto-D-mannonate, was observed. This same stereochemistry is observed for the reversible interconversion of hydroxypyruvate and D-glycerate.

The broad specificity observed for 2-KR was not totally unexpected. 2-Keto-D-gluconate reductases purified from a variety of ketogenic bacteria exhibit such nonspecificity (Ameyama and Adachi 1982). It has been postulated that these reductases are present in order to utilize oxidized carbohydrates for the production of gluconate, which can enter into central metabolism, and also to regenerate $NADP^+$. Based on our data as well as that reported in the literature (Ameyama and Adachi 1982), the minimum requirements for substrate activity for

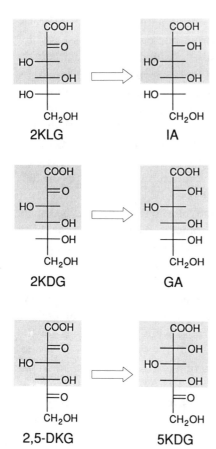

Figure 6.15. Product identification of *E. herbicola* 2-KR catalyzed reactions of NAD(P)H and different 2-ketoaldonate substrates.

2-KR include a carboxylate moiety at C1, a keto group at C2, and a hydroxyl group at C3, because hydroxypyruvate, but not pyruvate, is active. The products of the reduction reaction all possess the (R)-epimeric configuration at C2, consistent with the idea that the different keto-substrates are all bound to the enzyme in an identical manner and that the stereochemical transfer of hydrogen from NAD(P)H is conserved (Fig. 6.15).

Membrane-Bound Dehydrogenases

The carbohydrate oxidizing enzymes in *E. herbicola* were determined to be membrane-bound dehydrogenases based on the following observations: (1) the activities sediment with the membrane fraction in an ultracentrifuge; (2) the activities are solubilized by a variety of detergents; and (3) artificial electron acceptors such as 2,6-dichloroindophenol or O_2 coupled to phenazine methosulfate are reduced whereas neither NAD^+ or $NADP^+$ serve as electron acceptors. Activities measured either with an O_2 electrode or spectrophotometrically give comparable results. Thus, activities for glucose dehydrogenase (GDH), gluconate dehydrogenase (GADH), 2-keto-D-gluconate dehydrogenase (2-KDGDH), and, most interestingly, idonate dehydrogenase (IADH) have been identified. These enzymes are most likely heme and flavin containing proteins linked to the cytochrome chain (Ameyama et al. 1987). The product of the idonate dehydrogenase catalyzed reaction with IA was found to be 2-KLG based on HPLC and GC–MS data (Lazarus and Seymour 1986); 2-KDG, the product of GA oxidation, is further oxidized to 2,5-DKG. Although the specific activities in the crude membrane fraction for IA and GA oxidation are comparable, it is likely that different enzymes catalyze each reaction because 2-KLG inhibits idonate oxidation but not gluconate oxidation.

CONCLUSIONS

We have demonstrated that rDNA technology can be used to create new metabolic pathways by combining metabolic traits from dissimilar bacteria into a single recombinant organism. This process represents one of the first examples of a "metabolically engineered" recombinant microorganism designed to produce a specialty chemical. Using this approach we have created a bacterial strain that carries out the one-step bioconversion of glucose to 2-KLG, an important intermediate in the synthesis of vitamin C. During the course of our work, we have made some interesting observations. In particular, 2-KLG, which is not known to be a natural metabolite of *E. herbicola*, can be further metabolized to L-idonic acid, catalyzed by a nonspecific cytosolic NAD(P)H linked reductase. In addition, L-idonic acid, also not known to be a natural metabolite of this strain, can be oxidized by a membrane-bound dehydrogenase to 2-KLG. A model for the pathways involved in the carbohydrate metabolism in the recombinant *Erwinia herbicola* is shown in Figure 6.16. Although the key pathways are indicated, there is a great deal yet to be studied, such as the utilization of cell energy to regenerate NADPH and the specific transport mecha-

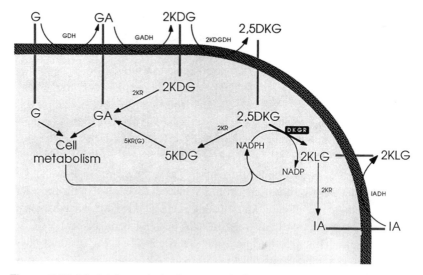

Figure 6.16. Model for carbohydrate metabolism in recombinant *E. herbicola* strain containing 2,5-DKG reductase.

nisms that are involved. An understanding of these phenomena should lead to further improvements in the overall process through the use of metabolic pathway engineering technology.

REFERENCES

Ameyama, M., and O. Adachi. 1982. 2-Keto-D-gluconate reductase from acetic acid bacteria. In *Methods in Enzymology, Vol. 89* (W. A. Wood, ed.), pp. 198–210. New York: Academic Press.

Ameyama, M., K. Matsushita, E. Shinagawa, and O. Adachi. 1987. Sugar-oxidizing respiratory chain of *Gluconobacter suboxydans. Agric. Biol. Chem.* **51:** 2943–2950.

Anderson, S., C. B. Marks, R. Lazarus, J. Miller, K. Stafford, J. Seymour, D. Light, W. Rastetter, and D. A. Estell 1985. Production of 2-keto-L-gulonate, an intermediate in L-ascorbate synthesis, by a genetically modified *Erwinia herbicola. Sci.* **230:** 144–149.

Asai, T. 1968. *Acetic Acid Bacteria: Classification and Chemical Activities.* Baltimore: University Park Press.

Crawford, T. C., and S. A. Crawford. 1980. Synthesis of L-ascorbic acid. *Adv. Carbohydrate Chem. Biochem.* **37:** 79–155.

DeBoer, H. A., L. J. Comstock, and M. Vasser. 1983. The tac promoter: a functional hybrid derived from the trp and lac promoters. *Proc. Natl. Acad. Sci. USA* **80:** 21–25.

Gaffe, G. M. 1984. Ascorbic acid. In *Kirk-Othmer Encyclopedia of Chemical Technology*, Vol. *24*, 3rd ed., pp. 8–40. New York: John Wiley & Sons.

Kulhanek, M. 1970. Fermentation processes employed in vitamin C synthesis. *Adv. Appl. Microbiol.* **12:** 11–33.

Lazarus, R. A., and J. L. Seymour. 1986. Determination of 2-keto-L-gulonic and other ketoaldonic and aldonic acids produced by ketogenic bacterial fermentation. *Anal. Biochem.* **157:** 360–366.

Lessie, T. G., and P. V. Phibbs Jr. 1984. Alternative pathways of carbohydrate utilization in *Pseudomonads*. In *Annu. Rev. Microbiol.* **38:** 359–387.

Makover, S., G. B. Ramsey, F. M. Vane, C. G. Witt, and R. B. Wright. 1975. New mechanisms for the biosynthesis and metabolism of 2-keto-L-gulonic acid in bacteria. *Biotechnol. Bioengin.* **17:** 1485–1514.

Miller, J. V., D. A. Estell, and R. A. Lazarus. 1987. Purification and characterization of 2,5-diketo-D-gluconate reductase from *Corynebacterium*. sp. *J. Biol. Chem.* **262:** 9016–9020.

Reichstein, T., and A. Grussner. 1934. A good synthesis of L-ascorbic acid (vitamin C). *Helv. Chim. Acta* **17:** 311–328.

Seymour, J. L., and R. A. Lazarus. 1989. Native gel activity stain and preparative electrophoretic method for the detection and purification of pyridine nucleotide-linked dehydrogenases. *Anal. Biochem.* **178:** 243–247.

Sonoyama, T., B. Kageyama, and S. Yagi. 1987. Distribution of Microorganisms capable of reducing 2,5-diketo-D-gluconate to 2-keto-L-gulonate. *Agric. Biol. Chem.* **51:** 2003–2004.

Sonoyama, T., and K. Kobayashi. 1987. Purification of two 2,5-diketo-D-gluconate reductases from a mutant strain derived from *Corynebacterium* sp. *J. Ferment. Technol.* **65:** 311–317.

Sonoyama, T., H. Tani, K. Matsuda, B. Kageyama, M. Tanimoto, K. Kobayashi, S. Yagi, H. Kyotani, K. Mitsushima. 1982. Production of 2-keto-L-gulonic acid from D-glucose by two-stage fermentation. *Appl. Environ. Microbiol.* **43:** 1064–1069.

Sonoyama, T., S. Yagi, and B. Kageyama. 1988. Facultative anaerobic bacteria showing high productivities of 2,5-diketo-D-gluconate from D-glucose. *Agric. Biol. Chem.* **52:** 667–674.

Takagi, Y. 1962. Reduction of 5-keto-D-gluconic acid to L-idonic acid by *Fusarium* species. *Agric. Biol. Chem.* **26:** 717–718.

Weenk, G., W. Olijve, and W. Harder. 1984. Ketogluconate formation by *Gluconobacter* species. *Appl. Microbiol. Biotechnol.* **20:** 400–405.

7
Chiral Synthons by Biocatalysis

STEPHEN C. TAYLOR

ICI Biological Products is a young and growing international business in industrial biotechnology. The application of biocatalysis to chemical synthesis is an important part of this business which is focused primarily on the synthesis of optically active molecules. It is widely recognized that this is an important area where biological synthesis can compete effectively with more traditional nonenzymic methods. This chapter will consider the novel technology that lies behind one of the new molecules in ICI's fine chemicals portfolio which provides an excellent example of the critical importance of working with a specialist multidisciplinary term to achieve success in this complex field. The technology referred to is based on a type of enzyme not previously used for biotransformations, a dehalogenase, and the product is (S)-2-chloropropanoic acid, a chiral molecule widely used in the synthesis of "effect products," particularly agrochemicals.

The major industrial requirement for (S)-2-chloropropanoic acid lies with the production of phenoxypropanoic acid herbicides of which there are many examples (Fig. 7.1). These herbicides all have the following characteristics:

1. All have a chiral center that can be derived from chloropropanoic acid.
2. The (R) isomer alone has significant herbicidal activity.
3. All have historically been produced as racemates.

There is increasing interest in producing the phenoxypropanoic acid herbicides as resolved, single-enantiomer compounds. This interest arises from the recognition that there are significant cost savings in producing the single active enantiomer alone. Furthermore, the envi-

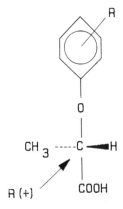

Figure 7.1. General structure of phenoxypropanoic acid herbicides.

ronmental impact of agrochemicals when produced as only the active enantiomer rather than as a racemate is also cut substantially (by half in the case of the phenoxypropanoics), providing a crucial and increasingly important motivation for this effort.

Although the final active herbicides could be resolved directly, the simplest and most economically attractive option is to start syntheses with optically active (S)-2-chloropropanoic acid rather than the racemate. Two basic synthetic approaches to (S)-2-chloropropanoic acid have been proposed (Fig. 7.2) and interestingly both require biotech-

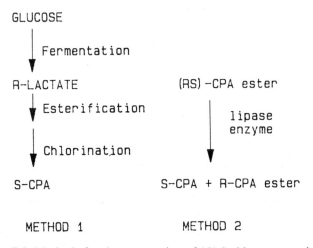

Figure 7.2. Methods for the preparation of (S)-2-chloropropanoic acid.

nology, indicative of the importance of enzymes in the production of single enantiomers. The first approach is based on fermentation to make (*R*)-lactic acid which is followed by nonenzymic esterification and chlorination. The second approach is based on enzymic resolution of an ester of racemic chloropropanoate with a suitable hydrolytic enzyme which shows differential rates of hydrolysis for the two enantiomers (US Patent 4613690, Stauffer Chemical Company). We considered these and other potential biological and chemical options and determined that a route based on novel enzymes called dehalogenases was potentially very attractive.

BACKGROUND OF DEHALOGENASE ENZYMES

Halogenated compounds are widespread in the environment and occur both naturally and as a result of man's activities. A substantial amount of academic work has involved the study of microbial growth on haloalkanoic acids, particularly Dalapon (dichloropropanoic acid), and its derivatives (Senior et al. 1976; Hardman and Slater 1981; Berry et al. 1979; Motosugi et al. 1982; Allison et al. 1983; Weightman et al. 1982). Many microorganisms have been found to degrade these compounds. Degradation is dependent on the induction of dehalogenase enzymes which apparently catalyze the hydrolytic removal of halogen from C2 yielding the corresponding hydroxy or oxocarboxylic acids from mono- and dihalogenated acids, respectively, both of which can be readily further metabolized (Fig. 7.3). This reaction does not require any additional enzyme cofactors, making it an attractive proposition for a commercial biotransformaton.

Dehalogenases show considerable variation in their catalytic activity and are often found in multiple forms in individual species of microor-

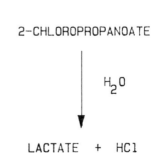

Figure 7.3. Dehalogenase reaction.

ganism (Hardman and Slater 1981). Where chloropropanoic acid has been studied as a substrate for dehalogenases, two specificities have been recorded. Enzymes have been found that are specific for the (S)-enantiomer and show no dehalogenase activity with the (R)-enantiomer. Nonspecific enzymes which have equal activity on both enantiomers have also been reported. Differences in reaction mechanism have also been seen. Some enzymes involve an inversion of configuration whereas others have a mechanism that maintains configuration (Weightman et al. 1982). However, despite the range of properties shown, no enzyme had hitherto been reported as being specific to the (R)-enantiomer of chloropropanoic acid.

DEHALOGENASE TECHNOLOGY

Incubation of an isomer specific dehalogenase enzyme with racemic chloropropanoic acid should yield lactic acid and leave one chloropropanoic acid isomer unreacted. Therefore to make (S)-2-chloropropanoic acid by this principle an (R)-2-chloropropanoic acid specific dehalogenase was required (Fig. 7.4). Because such an enzyme had not been reported, we established a program to isolate and identify microorganisms that we considered were likely to contain a dehalogenase with this property.

Using traditional microbiological methods, a wide range of microorgranisms were isolated that could use chloropropanoic acid as sole source of carbon and energy for growth. Thirteen of these isolates were found to grow only on (R)-2-chloropropanoic acid and not on (S)-2-chloropropanoic acid. Other organisms were found to utilize both enantiomers or only the (S)-enantiomer (Table 7.1). All 13 isolates were examined for dehalogenase enzyme activity which was found in all

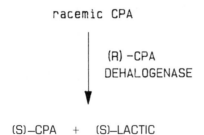

Figure 7.4. The proposed basis of dehalogenase technology for (S)-2-chloropropanoic acid production.

Table 7.1 Growth specificity of microorganisms isolated with chloropropanoic acid (CPA) as sole source of carbon and energy.

NUMBER OF ISOLATES	GROWTH SPECIFICITY
13	(R)-CPA$^+$, (S)-CPA$^-$
1	(R)-CPA$^-$, (S)-CPA$^+$
8	(R)-CPA$^+$, (S)-CPA$^+$

cases. However, the inability to grow on (S)-2-chloropropanoic acid was not due to the presence of an (R)-2-chloropropanoic acid dehalogenase as we expected. When screened, enzymes capable of dehalogenating both enantiomers of chloropropanoic acid were found. On further examination it was discovered that although these 13 strains had enzymes for dehalogenating both enantiomers of chloropropanoic acid, only one of the lactic acid isomers formed as products could be further metabolized. Thus these strains were unable to metabolize (R)-lactic acid, the product of (S)-2-chloropropanoic acid dehalogenase activity (Table 7.2), because no (R)-lactic dehydrogenase was present hence the lack of growth on (S)-2-chloropropanoic acid (1a, 2a, 3a). In all 13 cases it was found that the opposite enantiomer of lactic acid was formed and could not be metabolized, indicating that dehalogenation of (S)-2-chloropropanoic acid involved complete inversion of configuration.

Table 7.2. Growth specificity of a selection of chloropropanoic acid (CPA) degrading isolates. [1a, 2a, 3a were all unable to grow on (S)-CPA].

	GROWTH SUBSTRATE		
	(R)-CPA	(S)-CPA	(R)-LACTIC ACID	(S)-LACTIC ACID
WC14	−	+	+	+
WF1	+	+	+	+
AJ1	+	+	+	+
1a	+	−	−	+
2a	+	−	−	+
3a	+	−	−	+

The (R)-2-chloropropanoic acid growth specific phenotype has also been observed by Professor Howard Slater and his colleagues at the University of Wales, but in this case it appears to be due to an inability of strains to transport (S)-2-chloropropanoic acid into the cell rather than to a deficiency at the level of chloropropanoic acid dehalogenase or lactic acid dehydrogenase. The inability to use growth specificity as a selection tool for the required (R)-specific dehalogenase made the task of finding this enzyme more difficult. However when those isolates that were able to use both enantiomers of chloropropanoic acid as growth substrates were examined for their complement of dehalogenase enzymes, an unexpected result was found. One isolate identified as a strain of *Pseudomonas putida* was found to contain two separate dehalogenase enzymes. One of these dehalogenases was a low molecular weight enzyme and showed total specificity for (S)-2-chloropropanoic acid. The second dehalogenase enzyme was of a higher molecular weight and was specific to (R)-2-chloropropanoic acid. Thus although the isolation of an organism containing multiple dehalogenases was not in itself surprising, this strain did contain a complement of enzymes not previously reported and most critically contained an enzyme with the desired dehalogenase specificity. Both enzymes were intracellular and found in the soluble fraction of cell-free extracts. The two enzymes were readily separated from extracts by ion-exchange chromatography (Fig. 7.5) and could thus be individually studied.

Apart from its isomer selectivity, the novel (R)-2-chloropropanoic acid dehalogenase had similar properties to those reported for other dehalogenase enzymes active on haloalkanoic acids. For example the pH optimum was found to be 8.5 (Fig. 7.6) and K_m and V_{max} values were broadly comparable to those of other dehalogenases. Furthermore the reaction product was (S)-lactic acid, indicating an inversion mechanism.

The possibility of using this new (R)-2-chloropropanoic acid dahalogenase to prepare (S)-2-chloropropanoic acid via the proposed route was tested using the isolated enzyme. As predicted, when incubated with racemic chloropropanoic acid the enzyme rapidly converted the (R)-enantiomer to (S)-lactic acid and the reaction ceased to work when 50% of the racemic chloropropanoic acid had been converted. This reaction was monitored by both chloride ion release, alkali uptake

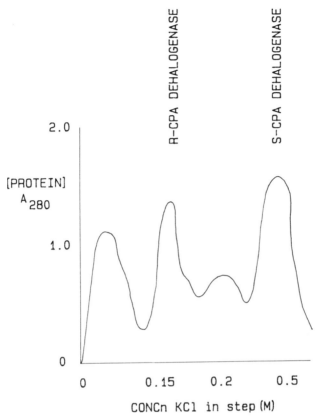

Figure 7.5. Separation of dehalogenases by ion-exchange chromatography on DEAE–Sephacel eluted stepwise with increasing concentrations of potassium chloride.

required for pH control, and by the accumulation of lactic acid in the reaction medium (Fig. 7.7). When hydrolysis ceased the reaction liquor was acidified, filtered to remove precipitated protein, and the chloropropanoic acid was solvent extracted with methylene chloride. Analysis showed this to be (S)-2-chloropropanoic acid with an enantiomer excess of >98% obtained in a yield of close to 50%.

In addition to exploring the use of the isolated enzyme we also considered the use of the intact microorganism as a catalyst. The first stage of testing involved mutation of the wild-type strain to inactivate

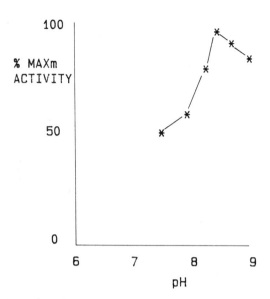

Figure 7.6. pH profile of (*R*)-2-chloropropanoic acid dehalogenase measured in Tris-sulfate buffer.

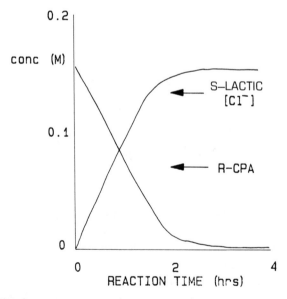

Figure 7.7. Profile of biotransformation using (*R*)-2-chloropropanoic acid de-halogenase to produce (*S*)-2-chloropropanoic acid. Initial concentration or racemic chloropropanoic acid was 0.3 *M*.

the (S)-2-chloropropanoic acid dehalogenase activity. A program of genetic work produced several new strains that were unable to utilize (S)-2-chloropropanoic as a growth substrate. This deficiency was confirmed to be due to an absence of an active (S)-2-chloropropanoic acid dehalogenase enzyme. The properties of the (R)-2-chloropropanoic acid dehalogenase within the intact microorganism were essentially the same as those of the isolated enzyme, and as with the isolated enzyme, the intact organism could also be used very effectively as a catalyst to produce (S)-2-chloropropanoic acid with the same protocol as previously described.

Thus it was successfully demonstrated in the laboratory that ICI's novel (R)-2-chloropropanoic acid dehalogenase could be used as either the free isolated enzyme or within intact organisms to effect kinetic resolution of racemic chloropropanoic acid yielding (S)-2-chloropropanoic acid in high enantiomer excess and in high yield. This work represents the first significant application of dehalogenases to the production of optically active molecules. Many more such applications are sure to follow in the future as these enzymes represent an important addition to the armory of catalysts for the production of single enantiomer products.

REFERENCES

Allison, L., A. J. Skinner, and R. A. Cooper. 1983. The dehalogenases of a 2,2′-dichloropropionate degrading bacterium. *J. Gen. Microbiol.* **129:** 1283–1293.

Berry, E. K. M., N. Allison, and A. J. Skinner. 1979. Degradation of the selective herbicide 2,2′-DCPA (Dalapon) by a soil bacterium. *J. Gen. Microbiol.* **110:** 39–45.

Hardman D. J., and J. H. Slater. 1981. Dehalogenases in soil bacteria. *J. Gen. Microbiol.* **123:** 117–128.

Motosugi, K., N. Esaki, and K. Soda. 1982. Bacterial assimilation of D- and L-2-chloropropionates and occurrence of a new dehalogenase. *Acta Microbiol.* **131:** 179–183.

Senior, E., A. T. Bull, and J. H. Slater. 1976. Enzyme evolution in a microbial community growing on the herbicide Dalapon. *Nature* **263:** 476–479.

US Patent 4613690 Assigned to Stauffer Chemical Company.

Weightman A. J., A. L. Weightman, and J. H. Salter. 1982. Stereospecificity of 2-monochloropropionate dehalogenation by the two dehalogenases of *Pseudomonas putida* PP3. *J. Gen. Microbiol.* **128:** 1755–1762.

8
Enzymatic Resolution of Ibuprofen in a Multiphase Membrane Reactor

F. X. MCCONVILLE, J. L. LOPEZ, AND S. A. WALD

The recent literature in the area of biocatalytic reactions on "nonphysiological or unnatural" substrates has been growing. More specifically, the use of enzymes of the kinetic resolution of optical isomers has been prevalent (Klibanov 1986; Whitesides and Wong 1983; Chen, Fujimoto, and Sih 1981; Jones and Beck 1976). A common problem with many of these systems is that the substrate, usually an organic ester, has low water solubility. The engineering challenge is encountered on scale-up of these systems. Often, an immobilized enzyme must be contacted with an emulsion of the substrate in an organic solvent/aqueous buffer system. This chapter is intended to demonstrate the use of membranes to address these problems.

This case study is focused on the enzymatic resolution of arylpropionic acid, ibuprofen. Ibuprofen, 2-(p-isobutylphenyl)propionic acid, is an important nonsteroidal antiinflammatory drug. It is currently sold as a racemic mixture; the (S)-enantiomer has 100 times the activity, in vitro, of the (R)-enantiomer (Adams, Bresloff, and Mason 1976).

Enzymatic resolution has been studied, resulting in the identification of an (S)-selective enzyme, a lipase from *Candida cylindracea* (Sih 1987; Cesti and Piccardi 1988). Figure 8.1 shows the general reaction scheme. A wide range of esterifying alcohols were described by the two groups. However, in all cases, the resultant esters had low water solubility (<1 mM).

This type of reaction system is well suited for a multiphase membrane reactor, that is, the contact of a sparingly soluble substrate with an immobilized enzyme. The use of a membrane as a phase contactor,

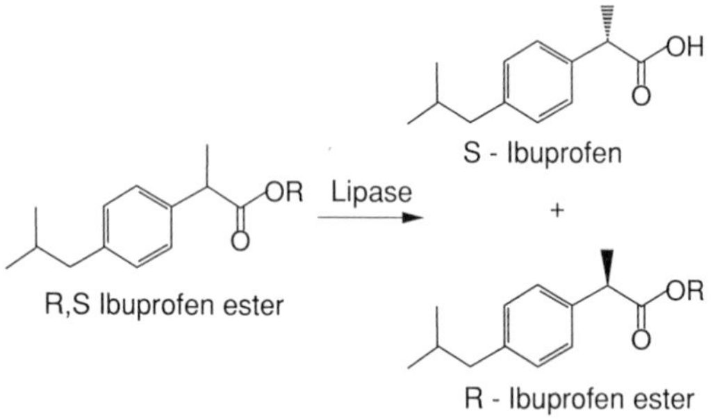

Figure 8.1. Enzymatic resolution of ibuprofen esters.

phase separator, and interfacial catalyst is the key feature of the multi-phase membrane reactor.

ENZYME–SUBSTRATE SCREENING

A range of ibuprofen esters were screened with regard to reactivity and selectivity. Reactivity was determined by an initial rate assay using a pH-stat technique. Lipase from *C. cylindracea* (Genzyme Corporation) was tested on an emulsion of the desired ester in 0.5 M phosphate buffer at 20°C. The pH was controlled at 7.8 with 20 mM sodium hydroxide. Table 8.1 shows that the reaction rate increases as the degree of electronic activation increases (Gu, Chen, and Sih 1986). It should be noted that all of the ibuprofen esters were liquids at room temperature. Based on these results the trifluoroethyl ester was chosen for the bioreactor studies.

Table 8.1. Ibuprofen ester screening.

ESTER	INITIAL RATE (μmol/h-mg)	ENANTIOMERIC RATIO
Methyl	4.2	>100
Methoxyethyl	16.8	65
Monochloroethyl	19.2	32
Cyanoethyl	47.2	21
Trifluoroethyl	51.6	26

C. cylindracea lipase (Genzyme Corporation). pH 7.8, 20°C.

Enantioselectivity was determined in a separate series of experiments. An emulsion of the ester and buffer was shaken overnight with the lipase. Ibuprofen acid was isolated by extraction and optical purity was determined by optical rotation. Selectivity was quantified as the enantiomeric ratio. This ratio between the pseudo-first-order rate constants (k_{cat}/K_m) for each isomer was developed by Sih (Chen et al. 1982). A simple mathematical expression can be derived for this enantiomeric ratio, E, to relate optical purity of either the residual reactant or product and degree of conversion in a stereoselective reaction. It is interesting to note that selectivity *decreases* with increasing degree of electronic activation (see Table 8.1).

MULTIPHASE MEMBRANE BIOREACTOR—THEORY OF OPERATION

Membrane bioreactors represent a powerful process engineering tool for performing biocatalytic reactions in multiphasic systems because membranes are fundamentally much more versatile than conventional, particulate enzyme support media (e.g., microporous particles, gel-type beads, and ion-exchange media). The key attribute of membranes that makes them particularly attractive for bioreactor applications is their ability to compartmentalize a multiphase reaction system. Thus, although there is only a single interface between an immobilized biocatalyst and the solution that surrounds it, there are two sides to every membrane. The second interface represents an additional degree of freedom that can be utilized in the following ways:

1. The enzyme-activated membrane functions as an interfacial catalyst, putting the supported enzyme in direct contact with the organic phase containing the substrate.
2. The membrane provides high surface area for contact between the two immiscible process streams on either side of it.
3. The membrane serves at the same time to separate the bulk phase, thus avoiding the need to disperse one phase within another.

Two major types of reactions that membrane bioreactors address are bioconversions of sparingly soluble reactants and bioconversions that require continuous removal of sparingly soluble products (due to inhibition, enzyme deactivation, or chemical instability in aqueous solutions). Both of these type of reactions can be conducted in this system (Matson 1989a).

Membrane as an Interfacial Catalyst

One of the major problems with conducting multiphase reactions in stirred tank or packed bed reactors is inefficient mass transfer. Figure 8.2 shows the nature of the productivity limitation. The organic substrate is typically dispersed in a continuous aqueous phase and the enzyme is immobilized on or within a porous support. For the reaction to proceed, the substrate must first partition into the aqueous phase and be transported to the catalyst. However, low water solubility of the substrate limits its diffusive flux to the catalyst. Consequently, reactor productivity suffers.

Figure 8.3 shows a cross-sectional view of the enzyme-activated membrane that is designed to address this limitation. It can be seen that the membrane serves as an interfacial catalyst in intimate contact with both the organic and aqueous phases on either side of it. This arrangement facilitates efficient mass transfer by putting the three phases in the correct order (i.e., organic/enzyme/aqueous as opposed to organic/aqueous/enzyme as in the conventional case) thereby minimizing the

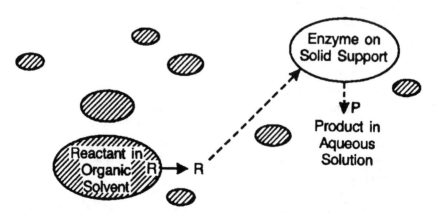

Figure 8.2. Bioconversions with sparingly soluble reactants.

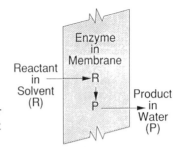

Figure 8.3. Multiphase membrane bioreactor cross-section. Efficient transport of reactant to enzyme.

diffusional distance of the poorly soluble substrate to the immobilized biocatalyst.

Enzyme Containment

An important feature of the membrane bioreactor is the ability to contain reversibly or "entrap" enzyme within the membrane (Matson 1989b). The reversibility allows convenient replacement of deactivated enzyme with fresh enzyme without having to replace its support (i.e., membrane). Figure 8.4 shows a cross-section of the type of asymmetric, hydrophilic hollow-fiber membrane used. The two most critical features are the finely microporous fiber walls and the existence of a "skin" layer at one of the membrane surfaces. This asymmetric mem-

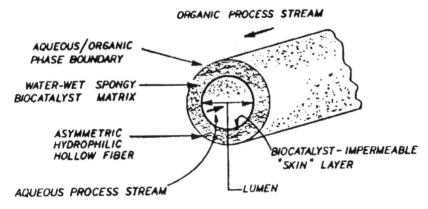

Figure 8.4. Enzyme entrapment within a hydrophilic, microporous membrane.

brane "immobilizes" the enzyme by entrapping it between two barriers: (1) the enzyme-impermeable skin layer and (2) the aqueous–organic interface maintained at the opposite membrane surface. The large size of the enzyme relative to the surface pores prevents it from diffusing across the skin layer of the membrane, while the poor solubility of the enzyme in organic solvents prevents it from partitioning across the aqueous/organic phase boundary. Enzyme activity is replenished by backflushing deactivated enzyme out of the membrane matrix followed by ultrafiltration of fresh enzyme back into the membrane.

It should be noted that "immobilization" of enzyme within a porous membrane results in reaction/diffusion tradeoffs similar to those seen with immobilization within porous particles or beads. These diffusional limitations can have the effect of reducing the enzyme effectiveness and hence decrease the "apparent" specific enzyme activity. However, membranes have the advantage over particulate supports in that a membrane has a short diffusional distance (in this case 50 μm wall thickness), thereby minimizing diffusional resistance. Particles of similar dimensions (diameter $\ll$ 100 μm) can result in severe pressure drop problems in packed beds.

Membrane as a Phase Contactor

Hollow fiber devices provide high interfacial area per unit volume (versus flat sheet and spiral membrane modules) for mass transfer. The aqueous/organic interface is located at the shell, or outer surface, of the hollow fiber membrane. Hence, the interfacial area is well defined, fixed (equal to the membrane surface area), and independent of both power input and the volume ratio between phases. This is contrary to the typical operation of a dispersed phase emulsion system. As a result, more flexibility is acquired. For example, the volume ratio (phase ratio) between the two immiscible phases can be established within a wide range from very low to very high. The phase ratio in an emulsion determines the desired continuous phase. For example, a high phase ratio (large volume of organic as compared to aqueous) will have the organic as the continuous phase. This type of water-in-oil emulsion will make pH control very difficult. This is not a problem in a membrane device.

Membrane as a Phase Separator

As stated above, the membrane is acting as a phase separator as well as a phase contactor. The membrane is designed to be water-wet within its interior, thus providing the preferred environment for the enzyme for most applications. The only operational requirement for complete phase separation is control over the fluid pressures. To separate the phases effectively, the organic phase (situated in the shell compartment as dictated by fiber structure) must be kept at a slightly greater pressure than the aqueous phase. If this positive shell-to-lumen trans-membrane pressure difference is not maintained, water will ultrafilter into the organic phase. Therefore, fluid pressures must be carefully controlled for proper operation of the membrane bioreactor.

RESULTS AND DISCUSSION

Membrane reactor experiments were conducted in hollow-fiber membrane modules. The modules are designed to be resistant to a wide range of solvents: ketones, esters, aromatic and aliphatic hydrocarbons, and alcohols. The membrane is an asymmetric, hydrophilic hollow fiber, having dimensions of 200 μm inner diameter and 300 μm outer diameter. Laboratory modules have 0.6 m^2 of membrane area.

Figure 8.5 shows a schematic of the batch reactor configuration. The *Candida* lipase (12 g, Genzyme Corporation) was loaded into the membrane. No measurable activity was detected in the filtrate. Phosphate buffer was circulated through the lumen, with the pH controlled at 7.8

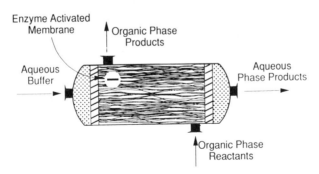

Figure 8.5. Multiphase hollow-fiber membrane bioreactor schematic.

with sodium hydroxide. Titration rate of the ibuprofen acid product corresponds closely to the hydrolysis rate. The racemic ibuprofen trifluoroethyl ester (neat) is circulated in the shell compartment. As the reaction proceeds, (S)-ibuprofen salt accumulates in the aqueous phase with the trifluoroethanol and the ester becomes enriched in the (R)-enantiomer.

Listed below are the reactor productivity and product quality for this set of reaction conditions. Productivity is expressed in two ways: on an enzyme basis and on a membrane basis. Both are important parameters for economic reasons.

Substrate: Ibuprofen trifluoroethyl ester
Enzyme: 12 g *Candida* lipase
                  pH 7.8, 20°C
Membrane: One 0.6 m² module
Productivity: 0.2 μmol/h/mg at x = 15%
                  4.7 kg acid/yr-m²
Optical purity: 81% ee

This reactor condition resulted in low enzyme effectiveness; that is, a low ratio of enzyme activity expressed to activity loaded. This is generally a result of diffusional limitations.

In parallel, the enzyme kinetics were further studied. While no significant product inhibition was observed, an enzyme poisoning effect by ibuprofen was identified. Table 8.2 shows results from enzyme incubation studies. An enzyme solution (10 mg/ml) was prepared and incubated at room temperature for 80 h with varying levels of ibuprofen. Initial rate assays were conducted to determine residual activity. Increasing the concentration of ibuprofen (as its sodium salt at pH 8) from 0 to 50 mM sharply decreased the residual enzyme activity down to <1% of the control level.

At lower pH, due to the low water solubility of the ibuprofen acid, the enzyme no longer "sees" the ibuprofen and the poisoning is eliminated. The mechanism of this poisoning is being further studied with various other organic acids. Nevertheless, minimizing the amount of ibuprofen salt in the enzyme environment was shown to significantly increase enzyme life. Enzyme kinetics were screened at low pH and the enzyme has a broad pH optimum from 5 to 8.

Table 8.2.   Effect of ibuprofen incubation
on *Candida* lipase.

IBUPROFEN LEVELS/pH DURING INCUBATION	RESIDUAL ACTIVITY AFTER INCUBATION (%)
0 mM/pH 8.0	100
20 mM/pH 8.0	37
50 mM/pH 8.0	<1
50 mM/pH 6.5	100
50 mM/pH 5.0	100

C. cylindracea lipase (Genzyme Corporation). Incubation: 10 mg/ml enzyme, 80 h, 20°C.

Membranes allowed the creation of two aqueous environments: one, at low pH for the enzyme and, two, at high pH for product removal. Two membrane devices were coupled together with a common organic phase shuttling the ibuprofen from one environment to the other.

Figure 8.6 shows a schematic of a multiphase membrane reactor coupled with a membrane extractor. The membrane reactor is the same as before, except the buffer pH is 5.0 and only 6 g of lipase was loaded. The product alcohol is collected in the buffer. However, as the ibupro-

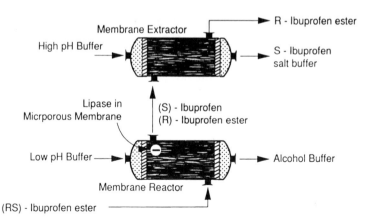

Figure 8.6. Ibuprofen resolution in a multiphase membrane bioreactor coupled with a membrane extractor.

fen remains protonated at pH 5, it is extracted into the water-immiscible ester phase. The ester with the (S)-ibuprofen acid is shuttled to the membrane extractor, where at high pH, the acid is stripped from the ester and neutralized as its salt. pH control monitors the extraction rate. Acid-free ester is returned to the membrane reactor for further enzymatic hydrolysis.

The productivity of this reactor design is shown below.

Substrate: Ibuprofen trifluoroethyl ester
Enzyme: 6 g *Candida* lipase
                 pH 5.0, 20°C
Membrane: Two 0.6 m² modules
Productivity: 1.2 μmol/h/mg at x = 15%
                 12.0 kg acid/yr-m²
Optical purity: 84% ee

The enzyme productivity or specific activity increased sixfold for the last run. While this table shows data only for short runs (1–3 days), long-term performance was also improved. However, the optical purity of the product was still unacceptably low.

Based on the substrate screening work, altering the choice of ester should allow greater enantiomeric purity (see Table 8.1). Reactivity must be sacrificed to gain the required selectivity. Condition II was repeated, except with the methoxyethyl ester. The reactor productivity dropped by about 50%, but the optical purity of the (S)-ibuprofen rose from 84% to an acceptable 96% enantiomeric excess:

Substrate: Ibuprofen methoxyethyl ester
Enzyme: 6 g *Candida* lipase
                 pH 5.0, 20°C
Membrane: Two 0.6 m² modules
Productivity: 0.7 μmol/h/mg at x = 15%
                 6.8 kg acid/yr-m²
Optical purity: 96% ee

Multiple kilograms of product were produced with this process by continuous operation over 3–4 months. The measured enzyme half-life in bioreactor operation was about 1 month.

CONCLUSIONS

A novel, multiphase membrane reactor, has been described, along with results of its use for conducting biocatalytic reactions in biphasic systems. The reactor was used to contact a sparingly soluble substrate, a simple ester of ibuprofen, with an immobilized enzyme, an enantioselective lipase from *Candida cylindracea*. The use of the membrane as a phase contactor, phase separator, and interfacial catalyst were demonstrated. A simple, reversible enzyme "immobilization" scheme was incorporated into the reactor for economical operation.

Significant process improvements were realized by taking advantage of the compartmentalization properties of a membrane; that is, creating two aqueous environments in the reactor system. One environment was maintained optimum for the enzyme (low pH to minimize poisoning) and one environment for the substrate (high pH to strip away the product). Continued study has further improved the ibuprofen process by increasing the water solubility of the ester. This has made the substrate more available to the immobilized enzyme, thus reducing diffusional limitations.

REFERENCES

Adams, S. S., P. Bresloff, and C. G. Mason. 1976. *J. Pharm. Pharmacol.* **28:** 256–257.

Cesti, P., and P. Piccardi. 1988. U.S. Patent 4,762,793.

Chen, C.-S., Y. Fujimoto, and C. J. Sih. 1981. *JACS* **103:** 3580–3582.

Chen, C.-S., Y. Fujimoto, G. Girdaukas, and C. J. Sih. 1982. *JACS* **104:** 7294–7299.

Gu, Q.-M., C.-S. Chen, and C. J. Sih. 1986. *Tetrahedron Lett.* **27:** 1763–1766.

Jones, J. B., and J. F. Beck. 1976. Asymmetric synthesis and resolutions using enzymes. In *Applications of Biochemical Systems in Organic Chemistry*, pp. 112–401. (J. B. Jones, C. J. Sih, and D. Perlman, eds.). New York: Wiley-Interscience.

Klibanov, A. M. 1986. *Chem. Tech.* **6:** 354–359.

Matson, S. L. 1989a. U.S. Patent 4,800,162.

Matson, S. L. 1989b. U.S. Patent 4,795,704.

Sih, C. J. 1987. European Patent Application 0 227 078.

Whitesides, G. M., and C. H. Wong. 1983. *Aldrichima Acta.* **16:** 27–34.

9
Microbial Reduction of Carbonyl Compounds
A Way to Pheromone Synthesis

ANNIE FAUVE

The control of stereochemistry during bioconversion reactions catalyzed by intact cells of various microbial species has been studied in our laboratory for the last decade. The definition of relative stereochemistry during microbiological reduction of C=C double bonds of α,β unsaturated ketones (Kergomard, Renard, and Veschambre 1982) and during reduction of carbonyl groups (Belan et al. 1987; Fauve and Veschambre 1988) has been studied extensively. Depending on the microbial cell involved, the stereochemical course of the reaction can lead to almost all of the possible stereoisomers of the reduced compounds. Chiral ketones and alcohols thus formed are valuable chiral building blocks in natural products synthesis.

Chiral molecules, natural compounds or chemicals, showing a biological activity (e.g., drugs, pesticides, pheromones) generally owe their biological properties to the well-defined geometry of their molecular structure. Chiral interactions are involved in molecular recognition processes. The use of the "wrong" enantiomer or diastereoisomer may have a different effect, sometimes with hazardous results. The biological tests on new drugs or biologically active compounds have to be carried out on all the purified enantiomeric forms of the molecule.

Enantiospecific synthesis of all enantiomers of natural compounds by use of biocatalysts is exemplified by our work on insect pheromones.

WHAT IS A PHEROMONE?

"A pheromone is a highly active chemical messenger secreted by a member of an animal species which elicits a definite behavior in other members of the same species" (Rossi 1977). Numerous insect behaviors are controlled by these communication substances: sex pheromones, aggregation pheromones, alarm pheromones (etc.). Insect pheromonal communication involves either one or several compounds, some of which are aliphatic, alicyclic, or heterocyclic, and most of which are chiral (Baker and Herbert 1984). Whereas one enantiomer attracts the insect species, the "wrong" enantiomer may act as a repellant. To a certain extent, some pesticides and fumigants are becoming less effective due to insect resistance; pheromones offer an alternative for insect control. By mass trapping with insect lures, crop or storage protection could be accomplished in one step in a "fatal attraction" procedure, initiated by the pheromones and called "Attract Them and Kill Them."

Because the quantity of pheromones isolated from the natural sources, i.e., from ground insects, is often less than a milligram, numerous stereospecific syntheses have been attempted. For a number of pheromones, though, the enantioselective synthesis of the active part remains an important challenge in organic chemistry. Examples of this difficulty are given by the enantioselective synthesis of simple compounds, chiral alcohols or ketones, active as pheromones on many pest insects. The described syntheses are numerous but generally lengthy and not so effective in preparation of all enantiomers in quantities required for accurate biological tests.

We reasoned that either of the enantiomers of biologically active alcohols or ketones could be obtained more easily by bioreduction of the corresponding saturated or unsaturated ketones, as already shown in our systematic studies. Enantiomeric alcohols, ketols, diols, and ketones and some bicyclo derivatives with pheromonal activity have been successfully obtained in this manner.

METHODOLOGY

There are different ways for obtaining all enantiomers of a desired compound by bioconversion. You may start a screening to find strains containing enzymes with opposite enantiotopic face specificities for the same substrate. Alternatively, you may act on the metabolic regulation of a given species. By modifications in culture conditions you may induce or repress the biosynthesis of such enzymes and study separately their stereospecificities. Modifications of the bioconversion conditions may inhibit or enhance some enzymatic activities and thus lead to different products. Another possibility is to exploit the structural substrate specificity of one enzyme by inverting sterically larger and smaller substituents of the substrate. All these possibilities have been exploited in our work on pheromone synthesis.

Experiments were carried out using whole cells of bacteria, yeasts, and fungi, either under growing or resting conditions. This procedure avoided any potential cofactor regeneration problems. All starting chemicals were carbonyl compounds commercially available or easily obtained in a one-vessel synthesis. After incubation in rotary shakers, under aerobic conditions unless otherwise stated, reduced compounds were extracted from the aqueous medium by organic solvents, purified, and analyzed. The absolute configurations and enantiomeric purities were established by chiropic comparisons and by analyses [gas chromatography (GC) on chiral phase, ^{1}H nuclear magnetic resonance, etc.] This brief description is not intended as an Experimental Section; the described syntheses are published and all details will be found in the corresponding articles (Belan et al. 1987; Fauve and Veschambre 1987; Bel-Rhlid, Fauve, and Veschambre 1989; Dauphin, Fauve, and Veschambre 1989).

CHIRAL ALCOHOLS OR DERIVATIVES
6-Methylhept-5-en-2-ol

This compound, commonly referred to as sulcatol, is the aggregation pheromone of ambrosia beetles, *Gnathotricus* spp., pest insects responsible for severe damage in coniferous forests of the North Pacific Coast (Byrne et al. 1974). The chirality of the molecule is involved in its biological activity. One species, *Gnathotricus sulcatus*, responds to a mixture of 65% of (2S) enantiomer and 35% of (2R) enantiomer (Fig. 9.1). Another species, *G. retusus*, sensitive only to the (S) enantiomer, appears to be inhibited by the (R) enantiomer.

As shown in Table 9.1, we were able to obtain both enantiomers of sulcatol by microbial reduction of the corresponding prochiral ketone,

Figure 9.1. Enantiomers of sulcatol.

6-methylhept-5-en-2-one. The best result for the (*S*) enantiomer was obtained with a thermophilic bacterium, *Thermoanaerobium brockii*, used in growing conditions at 70°C. The optical purity obtained is better than that obtained with bakers' yeast. Resting mycelium of the fungus *Aspergillus niger* afforded the (*R*) enantiomer with the highest optical purity.

In Table 9.1 are also displayed results obtained with one particular anaerobic bacterium, *Clostridium tyrobutyricum*, grown in different media. These experiments illustrate the possibility for obtaining different stereospecific enzymes from a given cell by acting on the metabolic regulation. According to the carbon source supplied to the growing

Table 9.1. Preparation of (*R*) and (*S*) enantiomers of sulcatol by microbiological reduction of 6-methylhept-5-en-2-one.

MICROORGANISM	EXPERIMENTAL CONDITIONS	REACTION TIME (h)	ABSOLUTE CONFIGURATION	ENANTIOMERIC EXCESS (%)
Bakers' yeast	Fermenting	120	(2*S*)	94
Thermoanaerobium brockii	Growing	24	(2*S*)	99
Clostridium tyrobutyricum grown on crotonic acid	Resting	48	(2*S*)	88
Clostridium tyrobutyricum grown on glucose	Resting	48	(2*R*)	80
Geotrichum candidum	Resting	17	(2*R*)	92
Aspergillus niger	Resting	24	(2*R*)	96

Adapted from Belan et al. 1987.

medium the reduction leads to either the (S) or the (R) enantiomer in satisfactory optical yields.

Other biocatalysts were used to obtain both enantiomers of sulcatol. Lipase resolution of racemic sulcatol as well as the use of pure enzymes in organic solvent are described in our previous article (Belan et al. 1987). All methodologies are discussed in terms of availability, cost, specific equipment, and chemical and optical yields.

This biocatalyzed synthesis of sulcatol enantiomers compares favorably with other approaches that start either from chiral synthons (Mori 1981) or from chiral building blocks of the so-called "chiral pool" (Takano, Gato, and Ogasawara 1982). Mori (1981) comments on the fact that bakers' yeast does not reduce 6-methylhept-5-en-2-one. Our results show that with some brands of bakers' yeast we could reduce this ketone. In general, a screening among various species, strains, and brands can rapidly answer one's expectations and it is important not to stop after one failure.

4-Methyl-5-hydroxyheptan-3-one

This compound has been identified as an aggregation-promoting pheromone secreted by the adult male of both rice weevil (*Sitophilus oryzae*) and maize weevil (*S. zeamais*) (Schmuff et al. 1984), and accordingly has been named sitophilure. Massive infestation of stored cereal grains by these pest insects, feeding directly on grain kernels, accounts for important economic losses throughout the world. The enantiomeric composition of the pheromone has been carefully studied by Burkholder (Walgenbach et al. 1987). This compound possesses two asymmetric centers and thus four stereoisomers have been synthesized for biological tests (Mori and Ebata 1986). Both *Sitophilus* spp. proved to respond to *syn* isomers (4R, 5S) and (4S, 5R) although most strongly to the latter. Maize weevils also showed a low but significant attraction to *anti* isomers (4R, 5R) and (4S, 5S).

Because the reported syntheses were long (13 steps) with low yields in chiral compounds (3% overall yield) we carefully studied the microbial reduction of the corresponding prochiral diketone, 4-methylheptane-3,5-dione. This β-diketone was obtained in a one-vessel synthesis by acylation of pentane-3-one by ethyl propionate.

The only microorganism found to reduce this compound was a yeast-like fungus, *Geotrichum candidum*. Reduction by other microbial cells (yeasts, fungi, and bacteria) proved to be either unsuccessful or poorly efficient. Thus, trying to obtain as many stereoisomers as possible, we studied the reducing ability of this species under various experimental conditions which can bring into play enzymes with different stereospecificities. Moreover, like other microbial species, *G. candidum* possesses a fermentative metabolism. It could be assumed that alcohol dehydrogenases involved in this metabolic pathway may have different enantiotopic face specificities than those involved in the oxidative metabolism.

Table 9.2 shows the results obtained after incubation of the β-diketone with *G. candidum* cells, harvested after 48 h of aerated growth, under aerobic and anaerobic bioconversion conditions. The major component of the pheromone, *syn* derivative (4*R*, 5*S*)-(−)-4-methyl-5-hydroxyheptan-3-one, is obtained under anaerobic conditions. GC analyses showed that whatever the sugar added for fermentation (dextrose, saccharose, lactose) the reaction took place and no starting ketone remained in solution, even in proof bottles where no sugar had been added. Under aerobic conditions, without any nutrient added to the mixture of fungal cells and diketone in water, the reaction also took place but led to an *anti* isomer, (4*S*, 5*S*)-(+)-4-methyl-5-hydroxyheptan-3-one. No diol has been detected in either experiment. In Figure 9.2 are shown the two stereoisomers of the pheromone obtained from the β-diketone.

The simplicity of this synthesis, described in full in our previous article (Fauve and Veschambre 1987), makes it very convenient for a large-scale synthesis of two stereoisomers of the pheromone sitophi-

Table 9.2. Reduction of 4-methylheptan-3,5-dione by *G. candidum* cells.

Experimental conditions	Aerobic	Anaerobic
Incubation time	48 h	48 h
Isolated yield	50%	70%
Absolute configuration of 4-methyl-5-hydroxyheptan-3-one	(4*S*, 5*S*)	(4*R*, 5*S*)
Enantiomeric excess	70%	100%

Adapted from Fauve and Veschambre 1987.

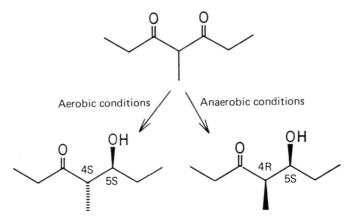

Figure 9.2. Reduction of 4-methylheptan-3,5-dione by *G. candidum* cells.

lure. Moreover, this work shows that a given species placed under different experimental conditions can afford enzymatic systems with dissimilar stereoselectivity toward the same substrate.

Octan-2,3-diol and 2-Hydroxyoctan-3-one

The male sex pheromone of an important pest of Japanese vineyards, the grape borer *Xylotrechus pyrrhoderus*, has been identified as a mixture of (2S, 3S)-(−)-octan-2,3-diol and (2S)-(+)-2-hydroxyoctan-3-one in a ratio of 80 : 20 to 95 : 5 (Sakai et al. 1984) (Fig. 9.3).

A number of methods have been reported for the synthesis of the two components of the pheromone. They require at least six or seven steps to obtain chiral molecules (Mori and Otsuka 1985 and references cited herein). Our predicted synthesis, microbial reduction of octan-2,3-

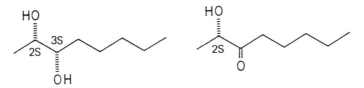

Figure 9.3. Pheromone components of *X. pyrrhoderus*.

dione, appeared much shorter and easier because α-diketones are rapidly reduced and can afford many stereoisomers of hydroxy compounds.

The starting dione has been obtained by hydrolysis of the oxime of 2-octanone. Resting cells able to reduce such a molecule were screened from our stock cultures and none of them failed. Either or both carbonyl groups can be reduced and thus, according to the choice of the microorganism, various stereoisomers of ketol and diols were obtained (Bel-Rhlid, Fauve, and Veschambre 1989).

To obtain the major component of the pheromone $(2S, 3S)$-octan-2,3-diol, we selected a fungus, *Beauveria sulfurescens*. Good yield and high enantiomeric excess were obtained after incubation of octan-2,3-dione with the resting mycelium of the fungus. In Table 9.3 are also displayed the microbial strains that led to other stereoisomers of the molecule, assumed to be biologically inactive on the grape borer *X. pyrrhoderus*.

The minor component of the pheromone $(2S)$-2-hydroxyoctan-3-one, biologically active when mixed to $(2S, 3S)$ diol, is more difficult to obtain. It is formed with all microorganisms but rapidly transformed into $(2S, 3R)$ or $(2S, 3S)$ diols. Only with *S. cerevisiae* used as commercial bakers' yeast has this compound been isolated. Under nonfermenting conditions, i.e., aerobically in distilled water without any nu-

Table 9.3. Microbiological reduction of octan-2,3-dione.

MICROORGANISM	INCUBATION TIME (h)	MAJOR PRODUCT	ABSOLUTE CONFIGURATION	ENANTIOMERIC EXCESS (%)
Bakers' yeast	1	2-OH, 3-keto	$(2S)$	90
	24	2,3-diol	$(2S, 3R)$	99
Beauveria sulfurescens	1	3-OH, 2-keto	$(3S)$	92
	24	2,3-diol	$(2S, 3S)$	99
Geotrichum candidum	7	2,3-diol	$(2R, 3S)$	97
Aspergillus niger	5	2,3-diol	$(2R, 3R)$[a]	92

Adapted from Bel-Rhlid, Fauve, and Veschambre 1989, with permission of *Journal of Organic Chemistry*. Copyright 1989, American Chemical Society.

[a] Obtained as a minor product, only with this fungus.

trient, this yeast reduces octanedione more slowly than the other microbial cells. After 1 h of incubation this compound is obtained as the major product along with isomeric (3 S) ketol. At present, we have not found a strain leading to the enantiomeric (2R) ketol.

For this pheromone again, provided the "right" microorganism be used for reduction, either of the two components has been obtained in a more convenient synthesis than previously reported.

1,3-Dimethyl-2,9-dioxabicyclo [3.3.1] nonane

This substance has been isolated from the bark of Norway spruce infested by a parasite, the ambrosia beetle *Trypodendron lineatum*. It has been suggested that it serves as a primary attractant for this species (Heemann and Francke 1976), and for that reason, it has been synthesized by various routes (Kongkathip and Kongkathip 1984 and references cited herein). It bears three asymmetric centers but due to the bicyclic form of the molecule there are only four stereoisomers, two *endo* and two *exo* molecules (Fig. 9.4). Most of the syntheses have given racemates and only one enantiospecific synthesis could provide the four stereoisomers after many steps (Redlich et al. 1983). Biological tests on racemates have shown that the parasite is attracted to the *endo* molecule. The natural enantiomeric composition of the molecule remains unknown.

As racemic pheromone was obtained by cyclization of racemic 8-nonen-2,4 diols (Kongkathip and Kongkathip 1984), we had good reason to believe that chiral bicyclo molecules could be obtained from

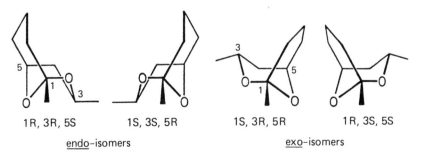

| 1R, 3R, 5S | 1S, 3S, 5R | 1S, 3R, 5R | 1R, 3S, 5S |

endo–isomers   exo–isomers

Figure 9.4. Stereoisomers of 1,3-dimethyl-2,9-dioxabicyclo [3.3.1.] nonane.

optically active diols. Starting from 8-nonen-2,4-dione, we were able to obtain two enantiomeric ketols, (2*S*) and (2*R*), as shown in our systematic investigation on reduction of β-diketones (Fauve and Veschambre 1988). Chemical reduction of each enantiomer afforded two diastereoisomeric diols easily separable and thus the four diastereoisomers have been obtained (Fig. 9.5). Subsequent cyclization of each diastereoisomer led to the four stereoisomers of the pheromone.

Biological and chemical steps are fully described in the corresponding article (Dauphin, Fauve, and Veschambre 1989). This chemoenzymatic synthesis is the first report that the four stereoisomers can be synthesized in only four steps. The previously reported synthesis, starting from a sugar of the "chiral pool," required about 10 steps (Redlich et al. 1983).

CHIRAL KETONES

Some pheromones are short-chain carbonyl compounds bearing substituents stereospecifically important for their biological activity. In our previous investigations we had shown that chiral molecules with a methyl group α to a carbonyl group can be obtained by microbial reduction of the corresponding α,β unsaturated ketone. We had shown that this reaction was a general property of living cells (Desrut et al. 1983). Moreover, working with one species, *B. sulfurescens*, we had proposed a rule to predict the absolute configuration of the reduced compound according to the size of the substituents of the C=C double bond, sterically smaller and larger (Kergomard, Renard, and Veschambre 1982). We intended to exploit this finding for the synthesis of some chiral ketones with pheromonal activity.

3-Methylhexan-2-one

This substance is the alarm pheromone of the triatomine bug *Dipetalogaster maximus* (Rossiter and Staddon 1983). This bloodsucking heteroptere is an insect vector important in the transmission of the pathogen *Trypanosoma cruzi*. No biological tests have been carried out on

Figure 9.5. Chemoenzymatic pathway to stereoisomers of 1,3-dimethyl-2,9-dioxabicyclo [3.3.1] nonane. (Dauphine, Fauve, and Veschambre, 1989. Reprinted with permission of *Journal of Organic Chemistry*. Copyright 1989, American Chemical Society.)

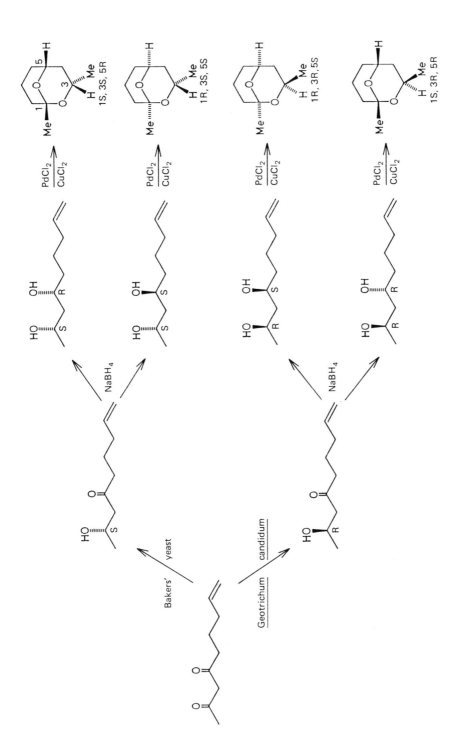

the two enantiomers (3R) and (3S) and the enantiomeric composition of the pheromone is unknown.

In 1982, we reported on the synthesis of one enantiomer (3S)-(+)-3-methylhexan-2-one by microbial reduction of 3-methylhex-3-en-2-one by growing mycelium of the fungus B. *sulfurescens* (Kergomard, Renard, and Veschambre 1982). Since then, we have shown that the enone-reductase system involved in this reaction is inducible (Fauve, Renard, and Veschambre 1987) and therefore, after induction, it is possible to work with resting mycelium which achieve higher yields by decreasing undesirable metabolic side reactions. With induced resting mycelium the reaction time has decreased from 48 h to 6 h.

To obtain the enantiomer, (3R)-(−)-3-methylhexan-2-one, the same fungus has been used, under the same experimental conditions, only the substrate was different. According to the stereochemical rule proposed by our group, reduction of 3-methylenhexan-2-one by B. *sulfurescens* leads to the (3R) enantiomer of 3-methylhexan-2-one (Fig. 9.6).

The biological activity of these two compounds is currently being investigated on *Dipetalogaster maximus* and their detailed synthesis will be published soon. Other pheromones with chiral methyl groups α to a carbonyl are being synthesized according to this procedure (e.g., 4-methylhexan-3-one, 4-methylheptan-3-one). Corresponding chiral alcohols are sometimes pheromones themselves (Kato and Mori 1985).

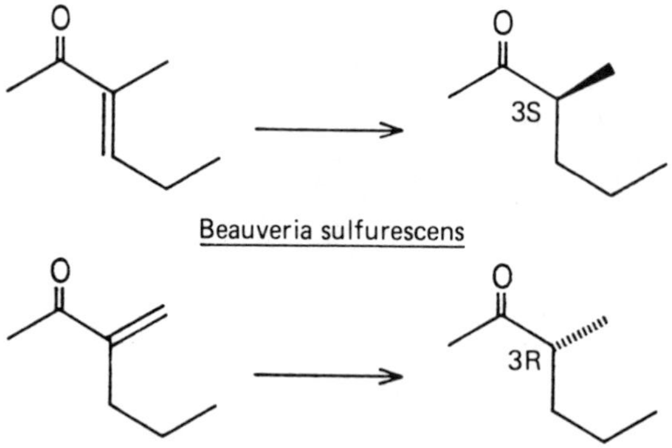

Figure 9.6. Preparation of enantiomers of 3-methylhexan-2-one by microbiological reduction of α,β-unsaturated carbonyl compounds.

Work on these substances is currently under investigation by our group.

CONCLUSION

The pheromones syntheses described above proved to be very efficient in terms of optical yields and number of steps. They all compete favorably with previously reported syntheses and illustrate the value of biocatalysts in enantioselective organic synthesis. Application of our systematic studies on microbial reduction as exemplified by these results may convince the synthetic practitioners of the potential of biocatalysis. This methodology is therefore competing with chemical catalysts and classical resolution. The recent breakthroughs made by chiral auxiliaries and chiral catalysts make it even more challenging.

ACKNOWLEDGMENTS

This work was supported by the CNRS within the framework of ATP "Messagers chimiques." It has been done in collaboration with Dr. M. F. Renard and Dr. H. Veschambre. I wish to express my gratitude for their help and friendship. We all gratefully acknowledge Professor A. Kergomard, who recognized very early the versatility of biocatalysts in organic synthesis and I personally dedicate this chapter to him on his retirement.

REFERENCES

Baker, R., and R. H. Herbert. 1984. Insect pheromones and related natural products *Natur. Prod. Rep.* 299–315.

Belan, A., J. Bolte, A. Fauve, J. G. Gourcy, and H. Veschambre. 1987. Use of biological systems for the preparation of chiral molecules 3. An application in pheromones synthesis: preparation of sulcatol enantiomers. *J. Org. Chem.* **52:** 256–260.

Bel-Rhlid, R., A. Fauve, and H. Veschambre. 1989. Use of biological systems for the preparation of chiral molecules 7. Synthesis of the pheromone components of the grape borer *Xylotrechus pyrrhoderus* by microbiological reduction of an α-diketone. *J. Org. Chem.* **54:** 3221–3223.

Byrne, K. J., A. A. Swigar, R. M. Silverstein, J. H. Borden, and E. Stokkink. 1974. Sulcatol, population aggregation pheromone of scolytic beetle *Gnathotricus sulcatus. J. Insect Physiol.* **20:** 1895.

Dauphin, G., A. Fauve, and H. Veschambre. 1989. Use of biological systems for the preparation of chiral molecules 6. Preparation of stereoisomeric-2,4 diols: synthesis and conformational study of bicyclo derivatives, isomeric components of a pheromone of *Trypodendron lineatum*. *J. Org. Chem.* **54:** 2238–2242.

Desrut, M., A. Kergomard, M. F. Renard, and H. Veschambre. 1983. Microbial reduction of α,β-unsaturated carbonyl compounds: a general property? *Biochem. Biophys. Res. Commun.* **110:** 908–912.

Fauve, A., and H. Veschambre. 1987. Use of biological systems for the preparation of chiral molecules 4. A two-step chemoenzymatic synthesis of a natural pheromone (4R, 5S)-(−)-4-methyl-5-hydroxyheptan-3-one, Sitophilure. *Tetrahedron Lett.* **28:** 5037–5040.

Fauve, A., and H. Veschambre. 1988. Use of biological systems for the preparation of chiral molecules 5. Microbiological reduction of acyclic β-diketones. *J. Org. Chem.* **53:** 5215–5219.

Fauve, A., M. F. Renard, and H. Veschambre. 1987. Inducibility of an enone-reductase system in the fungus *Beauveria sulfurescens*: application in enantioselective organic synthesis. *J. Org. Chem.* **52:** 4893–4897.

Heemann, V., and W. Francke. 1976. 1,3-Dimethyl-2,9-dioxabicyclo [3.3.1.] nonane; a host-specific substance in Norway spruce under attack by *Trypodendron lineatum*. *Oliv. Naturwissensch.* **63:** 344.

Kato, M., and K. Mori. 1985. Synthesis of (3R, 4S)-4-methyl-3-hexanol, the pheromone of the ant *Tetramorium impurum*. *Agric. Biol. Chem.* **49:** 3073–3075.

Kergomard, A., M. F. Renard, and H. Veschambre. 1982. Microbiological reduction of α,β-unsaturated ketones by *Beauveria sulfurescens*. *J. Org. Chem.* **47:** 792–798.

Kongkathip, B., and N. Kongkathip. 1984. Short-step synthesis of 1,3-dimethyl-2,9-dioxabicyclo [3.3.1.] nonane: an insect attractant. *Tetrahedron Lett.* **25:** 2175–2176.

Mori, K. 1981. A simple synthesis of (S)-(+)-sulcatol, the pheromone of *Gnathotricus retusus*, employing bakers' yeast for asymmetric reduction. *Tetrahedron* **37:** 1341–1342.

Mori, K., and T. Ebata. 1986. Synthesis of the four possible stereoisomers of 5-hydroxy-4-methyl-3 heptanone (sitophilure), the aggregation pheromone of the rice weevil and the maize weevil. *Tetrahedron* **42:** 4421–4426.

Mori, K., and T. Otsuka. 1985. Synthesis of (2S, 3S)-2,3-octanediol and (S)-2-hydroxy-3-octanone, the male sex pheromone of the grape borer *Xylotrechus pyrrhoderus*. *Tetrahedron* **41:** 553–556.

Redlich, H., B. Schneider, R. W. Hoffmann, and K.-J. Geueke. 1983. Synthese der vier isomeren 1,3-Dimethyl-2,9-dioxabicyclo [3.3.1.] nonane. *Lieb. Ann. Chem.* 393–411.

Rossi, R. 1977. Insect pheromones. 1. Synthesis of achiral components of insect pheromones. *Synthesis* 817–836.

Rossiter, M., and B. W. Staddon. 1983. 3-Methyl-2-hexanone from the triatomine bug *Dipetalogaster maximus* (Uhler) (Heteroptera; Reduviidae). *Experientia* **39:** 380–381.

Sakai, T., Y. Nakagawa, J. Takahashi, K. Iwabuchi, and K. Ischii. 1984. Isolation and identification of the male sex pheromone of the grape borer *Xylotrechus pyrrhoderus* Bates (Coleoptera; Carambycidae). *Chem. Lett.* 263–264.

Schmuff, N. R., J. K. Phillips, W. E. Burkholder, H. M. Fales, C.-W. Chen, P. P. Roller, and M. Ma. 1984. The chemical identification of the rice weevil and maize weevil aggregation pheromone. *Tetrahedron Lett.* **25:** 1533–1534.

Takano, S., E. Gato, and K. Ogasawara. 1982. A simple enantioselective synthesis of both enantiomers of sulcatol using a single chiral precursor. *Chem. Lett.* 1913–1914.

Walgenbach, C., J. K. Phillips, W. E. Burkholder, G. G. S. King, K. N. Slessor, and K. Mori. 1987. Determination of chirality in 5-hydroxy-4-methyl-3-heptanone, the aggregation pheromone of *Sitophilure oryzae* (L.) and *S. zeamais* Motschulsky. *J. Chem. Ecol.* **13:** 2159–2169.

10
Resolution of Binaphthols and Spirobiindanols Using Pancreas Extracts

Romas J. Kazlauskas

Pancreatic cholesterol esterase (EC 3.1.1.13), also known as carboxyl ester lipase (for reviews see Brockerhoff and Jensen 1974; Rudd and Brockman 1984), has been used to resolve 2-bromo alcohols (Chenault et al. 1987) and intermediates for the synthesis of chorismic and shikimic acids (Pawlak and Berchtold 1987). This chapter reports that cholesterol esterase (CE) can also be used to resolve binaphthols and spirobiindanols.

The interest in enantiomerically pure binaphthols stems from their usefulness as chiral auxiliaries (Miyano and Hashimoto 1986). For example, binaphthol, 1, has been used in chiral aluminum hydrides for reduction of carbonyl compounds (Noyori et al. 1984), in chiral titanium catalysts for the ene reaction (Mikami, Terada, and Nakai 1989), and in chiral crown ethers for complexation of amino acids (Cram 1988).

The best current preparation of enantiomerically pure 1 involves fractional crystallization of the diastereomic cinchonine salts of 1-cyclic phosphate ester (Jacques and Fouquay 1988). The phosphate ester is subsequently methylated with dimethyl sulfate and cleaved by reduction with an aluminum hydride (Truesdale 1988). This method is laborious, but is preferred to other methods because it can be carried out on a synthetic scale (~10 g) and yields material of >99.5% enantiomeric purity. The current standards of enantioselective organic synthesis require at least these levels of enantiomeric purity for a chiral auxiliary.

A number of other methods for resolution of 1 have been reported including chromatography using chiral stationary phase (Pirkle and

Schreiner 1981; Okamoto et al. 1981) enantiospecific hydrolyses using microorganisms (Fujimoto, Iwadate, and Ikewawa 1985; Chen, Girdaukas, and Sih 1985) or enzymes (Miyano et al. 1987), fractional crystallizations of copper–amphetamine complexes (Brussee et al. 1985), and complexation with chiral amides (Toda and Tanaka 1988). Enantioselective syntheses have also been reported (Miyano et al. 1985). The disadvantages of these methods are that they have not been demonstrated on a synthetic scale, are difficult or expensive to carry out, or yield material of insufficient enantiomeric purity.

The CE method described in this chapter meets the requirements of simplicity, ability to run on a synthetic scale, and >99% enantiomeric excess product. Further, it is simpler and involves fewer manipulations than the cinchonine method.

Spirobiindanols, like binaphthols, are chiral molecules. Enantiomerically pure 6 has been prepared previously in five steps after resolution of the corresponding diacid by fractional crystallization of the dibrucine salt (Hagishita et al. 1971). The substituted spirobiindanol 8 has been resolved by fractional crystallization of the diastereomeric carbamate derivative. Resolution of these compounds using CE is substantially simpler and may lead to the use of these compounds as chiral auxiliaries. Because the hydroxyl groups are further apart in the spirobiindanols (~7.4 Å) than in binaphthols (~4.1 Å), the spirobiindanols cannot chelate a single metal center. Nevertheless, they may be useful auxiliaries in bimetallic systems and in chiral cavities.

RESULTS
CE Has a Broad Substrate Specificity

Pancreatic CE catalyzes the hydrolysis of binaphthol, spirobiindanol, and sterol esters (Tables 10.1 and 10.2). Esters of the binaphthol, 1, and the octahydrobinaphthol, 3, are hydrolyzed at the same rate as cholesterol acetate. Hydrolysis is completely specific for the (S)-enantiomer. Esters of the bromobinaphthols 2 and 4 were not substrates for CE.

The butanoate of the benzindanol, 5, was hydrolyzed with moderate enantiospecificity for the (R)-enantiomer. Esters of the spirobiindanols 6, 7, and 8 were hydrolyzed with good enantiospecificity for the (R)-enantiomer. The diacetate of the spirobenzpyranol 9 was hydrolyzed with moderate enantiospecificity.

The sterol specificity (Table 10.2) indicates that equatorial acetates were good substrates while axial acetates were not hydrolyzed. Pancreatic CE, not an impurity, catalyzes the hydrolysis of binaphthol and spirobiindanol esters. The highest specific activity (2× partially purified CE) was observed using a highly purified sample of bovine CE. This sample was >90% pure by gel filtration and migrated as a single band on sodium dodecyl sulfate-(SDS)-polyacrylamide gel electrophoresis with an apparent molecular weight of 59K, consistent with the reported molecular weight of 63–67K (Van den Bosch et al. 1973). Partially purified bovine CE showed additional proteins migrating with apparent molecular weights of 43K, 26K, and 15K.

Cholesterol esterase is activated by taurocholate, a bile salt that induces association of monomers into more active dimers (Momsen and Brockman 1977) and hexamers (Hyun et al. 1971) and aids formation of an emulsion (Lynn, Chuaqui, and Clevette-Radford 1982). Omission of taurocholate in a hydrolysis of 6-dipentanoate reduced the rate of hydrolysis by a factor of two, consistent with the notion that hydrolytic activity is due to CE. Competitive hydrolysis of a 1 : 1 mixture of cholesterol acetate and binaphthol acetate proceeded with the same overall rate as a single compound and yielded equal amounts of each alcohol, consistent with the notion that the same catalyst was responsible for both hydrolyses. Neither porcine liver esterase nor purified pancreatic lipase catalyzed the hydrolysis of 6-dibutanoate (<0.01 U/mg). Crude porcine pancreatic lipase showed traces of CE activity (~0.002 U/mg) probably due to impurities of CE.

Substrates Were Hydrolyzed in Water–Ether Emulsions

Taurocholate-stabilized emulsions of ethyl ether and phosphate buffer were used to dissolve these water-insoluble substrates. Toluene or methylisobutyl ketone was also used as the organic phase, but no hydrolysis (relative rate <0.01) was observed using methylene chloride. While the initial rates of hydrolysis, as determined by pH-stat or HPLC, were similar for both cholesterol acetate and the unnatural esters, these two-phase conditions were not optimal. Cholesterol esterase was less active toward cholesterol acetate emulsified in ethyl ether–water than toward cholesterol acetate emulsified in water with a nonionic surfactant: $V_{max} = 0.5$ versus 2.0 U/mg for a partly purified

Table 10.1. Substrate specificity of CE.

SUBSTRATE		ESTER	RELATIVE[a] RATE	% ee OF[b] DIOL	E_1^c	E_2^c
	R_1 = H, 1	Diacetate	1.1	>95 S	ND	3.5 ± 0.1 S
		Dipropanoate	1.1	>95 S	ND	5.2 ± 0.3 S
		Dibutanoate	0.5	>95 S	ND	3.5 ± 0.4 S
		Dipentanoate	1.1	>95 S	>400 S^d	4.9 ± 0.2 S
		Dihexanoate	0.5	>95 S	ND	ND
		Diheptanoate	0.3	>95 S	ND	ND
		Dioctanoate	0.4	>95 S	ND	ND
	R_1 = Br, 2	Diacetate	<0.03[e]	—	—	—
		Dibutanoate	<0.001	—	—	—
	R_1 = H, 3	Diacetate	0.4[f]	ND	ND	ND
		Dipentanoate	0.2	>90 S^g	ND	ND
	R_1 = Br, 4	Dipentanoate	<0.004	—	—	—

Compound	Ester				
5	Butyrate	0.12[f]	39 R[h]	2.9 R	1.7 ± 0.2 S
$R_1 = R_2 = H$, 6	Diacetate	1.0	61 R	ND	ND
	Dipropanoate	0.8[i]	ND	ND	ND
	Dibutanoate	0.6	61 R	ND	1.6 ± 0.1 S
	Di-2-methyl propanoate	0.08	37 R	ND	ND
	Dipentanoate	1.3	79 R	ND	1.1 ± 0.1 S
	Dihexanoate	0.3	90 R	9.4 ± 2.2 R	1.2 ± 0.1 S
	Diheptanoate	0.2	88 R	ND	ND
	Dioctanoate	0.1	85 R	ND	ND
	Dinonanoate	0.07	90 R	ND	ND
$R_1 = CH_3, R_2 = H$, 7	Dihexanoate	0.1	>80[j]	ND	ND
$R_1 = CH_3, R_2 = Br$, 8	Dibutanoate	0.007	91 R[k]	ND	ND
9	Diacetate	1.0	43[l]	2.6 ± 0.6[m]	ND

Table 10.1. Substrate specificity of CE. (*continued*)

SUBSTRATE	ESTER	RELATIVE[a] RATE	% ee OF[b] DIOL	E_1^c	E_2^c
10	Diethyl ester	<0.001	—	—	—

[a] Rate of hydrolysis (relative to cholesterol acetate) was measured using a pH-stat which measured the amount of base (0.1 *M* NaOH) required to maintain the pH at 7.00 for an emulsion of ester (1.0 mmol) dissolved in ethyl ether (10 ml) and aqueous buffer (10 ml of 10 m*M* phosphate, pH 7 containing 30 mg sodium taurocholate monohydrate). The specific activity of CE from bovine pancreas (Genzyme lot. no. 3846) under these conditions was 0.14 U/mg with cholesterol acetate. ND, not determined. unit = μmol of ester hydrolyzed/min at 25°C.

[b] Enantiomeric purity of the product after ~15% hydrolysis.

[c] E_1 = enantiomeric ratio for the first hydrolysis (diester to monoester), E_2 = enantiomeric ratio for the second hydrolysis (monoester to diol).

[d] Based on a limit of >99% ee for the monoester product at 40% conversion.

[e] Poor solubility limits the concentration of ester in the ether to 14 m*M*. The rate of hydrolysis of 1-diacetate is ~4× slower at 14 m*M* than at 100 m*M*. The limit of 0.03 was calculated based on this slower rate.

[f] The concentration of the ester in the ether phase was 50 m*M*.

[g] Determined by ¹H-NMR using Eu[(+)-hfc]₃.

[h] Enantiomeric purity of alcohol was calculated based on a measured value of 26% ee *S* for the ester recovered after 40% conversion.

[i] Substrate was not completely dissolved in the ether.

[j] The diastereomeric camphanate diesters are incompletely resolved on HPLC. The order of elution suggests that the preferred enantiomer is the same as with the unsubstituted spirobiindane.

[k] Enantiomeric purity was determined from the optical rotation of the diol isolated by TLC after 58% hydrolysis.

[l] Enantiomeric purity of combined diol and monoester after 31% hydrolysis.

[m] Average of the values determined from the enantiomeric purity of the diester fraction and the diol/monoester fraction.

Reprinted with permission from *J. Am. Chem. Soc.* **111**, 1989, 4953–4959. Copyright 1989 American Chemical Society.

Table 10.2. Steroid specificity of CE.

SUBSTRATE	RELATIVE RATE[a]
C_8H_{17} (AcO structure)	1.0
C_8H_{17} (AcO structure)	1.5
C_8H_{17} (AcO structure)	1.2
C_8H_{17} (AcO structure)	<0.0005[b]
C_8H_{17} (RCOO structure)	<0.01[c]

[a] Determined as in Table 10.1, note a.
[b] The concentration of the ester in the ether phase was 55 mM.
[c] For synthesis of the oleate (Hernadez and Chaikoff, 1957).

sample of CE. The apparent K_m was also higher in the water–ether emulsions: 200 mM for cholesterol acetate in the ether phase while a K_m of 0.5 mM was reported for water-soluble phenyl acetates (Lynn, Chuaqui, and Clevette-Radford, et al. 1982). This lower activity and increased K_m is probably due to reduced availability of substrate when it partitions into the organic phase. In spite of the lower activity, water–ether emulsions were used to survey the substrate specificity of CE because these conditions mimic conditions used for synthesis.

Each Step of the Hydrolysis Is Enantiospecific

The enantiomeric ratio, E, is a measure of the enzyme's ability to discriminate between two enantiomers and is independent of substrate concentration (Chen et al. 1982). The equations for a simple, single-step reaction, Eqs. (1) and (2), were applied to the two-step hydrolysis by treating each step separately (Fig. 10.1).

$$E = \frac{\ln\{(1 - c)[1 - ee(S)]\}}{\ln\{(1 - c)[1 + ee(S)]\}}, \tag{1}$$

$$E = \frac{\ln\{1 - c[1 + ee(P)]\}}{\ln\{1 - c[1 - ee(P)]\}}, \tag{2}$$

where c = extent of conversion, ee (S) = enantiomeric excess of the recovered substrate fraction, and ee (P) = enantiomeric excess of the product fraction. The disappearance of diester is affected only by the first hydrolysis, Eq. (3); therefore, the value of E_1 was

$$\text{diester} \xrightarrow{E_1} \text{monoester} \tag{3}$$

$$\text{monoester} \xrightarrow{E_2} \text{diol} \tag{4}$$

determined by measuring the percent ee of the recovered diester and applying Eq. (1). Alternatively, the enantiomeric purity of the combined monoester and diol was determined and Eq. (2) was used. This alternate approach set a lower limit for E_1 during the hydrolysis of 1-dipentanoate. The second step, hydrolysis of the monoester to diol [eq. (4)], is a simple, single-step reaction when racemic monoester is used as the substrate. Either Eq. (1) or (2) was applied in a straightforward manner to determine E_2.

The enantiomeric ratios for 1-diesters show that CE is specific for the (S)-enantiomer in both steps of the hydrolysis; the first hydrolysis is considerably more specific than the second. On the other hand, the

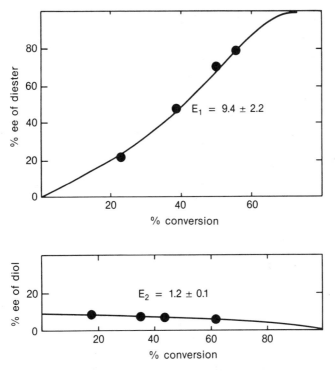

Figure 10.1. Determination of the enantiospecificity of each step in the hydrolysis of **6**-dihexanoate to the diol. The top curve shows data for the determination of the enantiomeric ratio for the first step: hydrolysis of **6**-dihexanoate to **6**-monohexanoate. The enantiomeric excess of (S)-enantiomer in the unreacted **6**-dihexanoate was measured as a function of conversion. The curve was calculated using Eq. (1) where $E = 9.4$. The bottom curve shows data for the determination of the enantiomeric ratio for the second step: hydrolysis of **6**-monohexanoate to **6**. The enantiomeric excess of the (S)-enantiomer in the product **6** was measured as a function of conversion. The curve shown was calculated using Eq. (2) where $E = 1.2$. (Reprinted with permission from *J. Am. Chem. Soc.* **111**, 1989, 4953–4959. Copyright 1989 American Chemical Society.)

two steps of the hydrolysis of **6**-dihexanoate show opposite enantiospecificity. The first step favors hydrolysis of the (R)-enantiomer, while the second step favors the (S)-enantiomer. The enantiospecificity is higher in the first step; thus the overall reaction yields an excess of the (R)-enantiometer.

Impure Cholesterol Esterase Can Also Be Used as a Catalyst

Crude extracts of pancreas also catalyzed the hydrolysis of binaphthol and spirobiindanol esters. Both homogenized porcine pancreas and bovine pancreas acetone powder catalyzed hydrolysis of 1-diacetate with enantiospecificities identical to purified CE [>95% ee (S)]. Pancreas acetone powder catalyzed hydrolysis of 6-diacetate, 6-dibutanoate, and 6-dihexanoate with identical enantiospecificities as purified CE [75% ee (R), 70% ee (R), 95% ee (R) at low conversion]. These crude preparations had lower specific activity than purified CE: ~2 U/g dry weight for porcine pancreas and ~14 U/g solid for pancreas acetone powder. Pancreas acetone powder is the least expensive source of CE activity; $50 worth (100 g) was enough to resolve 200 g of 1 (see below).

The stability of CE activity in bovine pancreas acetone powder was lower than in purified CE. The crude material had a half-life of ~2 days (phosphate buffer, 0.1 M, pH 7, 25°C, 5 mM taurocholate, 0.1% sodium azide) while the purified material showed <10% loss after 5 days under the same conditions. The lowered stability was presumably due to degradation by trypsin because addition of trypsin inhibitors (2 mM benzamidine or 0.25 mg/ml soybean trypsin inhibitor) to pancreas acetone powder increased the stability (~10% loss after 5 days).

Binaphthol and Spirobiindanol Were Resolved on a Synthetic Scale

Bovine pancreas acetone powder was used as the chiral catalyst to resolve 1 on a 200-g scale (Fig. 10.2).

The dipentanoyl ester was chosen for the synthetic-scale resolution because it is a good substrate and differs sufficiently from binaphthol to simplify separation after hydrolysis. Hydrolysis slowed and stopped when approximately half of the 1-dipentanoate had been hydrolyzed. Binaphthol and the diester were separated by crystallization, yielding 67 g (66% of theory) of (S)-1 having an enantiomeric purity of >99.95% ee (Fig. 10.3) and after hydrolysis of the diester, 64 g, 63% of theory, of (R)-1 having 98.8% ee.

High ee was expected for the (S)-enantiomer due to the high degree of specificity of CE for this enantiomer. The unexpected result was that

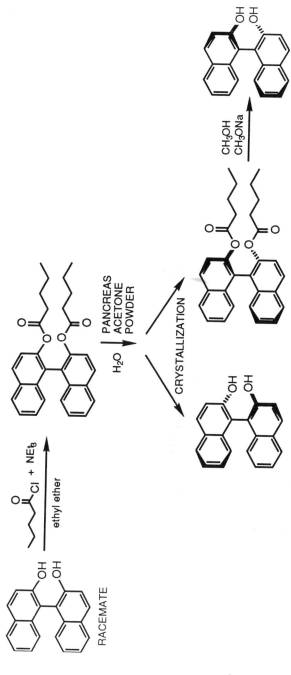

Figure 10.2. Resolution of binaphthol by CE-catalyzed hydrolysis of binaphthol dipentanoate. Bovine pancreas acetone powder is used as a crude source of cholesterol esterase activity.

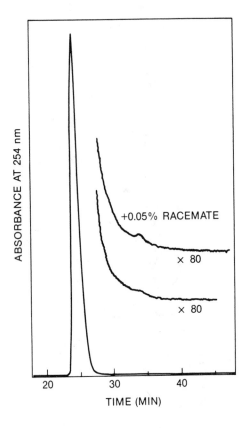

Figure 10.3. HPLC analysis establishes the enantiomeric purity of (S)-$(-)$-**1** to be $>99.95\%$. The upper traces show the region containing the peak for (R)-$(+)$-**1** amplified both before and after addition of racemate.

the (R)-enantiomer could also be isolated in high ee. After enzymic reaction a maximum of 96% ee (R) was expected for the unhydrolyzed diester because 52% of the original diester remained. Fortunately, **1**-dipentanoate crystallizes as a conglomerate; thus the crystallization used to isolate it also raised the enantiomeric purity to 99%.

In a similar manner the spirobiindanol **6** was resolved by hydrolyzing the dihexanoate using partially purified CE as the chiral catalyst. Because this hydrolysis was less enantioselective than the hydrolysis of **1**-dipentanoate, the enantiomeric purity of the diol after 35% hydrolysis was only 82% ee (R). After separation of the dihexanoate from the diol by column chromatography, the enantiomeric purity of the diol was increased to $>95\%$ by selective precipitation of racemic diol in CH_2Cl_2. Racemic **6** forms high melting point (213.5–216.5°C), slightly soluble crystals (32 mM, CH$_2$Cl$_2$; 15 mM, toluene, 25°C) while the enantiomerically pure material forms low melting point (125–127°C), very soluble crystals (1,900 mM, CH$_2$Cl$_2$, 760 mM, toluene, 25°C).

DISCUSSION

The Three-Dimensional Shape of the Substrates Determines Enantiospecificity

It is surprising that CE, whose presumed function is to hydrolyze steroid esters, can also enantiospecifically hydrolyze such apparently different molecules as binaphthol and spirobiindanol esters. Some similarities in the three-dimensional shapes of the preferred substrates can rationalize this result. Comparison of the structures of the two enantiomers of **1**-diacetate with the structure of cholesterol acetate (Fig. 10.4), shows a slightly similar front view as the A and B rings of cholesterol are aligned parallel to one of the naphthol rings in binaphthol. The carbonyl carbon of each ester might adopt a similar orientation for attack by the serine —OH. In the top view, all three structures look different and lopsided. It is satisfying to find that both cholesterol acetate and the preferred enantiomer of 1-diacetate are lopsided in the same direction.

This comparison can be extended to the spirobiindanols (Fig. 10.4). The preferred enantiomer of spirobiindanol, (*R*), has a similar three-dimensional shape to the preferred enantiomer of binaphthol, (*S*), in spite of the difference in nomenclature. Further, the opposite enantiospecificity of 6-di- and monohexanoate can be rationalized because the shape of the two are different. The ancillary ester group in the dihexanonate contributes to the bulk of group pointing in the page in Figure 10.5, while in the monoester the group pointing toward the back (phenol ring) and the one pointing toward the front (part of the dimethyl pentane ring) are more similar in size. Thus, for the monohexanoate a reduced, even reversed, enantiospecificity is predicted from the three-dimensional shape.

The relative rates of hydrolysis of steroid esters by CE show that the equatorial diastereomer is preferentially hydrolyzed. This preference parallels the increased chemical reactivity of equatorial over axial diastereomers (Barton 1950). Because diastereomers differ in physical and chemical properties as well as shape, the diastereospecificity of an

FRONT VIEW

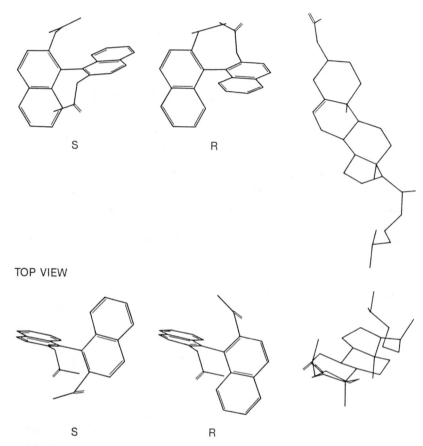

TOP VIEW

Figure 10.4. A comparison of the structures of cholesterol acetate and the two enantiomers of 1-diacetate. In the front view the A and B rings of cholesterol have been aligned with one of the naphthyl rings to emphasize their similarity. The two enantiomers differ in the orientation of the out-of-plane naphthyl ring. The top view shows that both cholesterol acetate and (S)-1-diacetate enantiomer have large groups pointing in the same direction. (Reprinted with permission from *J. Am. Chem. Soc.* **111**, 1989, 4953–4959. Copyright 1989 American Chemical Society.)

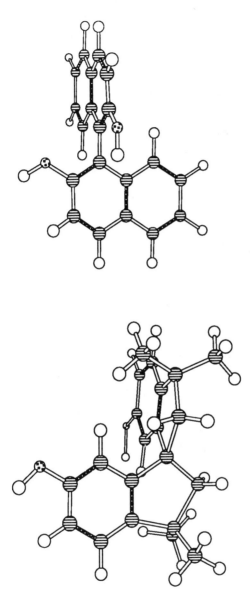

Figure 10.5. Three-dimensional models of (R)-(+)-spirobiindanol and (S)-(−)-binaphthol both show a similar shape. One ring is oriented in the plane of the paper while the bulk of the other ring extends into the paper. Structures were drawn using ChemDraw, Cambridge Scientific Computing, Cambridge, MA.

enzyme may be determined by factors other than the space available at the active site. Because enantiomers differ only in shape, the enantiospecificity of an enzyme gives reliable information about the space available at the active site.

Resolutions That Use Multiple Enzymic Steps Can Give Higher ee

A disadvantage of enzymes with a broad substrate specificity is that they often show imperfect enantiospecificity. When the substrate is a *meso* or prochiral compound, an enzyme-catalyzed enantioselective synthesis can be coupled to a kinetic resolution to enhance enantiomeric purity (Wang et al. 1984; Schreiber, Schreiber and Smith 1987). Another class of substrates that may show enhanced enantiomeric purity are multifunctional substrates, for example, diesters. Resolution of a diester can give higher enantiomeric purity than resolution of a monoester because the diester interacts with the enzyme twice (Fig. 10.6). A theoretical comparison of the enantiomeric purity of the product obtained from a single-step resolution and from a two-step resolution is given in Figure 10.7 for the special case where the enantiospecificity of each step is the same. These calculations show that for an enzyme of a given enantiospecificity, the two-step resolution gives higher enantiomeric purity than a single-step resolution. For example, an enzyme having an enantiospecificity of $E > 9$ yields product with $>90\%$ ee at 40% conversion during a two-step resolution, while a single-step resolution requires an enzyme with an enantiospecificity of $E > 35$ for the same purity product. Thus, substrates that require two enzymic reactions during resolutions can yield product of higher enantiomeric purity than similar resolutions involving only a single reaction.

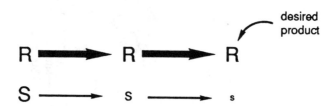

Figure 10.6. Multistep resolutions can enhance enantiomeric purity.

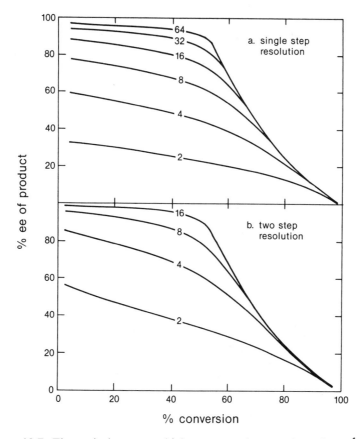

Figure 10.7. Theoretical curves which compare the enantiomeric purity of product as a function of conversion for a single- and two-step resolution. The different lines represent enzymes with different enantiospecificities and show that for a given enantiospecificity the two-step resolution yields product of higher enantiomeric purity. (Reprinted with permission from *J. Am. Chem. Soc.* **111**, 1989, 4953–4959. Copyright 1989 American Chemical Society.)

Unfortunately, the data in this chapter are not sufficient to test these calculations. The specificity of the first step of the resolution of **1** is so high that it is experimentally difficult to determine whether the second step affects enantiomeric purity. For spirobiindanol, resolution is probably worsened by the two-step process because the two steps have opposite enantiospecificity. Quantitative analysis indicated that the enantiomeric purity of **6** during a resolution was slightly higher than expected based on the measured values of E_1 and E_2. This suggests that the irreversible two-step model may be an oversimplification for this reaction. Nevertheless, the suggested strategy of using multistep resolutions to enhance the enantiomeric purities appears theoretically valid and may find use in other kinetic resolutions.

EXPERIMENTAL SECTION

Enantiomeric Purity

Enantiomers of **1** and 1-monoesters were separated using an HPLC column containing ionically bonded N-3,5-dinitrobenzoyl-D-phenylglycine (Regis Chemical Company, Morton Grove, IL, U.S.A.) eluted with hexane containing 5% (vol/vol) isopropanol (Pirkle and Schreiner 1981). The enantiomeric purity of the dipentanoate was determined after cleavage to binaphthol. A sample of dipentanoate was dissolved in methanol containing equimolar amount of sodium methoxide. After 30 min the solution was neutralized with excess acetic acid and analyzed by HPLC. To determine enantiomeric purities >99% ee, chromatograms of unknown and unknown with 0.1% deliberately added racemate were compared.

Enantiomeric purities of **5** and **8** were determined by optical rotation. Enantiomeric purity of **6** was determined by separation of diastereomeric camphanate derivatives by HPLC. To a sample of **6** (2 μmol, 50 μl of a 0.04 M solution) was added (1R)-(−)-camphanic acid chloride (20 μmol, 100 μl of a 0.2 M solution in CH_2Cl_2) and triethylamine (36 μmol, 5 μl). After 10 min at room temperature this mixture was diluted with acetonitrile and the diastereomers were separated by HPLC on a reverse-phase C18 column eluted with a gradient of 70–100% acetonitrile over 20 min. The extinction coefficients of the two diastereomers were assumed to be equal because racemic material gave equal peak areas to within integration errors (2%). To determine the enantiomeric purity of **6**-dihexanoate, samples were isolated by preparative TLC (silica gel eluted with methylene chloride), hydrolyzed to the diol with base, and derivatized with camphanic acid chloride as described above.

Enantiomeric purities of all other compounds were determined by ^{1}H-nuclear magnetic resonance using Eu[(+)-hfc]$_3$ as the chiral shift reagent. In a typical procedure a sample (20 mg) of the alcohol was isolated by preparative TLC (silica gel eluted with CH_2Cl_2) and acetylated with excess acetyl chloride (30 μl) and triethylamine (56 μl) in ethyl ether (1.0 ml) for 15 min. After washing with aqueous $KHCO_3$ and drying over $MgSO_4$ the ether was evaporated and a portion of the residue (~7 mg) was dissolved in 0.6 ml of C_6D_6. Shift reagent dissolved in C_6D_6 (100 mg/ml) was added in 50-μl portions until separation of the acetyl methyl resonances was observed, typically 200–300 μl. Baseline separation was not achieved; areas were estimated by cut and weigh.

Resolution of 1,1'-Binaphthalene-2,2'diol, 1

Pentanoyl chloride (185 ml, 1.56 mol) was added over 20 min to a suspension of 1,1'-binaphthalene-2,2'-diol (203 g, 0.71 mol) in ethyl ether (2 L) containing triethylamine (215 ml, 1.54 mol). After stirring for an additional hour the mixture was washed twice with aqueous sodium bicarbonate (2 L of 1 M) and once with water (2 L). Analysis by reverse-phase HPLC shows no detectable binaphthol and only 0.2% monopentanoate. For high enantiomeric purity of product it is important that no unreacted binaphthol remain at this point. The solution was diluted to 4 L with ethyl ether and stirred with phosphate buffer (pH 7.5, 0.1 M, 4 L) containing sodium taurocholate (60 g of crude material from ox bile, Sigma) to form an emulsion. Efficient stirring to emulsify the mixture completely is essential to get high conversion. Bovine pancreas acetone powder (125 g) was added and the pH of the aqueous phase was measured several times a day and readjusted to at 7.2 ± 0.2 with sodium hydroxide (1 M, 650 ml were required). The hydrolysis stopped after 3 days; HPLC showed 45.0% diol, 3% monopentanoate, and 52% dipentanoate. The mixture was transferred to a separatory funnel, ethanol (400 ml) was added, and the two phases were allowed to settle for 4 h. The brown aqueous phase was discarded and the remaining emulsion in the yellow ether phase was broken up by addition of magnesium sulfate (200 g). The ether phase was dried over additional magnesium sulfate, filtered, and concentrated to 300 ml. Toluene (500 ml) were added and the solution was cooled to 4°C overnight. The fine white crystals (71 g) were collected by filtration and washed twice with 20 ml portions of cold toluene. The filtrate was

concentrated to ~600 ml by rotary evaporation and cooled again to
4°C. The additional 10 g of crystals that form were collected by filtration and washed twice with 20 ml portions of toluene. Recrystallization of the combined crystals from 375–400 ml of toluene yielded 67 g
of white crystals, mp 211–213.5°C; >99% diol, >99.9% enantiomeric
purity.

Toluene in the filtrate above was removed by rotary evaporation, 400
mL of hexane were added and the solution cooled to 4°C overnight.
Yellow crystals (133 g) formed and were collected by filtration and
washed twice with 20 mL of hexane. Recrystallization from 500 mL of
methanol yielded pure binaphthol dipentanoate, mp 63–64°C, 96 g,
>99% ee R. This crystallization is essential to obtain the (R)-enantio-
mer with high ee. Sodium methoxide (0.15 mol) was added to the
combined dipentanoate crystals (96 g) in methanol (1 L). After 4 h at
room temperature the solution was neutralized with concentrated HCl
(~10 ml), and ethyl ether (1 L), toluene (0.5 L), and phosphate buffer (1
L of 0.1 M, pH 7) were added. The organic layer was separated,
washed once with water (1 L), dried over magnesium sulfate, and
concentrated to 300 ml. The white crystals which separated were collected by filtration, washed with cold toluene, and dried to yield (R)-
(+)-1,1'-binaphthalene-2,2'-diol (52 g, >99% diol, 98.8% ee R).

ACKNOWLEDGMENTS

I thank Dr. Gary Faler, Dr. Bruce F. Johnson, Dr. Thomas Guggen-
heim, and Professor Thomas Katz for gifts of compounds, and Dr. Ivar
Giaever for his help in calculating the curves in Figure 10.5.

REFERENCES

Barton, D. H. R. 1950. The conformations of the steroid nucleus. *Experientia*
6: 316–320.

Brockerhoff, H., and R. G. Jensen. 1974. *Lipolytic Enzymes*, pp. 176–193.
New York: Academic Press.

Brussee, J., J. L. G. Groemendijk, J. M. te Kopple, and A. C. A. Janse. 1985.
On the mechanism of the formation of S(−)-(1,1'-binaphthalene)-2,2'-diol
via copper(II) amine complexes. *Tetrahedron* 41: 3313–3319.

Chen, C.-S., Y. Fujimoto, G. Girdaukas, and C. J. Sih. 1982. Quantitative
analyses of biochemical kinetic resolutions of enantiomers. *J. Am. Chem.
Soc.* 104: 7294–7299.

Chen, C.-S., G. Girdaukas, and C. J. Sih. 1985. Bifunctional chiral synthons via biochemical methods. VII. Optically-active 2,2'-dihydroxy-1,1'-binaphthyl. *Tetrahedron Lett.* **26:** 4323–4326.

Chenault, H. K., M.-J., Kim, A. Akiyama, T. Miyazawa, E. S. Simon, and G. M. Whitesides. 1987. Enzymatic routes to enantiomerically enriched 1-butene oxide. *J. Org. Chem.* **52:** 2608–2611.

Cram, D. J. 1988. The design of molecular hosts, guests, and their complexes. *Science* **240:** 760–767.

Fujimoto, Y., H. Iwadate, and N. Ikewawa. 1985. Preparation of optically active 2,2'-dihydroxy-1,1'-binaphthyl via microbial resolution of the corresponding racemic diester. *J. Chem. Soc. Chem. Commun.* 1333–1334.

Hagishita, S., K. Kuriyama, M. Hayashi, Y. Nakano, K. Shingu, and M. Nakagawa. 1971. Optical activity of bis-1,1'-spiroindanes. I. Optical resolution and absolute configuration. *Bull. Chem. Soc. Jpn.* **44:** 496–505.

Hernandez, H. H., and I. L. Chaikoff. 1957. Purification and properties of pancreatic cholesterol esterase. *J. Biol. Chem.* **228:** 447.

Hyun, J., M. Steinberg, C. R. Treadwell, and G. V. Vahouny. 1971. Cholesterol esterase—a polymeric enzyme. *Biochem. Biophys. Res. Commun.* **44:** 819–825.

Jacques, J., and C. Fouquay. 1988. Enantiomeric (S)-(+)- and (R)-(−)-1,1'-binaphthyl-2,2'-diyl hydrogen phosphate. *Org. Synth.* **67:** 1–12.

Kazlauskas, R. J. 1989. Resolution of binaphthols and spirobiindanols using cholesterol esterase. *J. Am. Chem. Soc.* **111:** 4953–4959; U.S. Patent #4,879,421.

Kazlauskas, R. J. (S)-(−)- and (R)-(+)-1,1'-Bi-2-naphthol. *Org. Synth.*, submitted.

Lynn, K. R., C. A. Chuaqui, and N. A. Clevette-Radford. 1982. Kinetic studies of mammalian and microbial cholesterol esterases in homogeneous aqueous solution. *Bioorg. Chem.* **11:** 19–23.

Maruoka, K., T. Itoh, T. Shirasaka, and H. Yamamoto. 1988. Asymmetric hetero-Diels–Alder reaction catalyzed by chiral organoaluminum reagent. *J. Am. Chem. Soc.* **110:** 310–312.

Mikami, K., M. Terada, and T. Nakai. 1989. Asymmetric glyoxylate-ene reaction catalyzed by chiral titanium complexes: a practical access to alpha-hydroxy esters in high enantiomeric purities. *J. Am. Chem. Soc.* **111:** 1940–1941.

Miyano, S., and H. Hashimoto. 1986. Preparation and application of axially chiral biaryls to asymmetric reactions. *Yuki Gosei Kagaku Kyokai* **44:** 713–725 (in Japanese).

Miyano, S., K. Shimizu, S. Sato, and H. Hashimoto. 1985. The Ullmann coupling reaction of axially chiral (S)-2,2'-bis-(1-iodo-2-naphthyloxycarbonyl)-1,1'-binaphthyl. *Bull. Chem. Soc. Jpn.* **58:** 1345–1346.

Miyano, S., K. Kawahara, Y. Imone, and H. Hashimoto. 1987. A convenient preparation of optically active 1,1'-binaphthyl-2,2'-diol via enzymatic hydrolysis of the racemic diester. *Chem. Lett.* 355–356.

Momsen, W. E., and H. L. Brockman. 1977. Purification and characterization of cholesterol esterase from porcine pancreas. *Biochim. Biophys. Acta* **486:** 103–113.

Noyori, R., I. Tomino, Y. Tanimoto, and M. Nishizawa. 1984. Rational designing of chiral reducing agents. Highly enantioselective reduction of aromatic ketones by binaphthol-modified lithium aluminum hydride reagents. *J. Am. Chem. Soc.* **106:** 6709–6716.

Okamoto, Y., S. Honda, I. Okamoto, and H. Yuki. 1981. Novel packing material for optical resolution: (+)-poly(triphenylmethyl methyacrylate) coated on macroporous silica gel. *J. Am. Chem. Soc.* **103:** 6971–6973.

Pawlak, J. L., and G. A. Berchtold. 1987. Total synthesis of (−)-chorismic acid and (−)-shikimic acid. *J. Org. Chem.* **52:** 1765–1771.

Pirkle, W. H., and J. L. Schreiner. 1981. Chiral high-pressure liquid chromatographic stationary phases. 4. Separatio of the enantiomers of bi-β-naphthols and analogues. *J. Org. Chem.* **46:** 4988–4991.

Rudd, E. A., and H. L. Brockman. 1984. In *Lipases*, pp. 185–204. (B. Borgstrom, and H. L. Brockman, eds.). Amsterdam: Elsevier.

Schreiber, S. L., T. S. Schreiber, and D. B. Smith. 1987. Reactions that proceed with a combination of enantiotopic group and diastereotopic face selectivity can deliver products with a very high enantiomeric excess: experimental support of a mathematical model. *J. Am. Chem. Soc.* **109:** 1525–1529.

Toda, F., and K. Tanaka. 1988. Efficient optical resolution of 2,2'-dihydroxy-1,1'-binaphthyl and related compounds by complex formation with novel chiral host compounds derived from tartaric acid. *J. Org. Chem.* **53:** 3607–3609.

Truesdale, L. K. 1988. (*R*)-(+)-1,1'-binaphthalene-2,2'-diol. *Org. Synth.* **67:** 13–19.

Van den Bosch, H., A. J. Aarsman, G. N. De Jong, and L. L. M. Van Deenen. 1973. The lysophospholipase described in these papers is likely identical to cholesterol esterase. *Biochim. Biophys. Acta* **296:** 94–104; 105–115.

Wang, Y.-F., C.-S. Chen, G. Girdaukas, and C. J. Sih. 1984. Bifunctional chiral synthons via biochemical methods. 3. Optical purity enhancement in enzymatic asymmetric catalysis. *J. Am. Chem. Soc.* **106:** 3695–3696.

11

Chiral Synthons by New Oxidoreductases and Methodologies

HELMUT SIMON

For decades it has been known that the preparation of certain chiral compounds can be accomplished with oxidoreductases. The overwhelming majority of reactions that have been carried out involve the reduction of keto groups to chiral secondary alcohols or derivatives thereof. Many substrates can be reduced by alcohol dehydrogenase from liver or microbial sources, and for the former and some others it can be predicted which compound may be a substrate and which stereochemical course may be expected. Also the stereoselective dehydrogenation of prochiral dialcohols is a valuable approach to many chiral compounds. Many groups contributed to the use of dehydrogenases or microorganisms containing such enzymes and a system for the regeneration of the pyridine nucleotides. Only a few can be mentioned here, with recent articles cited from which their earlier work can be followed. Jones et al. mainly worked with horse liver alcohol dehydrogenase and Sih et al. with yeasts (Dodds and Jones 1988; Vanmiddlesworth and Sih 1987). Wong and Whitesides worked out important kinetic and technical aspects of coenzyme regeneration on a preparative scale for many systems (Chenault, Simon, and Whitesides 1988). Recently they showed which 2-oxocarboxylates can be stereospecifically reduced to the corresponding (2S)-hydroxycarboxylates with lactate dehydrogenases from various sources (Kim and Whitesides 1988). Kula and Wandrey established the feasibility of reductions, especially reductive aminations of 2-oxoacids and the formation of 2-hydroxycarboxylates with regeneration of NADH on a technical scale (Hummel et al. 1987).

So far, pyridine nucleotide-dependent reductases that have been completely or partially purified can be used only with enzymatic coenzyme regeneration. That means, in general, that two enzymes are necessary.

Glucose is usually the electron donor if yeasts or similar organisms are applied. For biochemical reasons this leads to rather low efficiencies, usually in the range of 5–200, if the productivity of cells or crude extracts is expressed by the ratio

$$PN = mmol\ product/biocatalyst\ (dry\ weight)\ kg \times time\ (h)$$

(Simon et al. 1985).

New and different approaches to the synthesis of chiral compounds will be described in the following section.

WHAT IS MEANT BY "NEW" ENZYMES AND METHODOLOGIES?

1. The results are unconventional because, by using whole cells, productivity numbers are obtained which are typically one and sometimes up to three orders of magnitude higher than those obtained using conventional methods. That means that 100–200 g of product can easily be prepared with normal laboratory equipment.

2. New types of reactions are catalyzed by newly detected enzymes. Yeasts are infrequently utilized and if they are applied, the method is modified (Thanos and Simon 1986). Resting cells of anaerobes such as clostridia or anaerobically grown *Proteus vulgaris* are employed.

3. As electron donors, hydrogen gas, formate, carbon monoxide, or the cathode of an electrochemical cell rather than glucose are used.

4. Most of the oxido-reductions that will be described are not pyridine nucleotide-dependent, but—and this is very important—the enzymes accept electrons from, or deliver electrons to, artificial mediators. Mostly these mediators are viologens of different redox potential as shown in Table 11.1.

Methylviologen (MV), which was used for many years as a total herbicide, and benzylviologen (BV) are commercially available. Their redox potential, E_0' is −443 and −360 mV, respectively. The tetramethylviologen (TMV) is about 225 mV more negative than carbamoyl

Table 11.1. Various useful viologens as artificial mediators.

STRUCTURE	E'_O (mV)	ABBREVIATION
$H_2NCOH_2C - N \overset{+}{\bigcirc} \overset{+}{\bigcirc} N - CH_2CONH_2$	−296	CAV
$\bigcirc - CH_2 - N \overset{+}{\bigcirc} \overset{+}{\bigcirc} N - CH_2 - \bigcirc$	−360	BV
$CH_3 - N \overset{+}{\bigcirc} \overset{+}{\bigcirc} N - (CH_2)_3 - \overset{+}{N}H_3$	−401	APMV
$CH_3 - N \overset{+}{\bigcirc} \overset{+}{\bigcirc} N - CH_3$	−443	MV
$H_3C \overset{CH_3}{\bigcirc} N+ \ +N \bigcirc$	−515	p-DMDQ
$H_3C - N \overset{+}{\bigcirc} \overset{+}{\bigcirc} \overset{CH_3}{N} - CH_3 \quad CH_3$	−553	TMV

methylviologen (CAV). Properties of the latter are described by Günther, Neumann, and Simon (1987).

5. By using mediators of different redox potentials, the equilibrium constants of redox reactions can be considerably affected. The aminopropyl group containing viologen (AMPV) can be bound to carboxyl group containing systems. Recently it turned out that the commercially available cobalt sepulchrate also can act as a mediator in microbial reductions using hydrogen gas or formate as electron donors or the cathode of an electrochemical cell (Günther and Simon unpublished).

For example, cobalt sepulchrate was known for its capability to reduce myoglobins and horseradish peroxidase (Balahura and Wilkins 1983).

PRACTICAL ASPECTS

Most of the reductions carried out with resting cells can be summarized by Reaction (1):

$$\rangle\!\!=\!\!X + H_2 \text{ or HCOOH or } (CO + H_2O) \rightarrow H\rangle\!\!-\!\!XH + (CO_2) \quad (1)$$

This in turn is the sum of the following two reactions if an artificial electron carrier such as a viologen is used:

$$2V^{2+} + H_2 \text{ or HCOOH or } (CO + H_2O) \rightarrow 2V^{+\cdot} + 2H^+ + (CO_2) \quad (2)$$

$$2V^{+\cdot} + \rangle\!\!=\!\!X + 2H^+ \rightarrow 2V^{2+} + H\rangle\!\!-\!\!XH \quad (3)$$

From a practical point of view, one can differentiate two cases of hydrogenations: (i) Reaction (2) (with hydrogen as electron donor) catalyzed by hydrogenase is rate limiting, or (ii) Reaction (3), the reductase reaction, is slower. In the latter case the solution shows the violet color of the viologen cation radicals.

In case (i) it is advisable to apply hydrogen gas under pressure. The K_m value of the hydrogenase for hydrogen in *Clostridium tyrobutyricum* is roughly 1.0 mM (Preiss, White, and Simon 1989). The solubility of H_2 in water at 35°C under 1 atm of H_2 is about 0.6 mM. That means the hydrogenase is far from saturation under normal pressure of hydrogen. The effect of increasing hydrogen pressure will be shown later.

Usually the described reactions have been conducted with microorganisms stored for months and even years under anaerobic conditions at −18°C. As lyophilized material, these cells can also be stored under anaerobic conditions at room temperature from 1 to 2 years with losses of enzyme activities of <10%.

C. tyrobutyricum, *P. vulgaris*, or *C. thermoaceticum* can also be used with hydrogen gas, formate, or CO as electron donors, respectively, without artificial mediators. This normally causes a decrease of the reaction rate by a factor of 1.5–3 and a lower long-term stability i.e., the half-lives of the systems may decrease from 40–80 to 20–30 h. Cobalt sepulchrate in 3–5 mM concentrations has about the same effect as 1 mM methylviologen. However, the color of the reduced viologens is in practice often an indicator of the stability of the system.

If the organisms or their crude extracts are used in preparative electrochemical cells, the mediators are necessary for carrying the electrons from the cathode to the enzymes or from there to the anode. Again cobalt sepulchrate can be used instead of viologens. The artificial mediators are also necessary if the enzymes are applied in a partially or completely purified form.

The reduced viologens and most of the enzymes discussed are sensitive to oxygen. In practice, this sensitivity is not a serious problem for preparative work. In such cases flushing the reaction vessels and solvent with nitrogen is sufficient. On the other hand, kinetic studies with these enzymes in cuvettes and bioconversions on a small scale of about 0.1 mmol requires special care. Although the aforenamed microbial cells are not commercially available yet, only materials available from Culture Collections have been used in this work. Culture conditions to obtain the cells for biocatalytical purposes have been described elsewhere (Thanos et al. 1987 and literature cited therein; White et al. 1989).

There are a few cases in which compounds react spontaneously with the reduced viologens. Examples are nitro groups or keto groups flanked by strongly electron-withdrawing groups, i.e., 4-acetylpyridine, esters of 4-chloroacetoacetate, phenylglyoxylic acid, or keto-isopheron. But these are the rare exceptions. In such cases the viologens must be omitted or cobalt sepulchrate may be used. For additional experimental details see Simon et al. (1985); Thanos and Simon (1986); Günther, Neumann, and Simon (1987); and Thanos et al. (1987).

REGENERATION OF VIOLOGENS

In contrast to pyridine nucleotides, viologens can be regenerated by chemical, electrochemical, as well as by enzymic means, as depicted by Figure 11.1.

The viologens which are in the oxidized form bis-cations (V^{2+}) take up one electron and are converted to the cation radicals ($V^{+\cdot}$). There

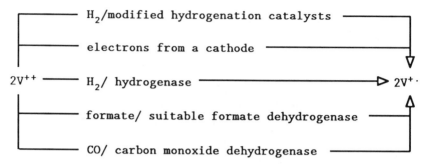

Figure 11.1. Various possibilities for the regeneration of reduced viologens.

are known reduction methods by hydrogenases, viologen-dependent formate dehydrogenases, or carbon monoxide dehydrogenases. To the best of our knowledge, the latter two have not been used for preparative work. Many anaerobic organisms contain the aforementioned enzymes. Recently modified hydrogenation catalysts were developed that reduce viologens selectively at the expense of hydrogen gas without attacking carbon–carbon or carbon–oxygen double bonds, or without destroying pyridine nucleotides (Thanos and Simon 1986). The latter effect is important since there are enzyme activities in many microbial organisms or other cells that catalyze one or both of the following reactions:

$$NAD^+ + H^+ + 2V^{+\cdot} \rightleftharpoons NADH + 2V^{2+} \qquad (4)$$

$$NADP^+ + H^+ + 2V^{+\cdot} \rightleftharpoons NADPH + 2V^{2+} \qquad (5)$$

Therefore, these modified catalysts can be used to couple hydrogen gas to reductions by yeasts or other cells possessing neither a hydrogenase, or viologen-dependent formate dehydrogenases, nor carbon monoxide dehydrogenases. The modification of the metal catalysts (Pt, Pd, Raney nickel) is achieved with highly fluorinated ZONYL® tensides (Thanos and Simon 1986). Using these catalysts or electrochemical cells leads to rather simple systems.

Figure 11.2 summarizes the flow of electrons which is probably valid for most cells. The aim is to reduce the double bond of the substrate

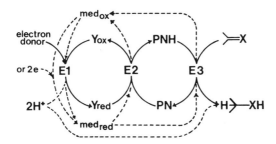

Figure 11.2. General scheme of the flow of electrons from various electron donors (hydrogen gas, formate, carbon monoxide, or an electrode). E1 may be a hydrogenase, viologen-dependent formate, or carbon monoxide dehydrogenase. Y symbolizes an endogeneous and med an artificial mediator. (For further explanations see text.)

$\rangle\!\!=\!\!X$ stereospecifically by enzyme E3 to a chiral product. If an E3 species is present that is able to accept electrons from artificial mediators that are reduced electrochemically or by the modified catalysts, only this enzyme (E3) is necessary. The need of only one enzyme can be an important advantage. One of the already mentioned enzymes E1 (Fig. 11.2) must be present if the artificial mediators shall be reduced enzymatically. Of course, there are many pyridine nucleotide-dependent reductases present in all cells. If E3 is pyridine nucleotide dependent and a pyridine nucleotide has to be regenerated, then E2 is necessary, too. Because many cells contain enzymes, often yet unknown, that are able to reduce pyridine nucleotides at the expense of reduced viologens [Reactions (4) and (5)] (Simon et al. 1985), such cells can be supplied with electrons by hydrogen gas or formate even when E3 is a pyridine nucleotide-dependent reductase (Thanos and Simon 1986). This shall be shown by one example.

The reduction of various alkyl esters of 4-chloro-3-oxobutyric acid with yeasts was studied carefully by Sih and Chen (1984). They observed productivity numbers of about 12. Application of *C. kluyveri* leads to productivity numbers of about 350 for the reduction of ethyl 4-chloro-3-oxobutyrate (Simon et al. 1985) and to about 1,800 in the presence of cobalt sepulchrate (Beer and Simon, unpublished). That means much smaller reaction volumes can be used if *C. kluyveri* is applied. Furthermore, the product has to be separated from <1% of that amount of biocatalyst that is present if yeasts are used. In other words, in this case 1 g of *C. kluyveri* with its pyridine nucleotide-dependent 3-oxocarboxylate reductase is as effective as 100–150 g of yeast.

HYDROGENATIONS WITH *C. TYROBUTYRICUM* CONTAINING HYDROGENASE AND ENOATE REDUCTASE

C. tyrobutyricum efficiently reduces 2-enoates without activation. It contains an enzyme referred to as enoate reductase (EC 1.3.1.31) (Kuno, Bacher, and Simon 1985, for a review see Simon, in press). This enzyme catalyzes various reactions as shown in Table 11.2, from which also the stereochemical course of the reaction can be seen. From a practical point of view, the reduction of enoates at the expense of $MV^{+\cdot}$ and this in turn with hydrogen gas is especially important. The substrate specificity of enoate reductase is astonishingly broad and the enantioselectivity very high (Simon et al. 1985). Table 11.3 shows a few types of substrates that have been studied kinetically with NADH. The use of $MV^{+\cdot}$ as electron donor leads to the same K_m values and relative rates for the substrates shown in Table 11.3. The reduction of an enoate with $MV^{+\cdot}$ is about 1.5 times faster than that with NADH [see Reactions (1) and (1a) in Table 11.2].

In addition to the substrates shown in Table 11.2, another 50 enoates have been hydrogenated with resting cells of *C. tyrobutyricum* on scales of up to 100 g.

Regarding the substrates, the following rules have been found to apply (Tables 11.2 and 11.3): ^1R should not be too large; if ^2R is a phenyl ring, $NHCOCH_3$ and OC_2H_5 residues are not accepted in place of NHCHO and OCH_3, respectively. The halogens F, Cl, Br, and I are tolerated in the α-position but are reductively eliminated in the β-position [Reaction (1b) of Table 11.2]. The choice of ^2R is subject to the least number of restrictions. If ^2R and ^3R are interchanged, i.e., if the E and Z isomers of substrates of the type $^3R^2RC{=}CHX$ (X = COO^- or CHO) are used, different enantiomers are produced. Therefore, E/Z mixtures of substrates in which the β-carbon atom becomes chiral by reduction cannot be employed if enantiomerically pure products are

Table 11.2. Reaction catalyzed by enoate reductase.

$$\overset{3}{R}\overset{^2R}{\underset{}{C}}=\overset{COO^-}{\underset{^2R}{C}}R^1 + NADH + H^+ \longrightarrow \begin{array}{c} H\;\;COO^- \\ {}^2R \diagup\diagdown \overset{}{R^1} \\ {}^3R\;\;H \end{array} + NAD^+ \qquad (1)$$

$$^3R^2RC{=}C^1RCOO^- + 2\,MV^{+\cdot} + 2\,H^+ \xrightarrow{\;\;\|\;\;} {}^3R^2RCH{-}CH^1RCOO^- + 2\,MV^{2+} \qquad (1a)$$

$$^2RXC{=}CHCOO^- + NADH + H^+ \longrightarrow {}^2RCH{=}CHCOO^- + NAD^+ + HX^a \qquad (1b)$$

$$^3R^2RC{=}C^1R\overset{H}{\underset{}{C}}{=}O + NADH + H^+ \longrightarrow {}^3R^2RCH{-}CH^1R\overset{H}{\underset{}{C}}{=}O + NAD^+ \qquad (2a)$$

$$+\;2\,V^{+\cdot} \qquad\qquad\qquad\qquad\qquad\qquad or \atop 2\,V^{2+} \qquad (2b)$$

$$2\,V^{+\cdot} + H^+ + NAD^+ \rightleftharpoons 2\,V^{2+} + NADH \qquad (4)$$

$$NADH + APAD^{+\,b} \longrightarrow NAD^+ + APADH \qquad (5)$$

$$^3R^2RCH{-}CH^1R\overset{H}{\underset{}{C}}{=}O + DCPIP_{ox}(or\;O_2) \longrightarrow {}^3R^2RC{=}C^1R\overset{H}{\underset{}{C}}{=}O + DCPIP_{red}\,(or\;H_2O) \qquad (6)$$

^{a} X = Cl or Br.
^{b} Acetylpyridine adenine dinucleotide

Table 11.3. Substrates ($^3R^2RC{=}C^1RCOO^-$) of enoate reductase that are reduced according to Reactions (1) or (2) of Table 11.2. (100% means a rate of 14 U/mg of purified enoate reductase at pH 6.0 and 0.25 mM NADH. The K_m value for NADH is 0.013 mM).

NO.	^1R	^2R	^3R	V_{max}	K_m (mM)
1	Me	Me	H	100	1.5
2	H	Me	H	280	0.8
3	H	Et	Me	11	—
4	Me	COO$^-$	H	26[a]	74
5	Me	COOMe	H	250	1.7
6	H	COOMe	Me	190	16
7	Me	C_6H_5	H	167	0.029
8	H	C_6H_5	Me	9	0.037
9	Me	C_6H_5	Me	8	0.078
10	NHCHO	C_6H_5	H	21	—
11	F	Me	H	150	6.2[b]
12	Cl	H	Me	180	5.2[b]
13	Br	Me	H	150[c]	28[b]
14	I	H	Me	180[d]	21[b]
15	F	C_6H_5	H	30	0.10
16	Br	C_6H_5	H	76	0.50
17	Cl	H	p-ClC$_6$H$_4$	0.03	—
18	Me	HOCH$_2$CH$_2$	H	120	—
19	H	HOCH$_2$CH$_2$	Me	10	10
20	H	Me	HOCH$_2$CH$_2$	9	15
21	H	(branched alkenyl group)	Me	19	0.014
22	H	Me	(branched alkenyl group)	1.1	0.19
23	H	(branched dienyl group)	H	16[a]	3.3

[a] Reaction could not be conducted under saturating conditions.
[b] 90% E and 10% Z isomer.
[c] Marked substrate inhibition above 40 mM.
[d] Marked substrate inhibition above 10 mM.

expected. An example for such a situation is the reduction of (E) and (Z) geraniate (Simon et al. 1985). As can be seen by comparison of substrates 7 and 8 or 18 and 19 or 20 (Table 11.3) branching in the β-position diminishes the rates severely without much effect on the K_m value in the former case. In the instance of monomethylesters of 2- or

3-methylfumarate (substrates 5 and 6) the V_{max} is diminished too. If the methyl group stands in the β- instead of the α-position, however, the K_m value for the β-branched substrate increases drastically. Polar ^{2}R or ^{3}R groups lead to relatively large K_m values as can be seen from substrates 4, 5, 19, and 20. Especially interesting is the comparison of substrates 4 and 5. The elimination of the negative charge increases V_{max} by a factor of 10 and decreases the K_m value roughly by a factor of 50. If a long chain in a 2-enoate is rigid due to additional double bonds, the K_m value increases and V_{max} decreases as can be seen by substrate 23.

Double and triple bonds in conjugation to the double bond in the 2-position do not prevent reaction (Görgen, Boland, Preiss, and Simon 1989; Preiss, White, and Simon 1989). The double bond can also be part of a ring as is shown by the fact that cyclohexene-1-carboxylate is a suitable substrate.

Especially interesting are allene carboxylates. As shown by Reactions (6) and (6'), three specificities can be observed with their reduction:

(i) Only the α,β double bond is reduced, (ii) in α-position a chiral center is created if a residue ^{1}R different from H is present in the starting material, (iii) because allene carboxylates with three different additional substituents exist in two enantiomeric forms, one of these is reduced to the (E)-3-eonate and the other to the (Z)-3-enoate.

The effect of the hydrogen pressure in the absence and presence of viologen is shown in Table 11.4. The use of hydrogen under pressure may also be necessary if larger volumes are applied by which the

Table 11.4. Effect of mediator and hydrogen pressure on the hydrogenation rate of (*E*)-2-methylbutenoate with *C. tyrobutyricum.*

HYDROGEN PRESSURE/MEDIATOR		PRODUCTIVITY NUMBER
Atmospheric pressure without mediator		725
80 bar without mediator		1,250
Atmospheric pressure, 1 mM methylviologen		1,540
20 bar	1 mM methylviologen	4,300
40 bar	1 mM methylviologen	6,625
80 bar	1 mM methylviologen	7,070

transfer of hydrogen from the gas phase into the liquid phase is rate limiting.

Figure 11.3 shows the reduction of 100 g of a substrate with very low solubility, proving that even such substrates can be converted. To avoid the use of hydrogen gas, in this case 10% of *P. vulgaris* cells were

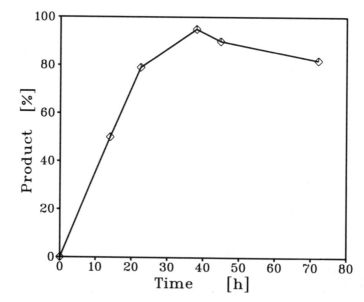

Figure 11.3. Stereospecific hydrogenation of an "unfavorable" enoate. Time course of the reduction of 286 mmol of an *E*-2-methyl-3-($C_{17}H_xN_y$)-propenate in 3 L of buffer with formate by 60 g (dry weight) cells. Solubility of the substrate ~30 μM; K_m ~130 μM. Enantiomeric excess of the product >98%.

added to *C. tyrobutyricum*. The former organism has a viologen-dependent formate dehydrogenase that regenerated $V^{+\cdot}$ at the expense of formate.

The use of free and immobilized enoate reductase in an electrochemical cell has also been described (Thanos and Simon 1987). Table 11.2 shows further reactions that are catalyzed by enoate reductase. Not only α,β-unsaturated carboxylates can be reduced at the expense of reduced viologens, but also α,β-unsaturated aldehydes. However, in the case of aldehydes due to racemization one has to take special measures to get high enantioselectivity (Thanos, Deffner, and Simon 1988).

2-HYDROXYCARBOXYLATE VIOLOGEN OXIDOREDUCTASE AND ITS USE FOR THE PREPARATION OF (2*R*)-HYDROXYCARBOXYLATES

A new and very versatile type of an oxidoreductase was found in the organism *P. vulgaris*. If grown under anaerobic conditions, such cells catalyze Reaction (7).

$$\underset{R}{\overset{O}{\|}}\!\!-\!\!COO^- \; + \; H_2 \; (or \; HCOO^- + H^+) \longrightarrow \overset{HO\;\;H}{\underset{R}{\diagdown}}\!\!-\!\!COO^- \; + \; (CO_2) \quad (7)$$

So far no limitation with respect to the residue R adjacent to the oxo group has been observed. Every compound tested is a substrate for this enzyme. Long carbon chains, dicarboxylates, and carbon–carbon double bonds in conjugation to the carbonyl group are accepted. Quarternary carbons in the β-position to the carboxyl group reveal the lowest rates, which means only 5–6% of that for phenylpyruvate (Simon et al. 1985; Schummer et al. 1989). The enzyme does not react with NADH or NADPH.

Figure 11.4 shows interesting substrates and products as well as only a few chemical reactions of polyfunctional chiral 2-hydroxycarboxylates which are possible. The resulting (2*R*)-hydroxy-3-enoates have been hydrogenated, converted to epoxides, or hydroxylated (Schummer, Yu, and Simon unpublished). Their formation has been studied under the influence of the inductive effect of the chiral carbon atom carrying the 2-hydroxy group. The (2*R*)-hydroxy acids have been ob-

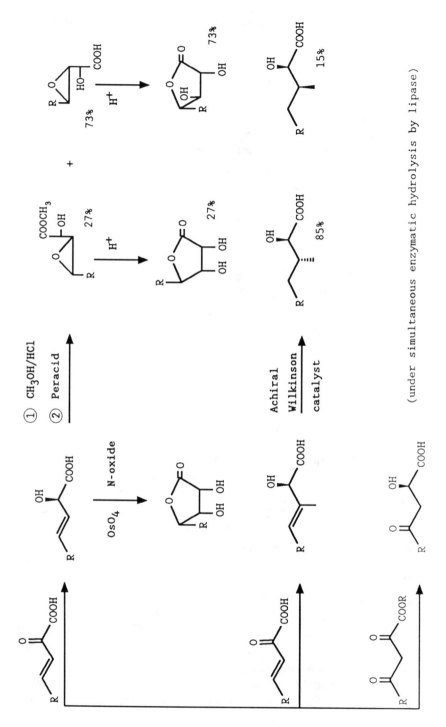

(under simultaneous enzymatic hydrolysis by lipase)

230

tained according to Reaction (7) in a scale of 100 mmol (Table 11.5). Figure 11.5 shows as an example the time course of the reduction of 100 mmol of an 3,4-unsaturated 2-oxocarboxylate. For the preparation of 1 mol of product in 20 h only 4–8 g of cells are necessary. That means under rather simple conditions productivity numbers of 10,000–25,000 can be achieved.

As already published, cells of clostridia as well as of *P. vulgaris* can be used in electrochemical cells (Simon et al. 1985; Günther, Neumann, and Simon 1987; Thanos et al. 1987; Thanos and Simon 1987). Figure 11.6 shows an experiment in which the very labile "keto-dopa" was reduced electromicrobially. The product is the chiral component of rosmarinic acid. Five millimoles of this substrate in 50 ml was converted by 12 mg of *P. vulgaris*. As soon as the reaction mixture in the electrochemical cell is complete, a current is observed. Due to a slight substrate inhibition, the current increases with time, and at the moment at which the substrate is completely converted the reaction stops and the current drops sharply. The beauty of such electromicrobial reductions is that one instantaneously monitors the reaction rate and the amount of product formation can be exactly calculated by determining the area on the recorder paper given by time × current. In a similar experiment, a productivity number of 130,000 was observed for phenylpyruvate (Günther, Neumann, and Simon 1987).

The enantiomeric purity has been determined for many of the products by chromatography of the copper complexes of 2-hydroxycarboxylates on a chiral column (CHIRAL 1, Schleicher and Schüll). Determinations were also done by different methods in other laboratories. In all cases, the enantiomeric excess (ee) value was >0.96–0.98.

DEHYDROGENATION OF (2R)-HYDROXYCARBOXYLATES FROM RACEMIC MIXTURES FOR THE PREPARATION OF (2S)-HYDROXYCARBOXYLATES

The 2-oxocarboxylate reductase present in *P. vulgaris*, which is always applied in the form of resting cells, also dehydrogenates (2R)-hydroxy

Figure 11.4. Reductions of various 2-oxocarboxylates to polyfunctional (2R)-hydroxycarboxylates by *Proteus vulgaris* and reactions carried out with the products of the bioconversion. (See also Table 11.5.)

Table 11.5. Preparative reductions of various 2-oxocarboxylates RCOCOO⁻ by *Proteus vulgaris*.

R	SCALE (mmol)	ELECTRON DONOR	YIELD[a] (%)	ENANTIOMERIC EXCESS[b] (%)	PN
CH_3—CH=CH—	10	HCOO⁻	84	ND	12,000
Ph—CH=CH—	100	HCOO⁻	95	>98	12,400
Ph—CH=C(CH_3)—	100	HCOO⁻	87	>98	3,500
Ph—CH=CH—CH=CH—	10	H_2	98	>98	3,800
p-Cl—Ph—CH=CH—	10	H_2	96	>98	7,500
	13	HCOO⁻	90	>98	27,000
	10	H_2	90	>98	3,500
Ph—CO—CH_2—	10[c]	HCOO⁻ + H_2	87	>96	3,700

[a] Isolated products; bioconversion >97%.
[b] Detection limit 96–98%.
[c] Ethyl ester combined with an esterase.

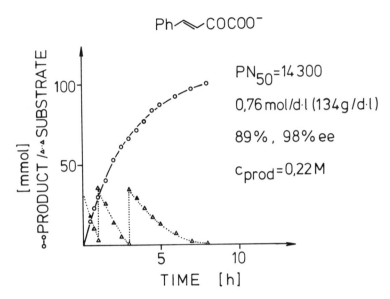

Figure 11.5. Time course of the hydrogenation of 2-oxo-4-phenyl-3-butenoate as a typical example. A total volume of 330 ml of 20 mM phosphate buffer containing 1 mM benzylviologen was stirred under a nitrogen atmosphere at 35°C. The pH value was automatically kept at 7.0. After 1 and 3 h each, 33 mmol of substrate and 50 mmol of formate were added. At the indicated times small aliquots were taken for analysis.

acids. Here the importance of using viologens as artificial electron mediators of different redox potential can be seen. If a 2-oxoacid is reduced, for instance with methylviologen, at pH 6, an equilibrium constant of almost 3×10^{10} is calculated. There is no chance to reverse this reaction. However, if CAV (Table 11.1) is applied at pH 8.5, the equilibrium constant decreases by nearly 10 orders of magnitude and therefore the reverse reaction can be carried out (Skopan, Günther, and Simon 1987). Note that in the general Reaction (3) two protons are involved in contrast to a pyridine nucleotide-dependent reaction which requires only one proton. That means that a change of the pH value affects the equilibrium constant of a viologen-dependent reaction by the second power.

Figure 11.7 shows the selective dehydrogenation of the (R)-enantiomer of a racemic 2-hydroxy acid by *P. vulgaris* cells under an atmo-

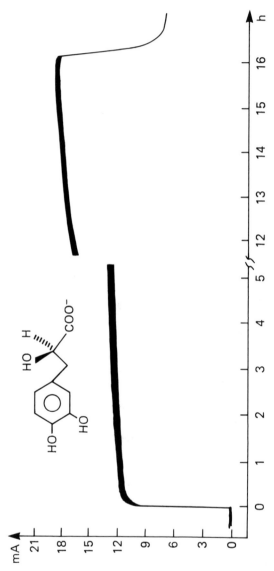

Figure 11.6. Time course of the current during an electromicrobial reduction of 2-oxo-3-(3,4-dihydroxy-phenyl)proprionate ("ketodopa") to the corresponding (R)-derivative which is the chiral constituent of rosmarinic acid. A volume of 50 ml of 0.1 M phosphate buffer, pH 7.0, contained: 5 mmol of substrate, 0.15 mmol of methylviologen, and 12 mg (dry weight) of cells of *Proteus vulgaris*. The potential of the mercury electrode was −800 mV versus standard calomel electrode. Yield of the isolated product about 95%.

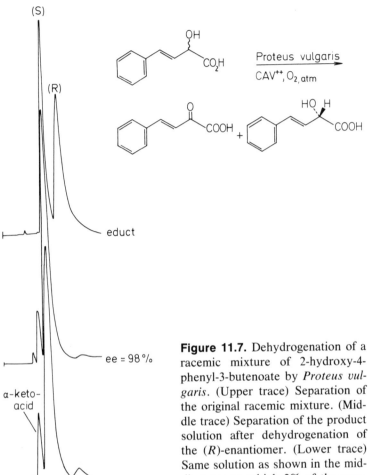

Figure 11.7. Dehydrogenation of a racemic mixture of 2-hydroxy-4-phenyl-3-butenoate by *Proteus vulgaris*. (Upper trace) Separation of the original racemic mixture. (Middle trace) Separation of the product solution after dehydrogenation of the (*R*)-enantiomer. (Lower trace) Same solution as shown in the middle trace to which 2% of the racemate has been added before.

sphere of air. The (*R*) enantiomer is quantitatively dehydrogenated and only (2*S*)-hydroxycarboxylate is left.

PREPARATION OF CHIRAL DEUTERATED COMPOUNDS

All reductions proceeding according to Figure 11.2 can be used for the preparation of stereospecifically deuterated products. These are often valuable research chemicals. The solvent water delivers the protons

Table 11.6. Various stereoselectively deuterium labeled compounds obtained by reductions in ^2H_2O buffer and freeze-dried microorganisms.

COMPOUND	AMOUNT (mmol)
$(2R, 3S)$-[2,3-^{2}H]2-Methylbutanoic acid	4.6
$(2R, 3R)$-[2,3-^{2}H]2-Methyl-3-phenylpropanoic acid	8.3
$(2S, 3R)$-[2,3-^{2}H]3-Phenylpropanoic acid	45
$(2S, 3R)$-[2,3-^{2}H]3-Hydroxyphenyl)propanoic acid	7.4
$(2S, 3R)$-[2,3-^{2}H]3-(4-Hydroxyphenyl)propanoic acid	18
(R)-[2,3-^{2}H]2-Methylpropanoic acid	7
$(2R)$-[2-^{2}H,1-^{13}C]4-Phenyl-Z-penten-3-oic acid	1.7
$(2S)$-[2-^{2}H]Succinic acid monomethyl ester	1.3
$(2S, 3S)$-[2-^{2}H, ^{3}H; 3-^{2}H]Succinic acid monomethyl ester	2.6
$(2S, 3R)$-[2,3-^{2}H,^{3}H]Propanoic acid	1
(R)-[2-^{2}H]12-Phenyldodecanoic acid	0.35
$(2S)$-[2,3-^{2}H]4-Phenylbuten-3-oic acid	1
(R)-[2-^{2}H]3-Dimethyl-2-hydroxypropanoic acid	2
(R)-[2-^{2}H]Hydroxybutanoic acid	2
(R)-[2-^{2}H]3-Phenyl-2-hydroxypropanoic acid	100

for the newly formed carbon hydrogen bonds. Therefore, the reductions have to be carried out in deuterium water. For this the cells are freeze dried under the exclusion of oxygen and resuspended in buffers made up from deuterium oxide. Especially interesting are compounds that are chiral only due to the stereospecific labeling of methylene groups by one deuterium or tritium. Table 11.6 lists some compounds that have been prepared (Simon and Günther 1988).

REDUCTIONS OF HYDROXYLAMINES

Another pyridine nucleotide-independent reduction that can be carried out with clostridia or methanogens is the selective reduction of dihydrooxazines (Klier et al. 1987) or hydroxylamines. Table 11.7 gives some examples. The dihydrooxazines are the products of hetero Diels–Alder reactions and the latter that of ene reactions. Both are intermediates on synthetic routes to polyfunctionalized chiral amines. It turned out that dihydrooxazines as well as the products of ene reactions show rather low yields by the chemical reductive splitting of the N–O bond without attacking carbon–carbon double bonds or chiral carbon atoms.

Table 11.7.   Examples of dihydrooxazines and hydroxylamines reductively converted to amines by various microorganisms.

SUBSTRATE FOR REDUCTION	PN	REMARKS[a]
(structure, bicyclic oxazine NH·HCl, R¹)	≤3,100	R^1 = H, OAc, OMe (Klier et al. 1987) *C. kluyveri* or *M. thermoautotrophicum*
(structure, Ph—C(=O)—NHOH)	800	*C. thermoaceticum* and formate (62% yield with LiAlH₄)
(structure, HNOH branched alkene)	1,100	*C. thermoaceticum* and formate (cannot be reduced with Zn/acetic acid or LiAlH₄)
(structure, cyclopentene NHOH)	8,100	*C. thermoaceticum* and CO

[a] All biocatalytic reductions proceed quantitatively.

C. thermoaceticum or other anaerobic microorganisms are very effective in this reduction.

EXAMPLE OF A VIOLOGEN-DEPENDENT PYRIDINE NUCLEOTIDE OXIDOREDUCTASE

From a thermophilic bacillus an enzyme has been isolated catalyzing Reactions (4) and (5) (Nagata et al. 1987). It is rather stable and useful for the regeneration of the four coenzymes NAD⁺, NADP⁺, NADH, and NADPH. For the reduction of the pyridine nucleotides, methylviologen can be used and for their reoxidation carbamoylmethyl viologen (Table 11.1) is suitable. Table 11.8 shows some kinetic parameters. One gram of wet-packed cells of *Bacillus* DSM 406 contains about 35 units for NADH as well as NAD⁺ regeneration and about 13–17 units for the regeneration of NADPH or NADP⁺, repsectively. One unit is the amount of enzyme catalyzing the formation of 1 μmol of NADH or NAD per minute under saturating conditions. *Bacillus* DSM 406 is easy to grow and the enzyme can easily be enriched. Another strain, *Bacil-*

Table 11.8. Kinetic parameters of Reactions (4) and (5) for both directions catalyzed by an enzyme from a thermophilic bacillus.

K_m VALUES		K_i VALUES		RELATIVE RATES[c] IN THE FORMATION	
$MV^{+\cdot}$	0.24	MV^{2+}	100	NADH	100
NAD	0.06	NADH	50	NAD	110[d]
NADP	1.82	NADPH	33	NADPH	50
CAV^{2+}	0.52[a]	$CAV^{+\cdot}$	1.8	NADH	37
CAV^{2+}	0.29[b]	NAD	19		
NADH	0.22	NADP	13.5		
NADPH	1.43				

For the formation of reduced pyridine nucleotide $MV^{+\cdot}$ and for their reoxidation CAV^{2+} were used. All values are given in mM. NAD(P)H formation was studied at pH 7.0 and that of NAD(P) at pH 8.5 at 27°C.
[a] For the reoxidation of NADH.
[b] For the reoxidation of NADPH.
[c] 100% corresponds to 58 U for the purified enzyme.
[d] At the pH optimum of the reoxidation of NADH the rate is about 260% of that of the NADH formation with $MV^{+\cdot}$ at pH 7.0.

lus DSM 466, has about twice as much of this enzyme for which the physiological role is not known, yet.

CARBOXYLATE REDUCTION WITH CLOSTRIDIA AND CARBON MONOXIDE

According to recent findings, *C. thermoaceticum* and *C. formicoaceticum* are able to reduce carboxylic acids in a nonactivated form in aqueous solution to alcohols via an intermediary aldehyde at the expense of formate or even more effectively with carbon monoxide (Simon et al. 1987; Fraisse and Simon 1988). That is surprising, because it is not possible to reduce an acylate in an aqueous medium by chemical means because the reduction of a proton at pH 7 occurs with a standard redox potential of only −420 mV. For the reduction of a carboxylate to an aldehyde about −560 mV are needed.

With *C. thermoaceticum* the reduction of carboxylic acids is rather general. Table 11.9 gives a series of examples and the relative initial rates by which various carboxylates are reduced. Unbranched and branched mono- and dicarboxylic acids are reduced to alcohols. Pro-

Table 11.9. Relative initial rates of the reduction of various carboxylates to the corresponding alcohols by *C. thermoaceticum* at the expense of carbon monoxide in the presence of 1 m*M* methylviologen.

CARBOXYLATE	RATE (%)	CARBOXYLATE	RATE (%)
Propionate	100	3-Methylglutarate monomethylester	75
Butyrate	100	Adipate	105
2-Methylbutyrate	90	3-Methyladipate	100
3-Methylbutyrate	75	Benzoate	115
Pentanoate	95	2-(*p*-Chlorphenoxy)propionate	50
2-Phenylpropionate	90	*p*-Methoxyphenylacetate	60
2-Methylpentanoate	75	(*R*)-Lactate	20
(*R*, *S*)-3-Methylpentanoate	85	Methyl propionate	0
Succinate	120	Butyrolacton	0
2-Methylsuccinate	25		
Glutarate	100		
3-Methylglutarate	25		
3-Hydroxy-3-methylglutarate	0		

pionate as a standard is reduced by productivity numbers of up to 1200. Branched dicarboxylates are relatively poor substrates. *C. thermoaceticum* and *C. formicoaceticum*, which are closely related species, show different substrate specificity.

Studies on the carboxylic acid reductase resulted in the assumption that this enzyme seems to be the first that is able to reduce nonactivated carboxylic acids to aldehydes at the expense of reduced viologens. There is no further reduction of the aldehydes to the corresponding alcohols with the isolated enzyme. In the presence of oxidized viologens, aldehydes can be dehydrogenated to carboxylates roughly 20 times faster as the latter are reduced. The specific enzyme activity in crude extracts is about 100 times greater if 1×10^{-5} M tungstate and a sulfur source such as dithionite or thiosulfate in addition to sulfate is given to the growth medium of *C. thermoaceticum*.

This labile enzyme has been enriched by a factor of about 145 by various chromatographic steps. After cell growth in the presence of [^{185}W]tungstate radioactivity coincides with the enzyme activity during all purification steps (White et al. 1989).

This enzyme activity may be useful for chemoselective bioconversions. The stereoselectivity seems to be rather low. The rate difference for the reduction of (*R*)- and (*S*)-lactate seems to be <1.5.

ACKNOWLEDGMENTS

The highly motivated cooperation of the co-workers mentioned in the references is gratefully acknowledged. The work was supported by Deutsche Forschungsgemeinschaft (SFB 145) and Fonds der Chemischen Industrie.

REFERENCES

Bader, J., H. Günther, S. Nagata, H. J. Schütz, M. L. Link, and H. Simon. 1984. Unconventional and effective methods for the regeneration of NAD(P)H in microorganisms or crude extracts of cells. *J. Biotechn.* **1:** 95–109.

Balahura, R. J., and R. G. Wilkins. 1983. Ligational effects on reduction of myoglobin and horseradish peroxidase by inorganic reagents. *Biochim. Biophys. Acta* **724:** 465–472.

Chenault, H. K., E. S. Simon, and G. M. Whitesides. 1988. Cofactor regeneration for enzyme-catalysed synthesis. *Biotechnol. Genet. Engineer. Rev.* **6:** 221–270.

Dodds, D. R., and B. J. Jones. 1988. Enzymes in organic synthesis. 38. Preparations of enantiomerically pure chiral hydroxydecalones via stereospecific horse liver alcohol dehydrogenase catalyzed reductions of decalindiones. *J. Am. Chem. Soc.* **110:** 577–583.

Fraisse, L., and H. Simon. 1988. Observations on the reduction of nonactivated carboxylates by *Clostridium formicoaceticum* with carbon monoxide or formate and the influence of various viologens. *Arch. Microbiol.* **150:** 381–386.

Görgen, G., W. Boland, U. Preiss, and H. Simon. 1989. Synthesis of chiral 12-phenyl(^{2}H)dodecanic acids: Useful metabolic probes for the biosynthesis of 1-alkenes from fatty acids. *Helv. Chim. Acta* **72:** 917–928.

Günther, H., S. Neumann, and H. Simon. 1987. 2-Oxocarboxylate reductase from *Proteus* species and its use in the preparation of (2R)-hydroxy acids. *J. Biotechnol.* **5:** 53–65.

Hummel, W., H. Schütte, E. Schmidt, C. Wandrey, and M.-R. Kula. 1987. Isolation of L-phenylalanine dehydrogenase from *Rhodococcus sp.* M4 and its application for the production of L-phenylalanine. *Appl. Microbiol. Biotechnol.* **26:** 409–416.

Kim, M. J., and G. M. Whitesides. 1988. L-Lactate dehydrogenase: substrate specificity and use as a catalyst in the synthesis of homochiral 2-hydroxy acids. *J. Am. Chem. Soc.* **110:** 2959–2964.

Klier, K., G. Kresse, O. Werbitzky, and H. Simon. 1987. The microbial splitting of the N–O bond of dihydrooxazines; an alternative to the chemical reduction. *Tetrahedron Lett.* **28:** 2677–2680.

Kuno, S., A. Bacher, and H. Simon. 1985. Structure of enoate reductase from a *Clostridium tyrobutyricum* (C. spec. La1). *Hoppe Seyler Z. Biol. Chem.* **366:** 463–472.

Nagata, S., R. Feicht, W. Bette, H. Günther, and H. Simon. 1987. On a viologen-dependent pyridine nucleotide oxidoreductase from a thermophilic *Bacillus. Appl. Microbiol. Biotechnol.* **26:** 263–267.

Preiss, U., H. White, and H. Simon. 1989. Additional enoates and other α,β-unsaturated carbonyl compounds as substrates for the enoate reductase from *Clostridium tyrobutyricum*, influence of elevated hydrogen pressure on the reduction rate. *DECHEMA Biotechnology Conferences, Vol. 3* (D. Behrens ed.), pp. 189–192. Weinheim: Verlag Chemie.

Schummer, A., H. Yu, C. Schinschel, P. Rauschenbach, and H. Simon. 1989. Preparation of multifunctional (2R)- or (2S)-hydroxy acids by *Proteus vulgaris* and their conversion to compounds with two or three chiral centers.

DECHEMA Biotechnology Conferences Vol. 3 (D. Behrens, ed.), pp. 271–274. Weinheim: Verlag Chemie.

Sih, C. J. and C.-S. Chen. 1984. Microbial asymmetric catalysis—enantioselective reductions of ketones. *Angew. Chem. Int. Ed. Engl.* **23**: 570–578.

Simon, H. 19xx. In *Chemistry and Biochemistry of Flavoenzymes*, Vol. III (F. Mueller, ed.). CRC Press, Boca Raton, FL, in press.

Simon, H., J. Bader, H. Günther, S. Neumann, and J. Thanos. 1985. Chiral compounds synthesized by biocatalytic reductions. *Angew. Chem. Int. Ed. Engl.* **24**: 539–553.

Simon, H., H. White, H. Lebertz, and I. Thanos. 1987. Reduktion von 2-Enoaten und Acylaten mit Kohlenmonoxid oder Formiat, Viologenen und *Clostridium thermoaceticum* zu gesättigten Säuren und ungesättigten bzw, gesättigten Alkoholen. *Angew. Chem.* **100**: 785–786.

Simon, H., and H. Günther. 1988. Microbial and enzymic production of labeled compounds. In *Biotechnology, Vol. 6b.* (H.-J. Rehm, and G. Reed, eds.), pp. 171–192. Weinheim: Verlag Chemie.

Skopan, H., H. Günther, and H. Simon. 1987. A biocatalyst for the preparation of (*R*)- and (*S*)-hydroxylic acids. *Angew. Chem. Int. Ed. Engl.* **26**: 128–130.

Thanos, J., and H. Simon. 1986. Stereospecific reductions with hydrogen gas, modified metal catalysts, methylviologen and enzymes or microorganisms. *Angew. Chem. Int. Ed. Engl.* **25**: 462–463.

Thanos, I., J. Bader, H. Günther, S. Neumann, F. Krauss, and H. Simon. 1987. Electroenzymatic and electromicrobial reductions for preparation of chiral compounds. *Methods Enzymol.* **136**: 302–317.

Thanos, I., and H. Simon. 1987. Electro-enzymic viologen-mediated stereospecific reduction of 2-enoates with free and immobilized enoate reductase on cellulose filters or modified carbon electrodes. *J. Biotechnol.* **6**: 13–29.

Thanos, I., A. Deffner, and H. Simon. 1988. Further reactions catalysed by enoate reductase; reductions of 2-enals. Dehydrogenation of saturated aldehydes and their racemisation. *Biol. Chem. Hoppe-Seyler* **369**: 263–267.

Vanmiddlesworth, F., and C. J. Sih. 1987. A model for predicting diastereoselectivity in yeast reductions. *Biocatalysis* **1**: 117–127.

White, H., G. Strobl, R. Feicht, and H. Simon. 1989. Carboxylic acid reductase. A new tungsten enzyme catalyses the reduction of nonactivated carboxylic acids to aldehydes. *Eur. J. Biochem.* **184**: 89–96.

12
Enzymes from Extreme Environments

Roy M. Daniel, Judy Bragger, and Hugh W. Morgan

We now know life is possible under conditions that were once thought to preclude it. The existence of life at very low pH, very high salinity, and very high temperatures contradicts what were once thought to be basic tenets of biochemistry. It is largely from environments such as these that members of a complete new kingdom of organisms, the archaebacteria (Woese and Fox 1977), have been isolated. Indeed, life at the uttermost extremes of pH, salt concentration, and high temperature is confined to archaebacteria. The designation of archaebacteria as belonging to a third kingdom has been based on a number of fundamental properties (see Fewson 1986) including ribosomal RNA sequences (e.g., Fox et al. 1980; Woese 1987).

The discovery of this third kingdom of organisms has forced a complete reassessment of our definition of the word "extreme". Evidence is now gathering that suggests that life may actually have evolved at very high temperatures. The archaebacteria are capable of growing at these very high temperatures, and seem to be among the more primitive forms of life in the sense that they originated nearest the phylogenetic separation point of the three Kingdoms (Woese and Olsen 1986), and the environments in which they are found match those thought to prevail at the time of the origin of life. There is also evidence that the most primitive members of the eubacteria are also the most thermophilic (Achenbach-Richter et al. 1987). This suggests that our current perception of "normal" life as being that at "room temperature" may be mistaken. Rather than marvel at how extreme thermophiles survive these extraordinary conditions, it may now be more rational to wonder how current life forms have managed to adapt to the disadvantages of life at temperatures as low as 35°C.

Out of three commonly encountered extreme environments, high salinity, extremes of pH, and extremes of temperature, only temperature *must* prevail inside the cell. Some acidophiles, for example, have intracellular pH values near neutrality maintained by proton pumping. But, generally speaking, if we have properly identified the environment in which an enzyme has evolved, we find that it has been well designed to withstand, and function in, this environment. Thus, enzymes from extreme halophiles are stable and function in high salt concentrations; enzymes from many extreme acidophiles function at low pH and are resistant to these extremes; and enzymes from extreme thermophiles withstand high temperatures. How has this stability been achieved? Relatively little work has been done on how enzyme structure is maintained in a functional form in the presence of very low pH, or very high salt concentrations (and low water activity), although some enzymes from halophiles have a large excess of acidic over basic amino acid residues (Werber, Sussman, and Eisenberg 1986). However, we now have quite a good picture of how thermal stability is maintained, thanks to the work of Langridge (1968), Matthews (1986, 1987) and his group, Perutz and Raidt (1975), and others, and this sheds light on protein stability generally.

PROTEIN STABILITY

We have good evidence that the stability of proteins from extreme thermophiles is intrinsic, i.e., that it is specified by the amino acid sequence and thus resides in the secondary and tertiary structure, rather than with external stabilizing factors. There are many pieces of evidence supporting this, but perhaps the most convincing and straightforward is simply that most proteins from thermophiles do not change in thermostability as they are purified from crude cell extracts to homogeneity. This is not to say that the presence or nature of a cofactor, for example, will not affect stability. Caldolysin, a protein from an extreme thermophile, depends on six calcium ions for its very high stability (Khoo et al. 1984; Cowan and Daniel 1982). Differently modified forms of caldolysin have half-lives ranging from about 1 h at 50°C to about 1 h at over 95°C (Table 12.1), corresponding to differences in the free energy of stabilization of the order of only 25 KJ/mol. If archaelysin (Cowan et al. 1987), the most stable protease we have

Table 12.1. Differences in thermal stability (half-lives) between different forms of caldolysin.

ENZYME	TEMPERATURE FOR HALF-LIFE OF 1 h
Apocaldolysin	50°C
Caldolysin	90°C
Immobilized caldolysin	97°C
Apocaldolysin plus lanthanum	99°C
[Archaelysin	100°C]

Data from Cowan and Daniel 1982; Khoo et al. 1984; Cowan et al. 1987.

isolated to date, possessed an extra free energy of stabilization of 25–30 KJ/mol, its half-life would be about 1 h at 150°C (but see Ahern and Klibanov, 1986).

It is now widely accepted that in their ''natural'' surroundings native proteins are poised on the brink of denaturation, their net free energy of stabilization being a small resultant of large stabilizing and destabilizing forces (Brandts 1967). Relatively small changes in the stabilizing forces, whether they be in hydrogen bonding, or in hydrophobic or ionic interactions, can cause large changes in the net free energy of stabilization. The net free energy of stabilization of most proteins from mesophiles is in the range 10–40 KJ mol^{-1}. Individual intramolecular stabilizing interactions (e.g., a salt bridge) can contribute of the order of 2–20 KJ mol so that a single interaction more or less can dramatically alter thermostability.

Thus the stabilization or destabilization of proteins can be accomplished by minor changes in the amino acid sequence giving rise to the elimination or addition of a few intramolecular interactions, without any obvious structural alterations (Matthews 1986, 1987). The difference in the free energy of stabilization between ''normal'' enzymes with a half-life of about an hour at 50°C, and those enzymes that are exceptionally stable (having a half-life of the same sort of period at above 90°C) is only a few tens of kilojoules per mole. A single salt bridge can probably contribute between 4 and 12 KJ per mole so a small number of extra salt bridges, for example, could confer this extra degree of stabilization (e.g., Perutz and Raidt 1975). Small numbers of additional hydrophobic interactions or hydrogen bonds could perform a similar function. As we might expect from this, no systematic structural differences between extremely stable and normal proteins have been found (e.g., Grutter, Hawkes, and Matthews 1979).

If life has evolved at high temperatures, and there is no intrinsic barrier to enzyme stability at, say 100°C, why are not all enzymes more stable? We now know that proteins are flexible, from the level of amino acid side chain rotation, to the movement of whole protein domains relative to one another. This flexibility is essential for catalytic function, but flexibility is the first step toward unfolding, and for thermostability a more rigid structure is necessary. So, the requirements for catalysis and thermostability are, to some extent at least, opposed. If the stabilizing forces within the molecule are too strong, it will be too rigid to work as a catalyst. If the forces are too weak, the enzyme will be unstable. A balance must be struck between these requirements, and that depends on the temperature at which the enzyme has evolved to function. In the case of enzymes from extreme thermophiles, the additional stabilizing forces are offset by the destabilizing effect of the high temperatures, so that we may expect the enzyme to be as flexible at its working temperature of, say, 85°C as its mesophilic counterpart is at 25°C. In support of this it has been found that mutations to stabilize enzymes do decrease specific activity (Matsumura, Yasumura, and Aiba 1986). The resistance of thermostable enzymes to proteolysis at room temperature has also been explained on this basis (Daniel et al. 1982; Daniel 1986). A corollary is that we should not expect, and do not find, enzymes from extreme thermophiles to work more rapidly at their "evolved" or "design" temperatures than those from mesophiles (Cowan, Daniel, and Morgan 1987).

THERMOSTABLE ENZYMES

A number of workers have pointed out that stable enzymes from extreme thermophiles could have important applications (e.g., Doig 1974; Daniel, Cowan, and Morgan 1981; Cowan, Daniel, and Morgan 1985). If our intention is to seek particularly robust enzymes for industrial applications, then an important question is, just how robust can enzymes actually be? Reports of microbial growth at 250°C (Baross and Deming 1983) have not been confirmed (to the bitter disappointment of many biochemists!), and the highest temperature for which microbial growth is well established is 110°C (Stetter 1982). An obvious implication of this is that a whole array of intracellular enzymes is present within these organisms that can withstand, and function at, 110°C. This implies, at the very least, a half-life of minutes at this temperature. Furthermore, because the enzymes are unlikely to have uniform stability, some of them may have stabilities significantly higher than this.

This expectation is certainly in accord with our data on the stability of archaebacterial extracellular enzymes (Bragger et al. 1989), albeit of organisms that grow only up to about 95°C. As part of a study of the enzymes of extreme thermophiles, we have screened the major groups of archaebacterial extreme thermophiles for extracellular enzymes. Table 12.2 shows the occurrence of these enzymes. Table 12.3 shows the stability of some of these extracellular enzymes, and includes comparable values for enzymes from the recently discovered, most extremely thermophilic eubacterium (Huber et al. 1986; Huser et al. 1986), *Thermotoga* sp.. It can be seen that enzyme half-lives of several minutes at 103°C are not exceptional for these thermophilic systems.

The method we used for assessing thermostability consisted simply of heating the enzyme in dilute buffer, removing samples at various time intervals, and assaying residual activity at a temperature known not to cause denaturation. This method is convenient, but allows for thermostability to be overestimated if renaturation occurs between the heat treatment and the assay. We have never in fact observed this to

Table 12.2. Production of extracellular enzymes by archaebacteria.

ARCHAEBACTERIA	PROTEASE	AMYLASE	CELLULASE	XYLANASE
Sulfolobus acidocaldarius	+	+	−	−
Sulfolobus solfataricus	+	+	−	−
Acidianus brierleyi	−	−	−	−
Pyrodictium brockii	−	−	−	−
Thermoplasma acidophilum	−	−	−	−
17 *Thermoproteus* strains	−	−	−	−
2 *Thermofilum* strains	−	+	+	+
8 *Desulfurococcus* strains	+	+	−	−
2 *Thermococcus* strains	+	+	−	−

Table 12.3. Thermal stability (half-lives) of some archaebacterial and eubacterial enzymes at various temperatures.

	95°C	103°C
^rchaebacteria		
AN1 (DSM 2770)		
α-Glucosidase	>60 min[a]	>20 min[a]
β-Glucosidase	10–15 min	
Pullulanase	>60 min[a]	15–20 min
Transglucosylase	10–20 min	
Desulfurococcus Tok12S.1 (NZ isolate)		
Archaelysin (Cowan et al 1987)	70–90 min	8–9 min
Amylase		15–25 min
Transglucosylase		4–9 min
Thermococcus celer (DSM 2476)		
Protease	34–45 min	
α-Glucosidase		>20 min[a]
β-Glucosidase		20–30 min
Sulfolobus solfataricus (DSM 1616)		
Protease	35–40 min	
Amylase	>60 min[a]	
Thermophilic eubacteria		
Thermotoga FjSS3B.1		
Cellulase		15 min
β-Glucosidase		15–20 min
Xylanase		>20 min[a]
Amylase	20–30 min	

[a] More than 80% of the initial activity remains after this time.

occur during detailed studies of more than 30 enzymes from extreme thermophiles. We do not believe it occurred in this work because more extensive studies on some of the enzymes have given Arrhenius plots consistent with the stabilities observed (Cowan et al. 1987) or have shown faster reaction rates when the enzymes are assayed for significant periods at higher temperatures; e.g., the reaction rates for the α-glucosidases from *Thermococcus celer* and AN1 were shown to be 8- to 10-fold faster at 103°C than at 75°C over a 10-min reaction period. Most of the enzymes listed are the most stable recorded for enzymes of that type, and it is clear that archaebacteria growing at these high temperatures are a potentially rich source of very stable enzymes.

These half-lives are relatively conservative in the sense that the enzymes were not selected for stability, and that all were derived using partially purified enzymes in dilute buffers. It is likely that stabilities could be increased by including, for example, substrate, or by immobilization.

These results suggest that enzyme reaction temperatures of 120°C are not hopelessly out of reach, although there is some evidence (e.g., Ahern and Klibanov 1986) from work on mesophilic enzymes such as lysozyme that irreversible inactivation processes due, for example, to hydrolysis of peptide bonds may set in at around this temperature. In any application, to obtain the full benefit of the stability of enzymes of this type may entail using them at temperatures rather lower than their evolved temperature. We may even have to use these enzymes at as low as 80°C to obtain long-term retention of activity, especially where relatively harsh conditions are involved. Going too far "down temperature" to obtain even greater stability may not be practical. At 20°C the activity of such enzymes is usually only a few percent of their activity inside the native organism.

RECOMBINANT THERMOSTABLE ENZYMES

An alternative way to obtain stable enzymes is to use site-directed mutagenesis to engineer stability into mesophilic enzymes that may, for example, already have FDA approval (e.g., Yutani et al. 1977; Matsumara et al. 1984; Matsumara, Yasumara, and Aiba 1986; Imanaka, Shibrayahi, and Takagi 1986; Bryon et al. 1986). This of course avoids both the cost of obtaining regulatory approval for enzymes from extreme thermophiles and also the cost of adjusting process variables to suit new enzymes. Although this type of work is of enormous importance in our efforts to understand the mechanism of protein stability, and other features such as pathways of folding and unfolding, it has not so far been very successful in terms of raising stability, the increases generally being of the order of 5–10°C.

In addition to their potential industrial applications (see Doig, 1974; Daniel, Cowan, and Morgan, 1981; Bergquist et al. 1987) enzymes from extreme thermophiles are very useful in research. They enable the use of harsh techniques and investigation of properties over very wide temperature ranges and (at low temperatures) at quite slow rates.

Table 12.4. Purification of cellulase and hemicellulase enzymes of "*C. saccharolyticum*", cloned into *E. coli*, by heat treatment.

	TREATMENT	SOLUBLE PROTEIN (mg ml^{-1})		SPECIFIC ACTIVITY (units mg^{-1})		PURIFICATION	RECOVERY (%)
		BEFORE	AFTER	BEFORE	AFTER		
CMCase PNZ1012/PB2477	70°C, 60 min	47.8	2.1	0.05	0.53	11.5	51
β-Glucosidase PNZ1001/PB2481	70°C, 65 min	14.4	0.68	2.26	42	18.6	88
Xylanase PNZ1076/PB2477	70°C, 30 min	5.24	0.80	3.01	22	7.2	109
β-Xylosidase PNZ1076/PB2477	70°C, 30 min	31.0	3.5	0.08	0.66	7.9	89

Stable enzymes from extreme thermophiles (or indeed any source) also have a number of advantages that are perhaps slightly less obvious. One of the reasons that intracellular enzymes are not more widely used in industry is the difficulty of purifying the enzyme concerned away from the hundreds of others within the cell. Our colleagues at the University of Auckland have cloned a number of genes from extreme thermophiles into mesophiles (Bergquist et al. 1987). This enables a major purification to be undertaken quickly and simply by heating (Tanaka, Kawano, and Oshima 1981; Iijima, Uozumi, and Beppu 1986; Daniel 1986; Patchett et al. 1989). Table 12.4 shows the use of this technique to purify thermostable cellulases and hemicellulases from a new extreme thermophile, *"Caldocellum saccharolyticum"*, which have been cloned into *E. coli* (Patchett et al. 1989; Reynolds et al. 1986; Sissons et al. 1987). Cloning is a very useful first step in the investigation of any thermostable enzyme because it makes purification so much easier. Although purification is not complete after the heat treatment there are not normally any contaminating enzyme activities present. We have not, so far, found any significant stability difference between the cloned form of an enzyme and its native form.

FUTURE APPLICATIONS

Perhaps the most exciting application of enzymes from extreme thermophiles is their potential to open up new areas of enzyme-catalyzed synthesis. Although these enzymes differ one from another in the particular nature and extent of their stability, they are in general very resistant to organic solvents (e.g., Owusu and Cowan 1989), detergents, and chaotropic agents. Because the diversity of extreme thermophiles is enormous, it follows that a diversity of stable enzymes is available. It should therefore be possible to select enzymes to carry out particular catalyses involving aggressive substrates or products, or under harsh conditions. Work in this area has begun (e.g., Gambacorta et al. 1988), but much remains to be achieved.

ACKNOWLEDGMENTS

We thank Pacific Enzymes Limited for financial support, and the New Zealand University Grants Committee for the award of a postgraduate scholarship to one of us (J.B.).

REFERENCES

Achenbach-Richter, L., R. Gupta, K. O. Stetter, and C. R. Woese. 1987. Were the original eubacteria thermophiles? *System. Appl. Microbiol.* **9:** 34–39.

Ahern, J. J., and A. M. Klibanov. 1986. Why do enzymes irreversibly inactivate at high temperatures? In *Protein Structure, Folding and Design: GENEX-UCLA Symposium Vol 39* (D. L. Oxender, ed.), pp. 283–289. New York: Allan R. Liss.

Baross, J. A., and J. W. Deming. 1983. Growth of black smoker bacteria at temperatures of at least 250°C. *Nature* **303:** 423–426.

Bergquist, P. L., D. R. Love, J. E. Croft, M. B. Streiff, R. M. Daniel, and H. W. Morgan. 1987. Genetics and potential biotechnological application of thermophilic and extremely thermophilic microorganisms. *Biotechnol. Genet. Engineer. Rev.* **5:** 199–244.

Bragger, J. M., R. M. Daniel, T. Coolbear, and H. W. Morgan. 1989. Very stable enzymes from extremely thermophilic archaebacteria and eubacteria. *Appl. Microbiol. Biotechnol.* **31:** 556–561.

Brandts, J. F. 1967. Heat effects on proteins and enzymes. In *Thermobiology* (A. H. Rose, ed.), pp. 25–72. New York: Academic Press.

Bryon, P. N., M. L. Rollence, M. W. Pantoliano, J. Wood, B. C. Finzel, G. L. Gilliland, A. J. Hamond, and T. L. Panlos. 1986. Proteases of enhanced thermostability: characterisation of a thermostable varient of thermolysin. *Proteins* **1:** 326–334.

Cowan, D. A., and R. M. Daniel. 1982. Purification and some properties of an extracellular protease (caldolysin) from an extreme thermophile. *Biochim. Biophys. Acta* **705:** 293–305.

Cowan, D., R. M. Daniel, and H. Morgan. 1985. Thermophilic proteases: properties and potential applications. *Trends Biotechnol.* **3:** 68–72.

Cowan, D. A., R. M. Daniel, and H. W. Morgan. 1987. The specific activities of mesophilic and thermophilic proteinases. *Int. J. Biochem.* **19:** 741–743.

Cowan, D. A., K. A. Smolenski, R. M. Daniel, and H. W. Morgan. 1987. An extremely thermostable extracellular proteinase from a strain of the archaebacterium *Desulfurococcus* growing at 88°C. *Biochem. J.* **247:** 121–133.

Daniel, R. M. 1986. The stability of proteins from extreme thermophiles. In *Protein Structure, Folding and Design: GENEX-UCLA Symposium, Volume 39* (D. L. Oxender, ed.), pp. 291–296. New York: Allan R. Liss.

Daniel, R. M., D. A. Cowan, and H. W. Morgan. 1981. The industrial potential of enzymes from extremely thermophilic bacteria. *Chem. Industr. N. Zeal.* **15:** 94–97.

Daniel, R. M., D. A. Cowan, H. W. Morgan, and M. P. Curran. 1982. A correlation between protein thermostability and resistance to proteolysis. *Biochem. J.* **207**: 641–644.

Doig, A. R. 1974. Stability of enzymes from thermophilic microorganisms. In *Enzyme Engineering, Vol. 2* (E. K. Pye, and L. B. Wingard, eds.), pp. 17–21. New York: Plenum Press.

Fewson, C. A. 1986. Archaebacteria. *Biochem. Educat.* **14**: 103–107.

Fox, G. E., E. Stackebrandt, R. B. Hespell, J. Gibson, J. Maniloff, T. A. Dyer, R. S. Wolfe, W. E. Balch, R. Tanner, L. Magrum, L. B. Zablen, R. Blakemore, R. Gupta, L. Bonen, B. J. Lewis, D. A. Stahl, K. R. Luehrsen, K. N. Che, C. R. Woese. 1980. The phylogeny of prokaryotes, *Science* **209**: 457–463.

Gambacorta, A., R. Rella, M. Rossi, and M. D. E. Rosa. 1988. Biotransformations with an NAD^+-dependent alcohol-aldehyde/ketone oxidoreductase activity from the extremely thermophilic archaebacterium *Sulfolobus solfataricus*. In *Enzyme Engineering, Vol. 9* (A. M. Klibanov, ed.). New York: Plenum Press.

Grutter, M. G., R. M. Hawkes, and B. W. Matthews. 1979. Molecular basis of thermostability in the lysozyme from bacteriophage T4. *Nature* **277**: 667–669.

Huber, R., T. A. Langworthy, H. Konig, M. Thomm, C. R. Woese, U. B. Sleytr, and K. O. Stetter. 1986. *Thermotoga maritima* sp. nov. represents a new genus of unique extremely thermophilic eubacteria growing up to 90°C. *Arch. Microbiol.* **144**: 324–333.

Huser, B. A., B. K. C. Patel, R. M. Daniel, and H. W. Morgan. 1986. Isolation and characterisation of a novel extremely thermophilic anaerobic chemoorganotrophic eubacterium. *FEMS Microbiol. Lett.* **37**: 121–127.

Iijima, S., T. Uozumi, and I. Beppu. 1986. Molecular cloning of *Thermus flavus* malate dehydrogenase gene. *Agric. Biol. Chem.* **50**: 589–592.

Imanaka, T., M. Shibrayahi, and M. Takagi. 1986. A new way of enhancing the thermostability of proteases. *Nature* **324**: 695–697.

Khoo, T. C., D. A. Cowan, R. M. Daniel, and H. W. Morgan. 1984. Interactions of calcium and other metal ions with caldolysin, the thermostable proteinase from *Thermus aquaticus* strain T351. *Biochem. J.* **221**: 407–413.

Langridge, J. 1968. Genetic and enzymatic experiments relating to the tertiary structure of β-galactosidase. *J. Bacteriol.* **96**: 1711–1717.

Matsumara, M., S. Yasumara, and S. Aiba. 1986. Cumulative effect of intragenic amino acid replacements on the thermostability of protein. *Nature* **323**: 356–358.

Matsumara, M., Y. Katakura, T. Imanaka, and S. Aiba. 1984. Enzymatic and nucleotide sequence studies of a kanamycin-inactivating enzyme encoded by a plasmid from thermophilic bacilli in comparison with that encoded by plasmid pUB110. *J. Bacteriol.* **160:** 413–420.

Matthews, B. W. 1986. Structural basis of protein stability and DNA-protein interaction. *Harvey Lect.* **81:** 33–51.

Matthews, B. W. 1987. Genetic and structural analysis of the protein stability problem. *Biochemistry* **26:** 6885–6888.

Owusu, R. K., and D. A. Cowan. 1989. Correlation between microbial protein thermostability and resistance to denaturation in aqueous: organic solvent two-phase systems. *Enzyme Microb. Technol.* (in press).

Patchett, M. L., T. L. Neal, L. R. Schofield, R. C. Strange, R. M. Daniel, and H. W. Morgan. 1989. Heat treatment purification of thermostable cellulase and hemi-cellulase enzymes expressed in *E. coli. Enzyme Microb. Technol.* **11:** 113–115.

Perutz, M. F., and H. Raidt. 1975. Stereochemical basis of heat stability in bacterial ferridoxins and in haemoglobin A2. *Nature* **244:** 256–259.

Reynolds, P. H. S., C. H. Sissons, R. M. Daniel, and H. W. Morgan. 1986. Comparison of cellulolytic activities in *Clostridium thermocellum* and three thermophilic cellulolytic anaerobes. *Appl. Environ. Microbiol.* **51:** 12–17.

Sissons C. H., K. R. Sharrock, R. M. Daniel, and H. W. Morgan. 1987. Isolation of cellulolytic anaerobic extreme thermophiles from New Zealand thermal sites. *Appl. Environ. Microbiol.* **53:** 832–838.

Stetter, K. O. 1982. Ultrathin mycelia-forming organisms from submarine volcanic areas having an optimum growth temperature of 105°C. *Nature* **300:** 258–260.

Tanaka, T., N. Kawano, and T. Oshima. 1981. Cloning of the 3-isopropylmalate dehydrogenase of an extreme thermophile and partial purification of the gene product. *J. Biochem.* **89:** 677–682.

Werber, M. M., J. L. Sussman, and H. Eisenberg. 1986. Molecular basis for the special properties of proteins and enzymes from *Halobacterium marismortuii. FEMS Microbiol. Rev.* **39:** 129–135.

Woese, C. R. 1987. Bacterial phylogeny. *Microbiol. Rev.* **51:** 221–300.

Woese, C. R., and G. E. Fox. 1977. Phylogenetic structure of the prokaryotic domain: the primary kingdom. *Proc. Natl. Acad. Sci. USA* **74:** 5088–5090.

Woese, C. R., and G. J. Olsen. 1986. Archaebacterial phylogeny: perspectives on the Urkingdoms. *System. Appl. Microbiol.* **7:** 161–177.

Yutani, K., K. Ogasahara, Y. Sugino, and A. Matsushiro. 1977. Effect of a single amino acid substitution on stability of conformation of a protein. *Nature* **267:** 274–275.

13
Biocatalysis in Anaerobic Extremophiles

J. GREGORY ZEIKUS, SUSAN E. LOWE, AND BADAL C. SAHA

By and large most studies on biocatalysis deal with enzymes derived from animal, plant, or microbial sources that grow under "normal physiological" conditions, that is, those environmental conditions that would describe the discovery of physiological biochemistry in the 1930–1950s when normal enzyme environmental conditions (e.g., pH 7.0, 1.5% NaCl, 37°C) were representative of an animal cell as the model system of the times. In the 1970–1980s the vast physiobiochemical diversity of microbes has been extended by microbiologists who can now recognize microbes as living in both normal environments (e.g., human body flora) or in environments (e.g., thermal springs, hypersaline lakes, acidic peat bogs) that represent extreme conditions in relation to the origins of physiological chemistry and a normal animal or plant cell. This chapter will report on our laboratory's efforts to understand biocatalysis in both obligate anaerobes and their enzymes that have adapted to extreme environmental conditions of pH, salinity, or temperature. Our general hypothesis is that anaerobic microbes have evolved different biocatalysis mechanisms for adaptation to extreme environmental conditions than aerobic microbes because they perform biocatalysis under conditions of limited chemical free energy and can afford neither enzyme instability and rapid turnover nor the costly synthesis of general protection mechanisms against enzyme denaturation. Because anaerobic microbes also evolved first on earth under assumed extreme conditions (e.g., high temperature), evolution may have stressed development of proteins with high stability and catalytic efficiency for successful environmental adaptation by these microbes.

We will describe the biocatalytic features of the model anaerobic extremophiles (i.e., acidoanaerobes, haloanaerobes, and thermo-anaerobes) shown below with their growth ranges, in terms of specific physiobiochemical adaptation mechanisms to stress and, specific enzymatic properties related to stability, activity, function, and biotechnological interest.

Thermophile
   Thermoanaerobium brockii $\geq 40-\leq 80°C$
Halophile
   Halobacteroides acetoethylicus $\geq 6-\leq 20\%$ NaCl
Acidophile
   Sarcina ventriculi $\geq 2-\leq 7.5$ pH

The intent of this chapter is to demonstrate our research progress, not to review the literature. Research on the biocatalytic features of thermoanaerobes is very active by others (see R. Daniels, *this volume*), whereas studies on haloanaerobes or acidoanaerobes are very limited.

ACIDOANAEROBES

Peat bogs contain acid sediments with high organic content but limited microbial activity as low pH is a general inhibitor of anaerobic biocatalysis and accounts in part for the accumulation of peat, coals, and lignites in the biosphere (Goodwin and Zeikus 1987a; Zeikus 1983). *Sarcina ventriculi* is found in peat bogs and proliferates at pH 2.0 but sporulates at neutral pH (Goodwin and Zeikus 1987a; Lowe, Pankratz, and Zeikus 1989); thus, it also serves as a model acidoanaerobe to study the effect of pH on biocatalysis. Table 13.1 shows the relationship between external growth pH and internal cytoplasmic pH of *S. ventriculi*. Notably when grown at acid pH (i.e., pH 3.0), the organism maintains an acidic cytoplasm (approximately pH 4.25). Thus the organism has a physiological biochemistry that is unlike that of aerobic acidophiles (e.g., *Thiobacillus* sp.) which expend metabolic energy to maintain a neutral cytoplasm when grown at extremely acid pH values (Goodwin and Zeikus 1987b). Growth at very acidic versus neutral conditions (Fig. 13.1) shows that a dynamic adaptive response occurs

Table 13.1. Effect of environmental pH on growth and electrochemical parameters of *S. ventriculi* in pH-controlled batch cultures.

MEDIUM pH	GROWTH[a]	INTERNAL pH	ΔpH[b]	$\Delta\psi$ (mV)[c]	PMF (mV)[d]
10.0	−				
9.0	+				
7.5	+	7.6	0.1	46	50 ± 4
7.0	+	7.1	0.1	63	67 ± 2
4.2	+	5.1	0.9	55	112 ± 0.4
3.0	+	4.3	1.3	38	114 ± 4
2.0	+				
1.0	−				

From: Goodwin S. and J. G. Zeikus. 1987b. Physiological adaptations of anaeroboic bacteria to low pH: metabolic control of proton motive force in *Sarcina ventriculi*. *J. Bacteriol.* **169:** 2150–2157.
[a] Positive growth indicates that the organism produced >200 μg of protein per liter.
[b] The ΔpH was determined from the distribution of [^{14}C]salicylic acid.
[c] The $\Delta\psi$ was determined from the distribution of [^{3}H]tetraphenylphosphonium bromide.
[d] The values for PMF are averages of triplicate analysis ± 1 standard deviation.

by the catabolic enzyme machinery of the organism in that it shifts from a pyruvate dehydrogenase-directed mixed acid fermentation (acetic, formic, H_2, and ethanol) at pH 7.0 to a pyruvate decarboxylase-directed ethanolic fermentation at pH 3.0 (Goodwin and Zeikus 1987b; Lowe and Zeikus, unpublished findings). Pyruvate decarboxylase activities notably increase at pH 3.0 versus pH 7.0. The inhibition of in vivo hydrogenase activity by growth under decreasing medium pH has also been shown to correlate with less H_2 production. Because acid production by acidoanaerobes growing at low pH will dissipate the protonmotive force when the buffering capacity of the cell is exceeded, this organism shifts to production of neutral products. This understanding can help explain why anaerobic solvent fermentations can take place at lower pH values than acid fermentation and may explain in part why biocatalysis is limited in acidic anaerobic environments. Acidoanaerobes such as *S. ventriculi* or other species can be considered as potential sources of intracellular enzymes (e.g., pyruvate decarboxylase, glucose isomerase, etc.) to industrial biocatalytic processes requiring acid stability.

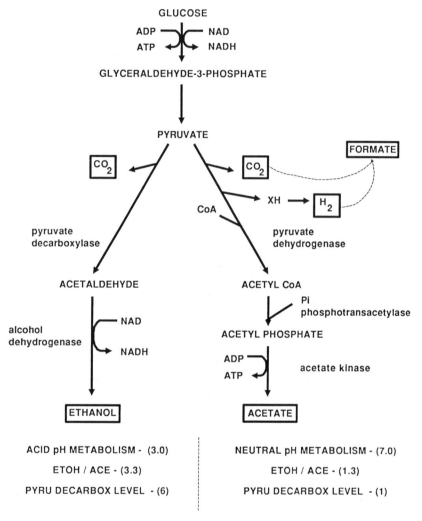

Figure 13.1. Comparison of glucose catabolism at a pH of 3.0 versus 7.0 in *S. ventriculi.*

HALOANAEROBES

Anoxic hypersaline sediments and waters of lakes and deep subsurface environments associated with oil and gas deposits have microbial biocatalytic activities that appear to be limited by extremely high salt. This inhibition of anaerobic biocatalysis may in part account for the accumulation of oils and methane (Zeikus 1983; unpublished lab findings). Nonetheless, haloanaerobic bacteria have evolved that are adapted to moderately high salt levels (i.e., 15–20%).

Halobacteroides acetoethylicus (Rengpipat, Langworthy, and Zeikus 1988) is a model haloanaerobe for biocatalysis studies and it readily grows on glucose at hypersalinities between 6 and 20% NaCl (Fig. 13.2). The organism produces ethanol, hydrogen, acetate, and CO_2 as fermentation products of glucose. Table 13.2 shows the relationship between internal and external salt concentration in *H. acetoethylicus*. As with other halophiles described, K^+ is accumulated in response to

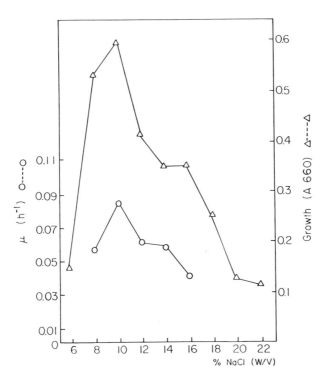

Figure 13.2. Relationship between NaCl concentration and growth of *H. acetoethylicus*. The organism was grown at 34°C in anaerobic pressure tubes that contained halo-YEG stock medium (10 ml) and the NaCl concentration indicated. (Reprinted by permission of VCH Publishers, Inc., 220 East 23 Street, New York, N.Y., 10010 from: Figure 3 of *Systematic and Applied Microbiology*, **11**, 28–35 (1988).)

increasing NaCl concentrations. However, unlike in aerobic halophiles, both Na^+ and Cl^- concentrations increase intracellularly in response to higher external salt and they are present in the cytoplasm at higher values than K^+ (Rengpipat, Lowe, and Zeikus 1988). Unlike halophilic phototrophs or chemoorganotrophic aerobes, this organism does not produce osmoregulants (e.g., glycerol or betaine) in the presence of high NaCl to maintain osmotic balance.

Table 13.2. Intracellular and supernatant concentrations of sodium, potassium, magnesium, and chloride in *H. acetoethylicus* cultures grown with different NaCl concentrations in the culture medium.[a]

ION AND SOURCE	ION CONCENTRATION (M) IN MEDIUM WITH NaCl CONCENTRATION (M) OF:	
	1.70	2.32
Na^+		
Supernatant	1.16	2.52
Intracellular	0.92	1.50
K^+		
Supernatant	0.032	0.034
Intracellular	0.240	0.780
Mg^{2+}		
Supernatant	0.006	0.006
Intracellular	0.020	0.040
Cl^-		
Supernatant	1.4	2.7
Intracellular	1.2	2.5

From: Rengpipat S., S. E. Lowe, and J. G. Zeikus. 1988. Effect of extreme salt concentrations on the physiology and biochemistry of *Halobacteroides acetoethylicus*. *J. Bacteriol.* **170:** 3065–3071.

[a] Cells were grown to midexponential phase in TYG medium with the [NaCl] indicated. Each culture was processed into a supernatant and an intracellular fraction prior to analysis of ions.

The question as to why moderate hypersalinity is required for and stimulates growth whereas extreme hypersalinity inhibits growth of this organism was pursued at the enzyme level. Figure 13.3 shows that in vivo hydrogenase activity is stimulated by moderately high external NaCl concentrations, whereas extremely high salt is inhibitory to activity. Figure 13.4 compares the effect of different salt concentrations on alcohol dehydrogenase activities of *H. acetoethylicus* versus *C. thermohydrosulfuricum* as control. Increasing [Na$^+$] and [K$^+$] in the assay mixture generally correlated with increasing ADH activity in *H. acetoethylicus* but inhibited activity in the nonhalophilic control. In general, it appears from enzymatic analysis studies that the salt concentration that appears optimal for growth correlates with the level of Na$^+$ required for optimal activity of enzymes operative in the glucose fermentation pathway. Interestingly, at extremely high salt concentrations (>15%) key catabolic enzyme activities (pyruvate dehydrogenase

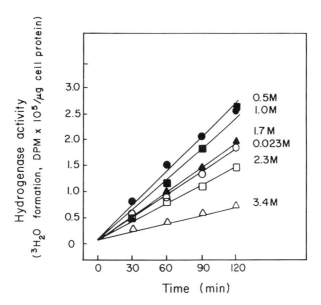

Figure 13.3. Relation of NaCl concentration (on the right) to in vivo hydrogenase activity in whole cells of *H. acetoethylicus*. Cells were grown to the midexponential phase at 34°C in 58-ml serum bottles. Before the assay, the headspace was evacuated and flushed with N$_2$ gas and 1 ml of ^{3}H$_2$ was added. The assays were performed at 34°C. (From: Rengpipat S., S. E. Lowe, and J. G. Zeikus. 1988. Effect of extreme salt concentrations on the physiology and biochemistry of *Halobacteroides acetoethylicus*. *J. Bacteriol.* **170:** 3065–3071.)

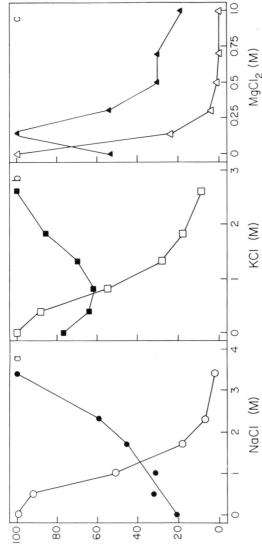

Figure 13.4. Comparison of the effects of different salt concentrations on the alcohol dehydrogenase activities of cell extracts prepared from either *H. acetoethylicus* (closed symbols) or *C. thermohydrosulfuricum* (open symbols). Highest (100%) activities in panels (a), (b), and (c) correspond to 0.133, 0.042, and 0.045 U/min per mg of protein, respectively, for *H. acetoethylicus* and to 4.45 U/min per mg of protein for *C. thermohydrosulfuricum*. Conditions: volume of cell extract of *H. acetoethylicus* used, 15 μl (0.72 mg of protein); volume of cell extract of *C. thermohydrosulfuricum* used, 5 μl (0.18 mg of protein); total volume, 1 ml; gas phase, nitrogen; temperature, 34°C for *H. acetoethylicus* and 60°C for *C. thermohydrosulfuricum*. (From: Rengpipat S., S. E. Lowe, and J. G. Zeikus. 1988. Effect of extreme salt concentrations on the physiology and biochemistry of *Halobacteroides acetoethylicus*. *J. Bacteriol*. **170:** 3065–3071.)

and glyceraldehyde 3-phosphate hydrogenase) are severely inhibited. The effect of increasing sodium concentration on activation of certain oxidoreductases in this organism bears further examination to understand how this is achieved at the enzyme molecular level.

Activity of the alcohol dehydrogenase of this haloanaerobe continues to increase even at extremely high salt concentrations. These enzymes could be viable sources for a study of biocatalysis under nonaqueous conditions. Certainly haloanaerobes should be considered as potential sources of biocatalysts for industrial processes that require high salt such as the fermentative production of calcium magnesium acetate-based road salts or where a specific haloactive and tolerant enzyme is required for a specific chemical process reaction.

THERMOANAEROBES

Hot spring environments such as those found in Yellowstone National Park contain a large diversity of thermophilic bacteria although extremely high temperature does place some constraints on certain biocatalytic processes (Zeikus 1979). For example, eucaryotes and procaryote phototrophs do not proliferate in these ecosystems above 50° and 70°C, respectively (Langworthy et al. 1979). At the hyperthermophilic region above 85–105°C, archaebacterial chemotrophs but not eubacterial chemotrophs thrive (see R. Daniels, *this volume*).

Our work on thermophiles has focused on understanding the biocatalytic features of the saccharidases and alcohol dehydrogenases of various thermoanaerobes isolated from Yellowstone National Park that grow on either starch, hemicellulose, pectin, or simple sugars and produce ethanol, acetate, and lactate as major end products (Zeikus, Hegge, and Anderson 1979; Zeikus, Ben-Bassat, and Hegge 1980; Schink and Zeikus 1983). In nature, this microbial group catalyzes the anaerobic degradation of biochemical components produced by the thermophilic phototrophic bacteria (Zeikus 1979, 1983). Like aerobic eubacterial thermophiles, thermoanaerobes have evolved thermostable and thermoactive enzymes to account for adaptation to high temperatures. Evidence is not yet available as to whether thermoanaerobes produce generalized enzyme protectants or cofactors such as thermine (Langworthy et al. 1979) in *T. aquaticus* or if they have rapid rates of protein synthesis and turnover as in *B. stearothermophilus*

Table 13.3. Physiobiochemical characteristics of selected saccharolytic fermentative thermoanaerobes isolated from Yellowstone National Park.

SPECIES STRAIN	GROWTH (TEMPERATURE OPTIMUM, °C)	ENERGY SOURCES	THERMOACTIVE ENZYMES (TEMPERATURE OPTIMUM, °C)
T. brockii strain HTD4	65	Hexose, starch, pentoses	Alcohol dehydrogenase—80
C. thermohydrosulfuricum strain 39E	65	Starch, pentose, hexoses	Amylopullulanase—90
			α-Glucosidase—75
C. thermosulfurogenes strain 4B	60	Pectin, starch, pentoses, hexoses	β-Amylase—75
			Xylose isomerase—80
Thermoanaerobacter Strain B6A	60	Hemicellulose, starch, hexoses, pentoses	Amylopullulanase—75
Strain LX11	60	Hemicellulose, starch, hexoses, pentoses	Endoxylanase—75

References: Zeikus et al., 1979; Zeikus et al., 1980; Lamed and Zeikus, 1981; Schink and Zeikus, 1983; Hyun and Zeikus, 1985a; Hyun and Zeikus, 1985b; Saha et al., 1988; Shen et al., 1988; and unpublished results of Saha et al., 1990.

(Langworthy et al. 1979). Table 13.3 illustrates temperature optima for growth and key catabolic enzymes of several thermoanaerobic species isolated from Yellowstone National Park. In general, the temperature activity and stability optima of catabolic enzymes characterized in thermoanaerobes is higher than the growth temperature optimum and longer than the species doubling time.

ALCOHOL DEHYDROGENASE

Both *T. brockii* (strain HTD4) and *C. thermohydrosulfuricum* (strain 39E) contain a novel, thermoactive NADP-linked alcohol dehydrogenase that is absent in other ethanol-producing thermoanaerobes such as *C. thermocellum* and *Thermobacteroides acetoethylicus* (Lamed and Zeikus 1981). Table 13.4 illustrates the very broad substrate speci-

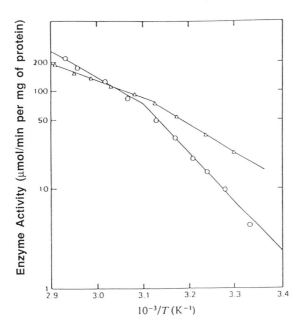

Figure 13.5. Temperature dependence of *T. brockii* secondary-alcohol dehydrogenase. The reaction mixture contained (1 ml total volume) 50 mM potassium phosphate buffer, pH 7.3, 0.5 mM NADP$^+$, 150 mM propan-2-ol ($\triangle$) or butan-2-ol ($\bigcirc$), and 0.5 μg of the purified enzyme; the temperature was as indicated. (From: Lamed, R. J., and J. G. Zeikus. 1981. Novel NADP-linked alcohol-aldehyde/ketone oxidoreductase in thermophilic ethanologenic bacteria. Biochem. J. **195:** 183–190.)

Table 13.4. Substrate specificity of NADP-linked alcohol-aldehyde/ketone oxidoreductase from *T. brockii*.

SUBSTRATE	APPARENT V_{max} (μMOL/MIN PER MG OF PROTEIN AT $40°C$)
Primary alcohols	
CH_3OH	0
CH_3—CH_2OH	3.2
CH_3—CH_2—CH_2OH	3.6
CH_3—CH_2—CH_2—CH_2OH	4.1
CH_3—CH_2—CH_2—CH_2—CH_2OH	0.9
Branched primary alcohols	
CH_3—$CH(CH_3)$—CH_2OH	3.1
CH_3—CH_2—$CH(CH_3)$—CH_2OH	3.0
Secondary alcohols	
CH_3—$CH(OH)$—CH_3	59.0
CH_3—CH_2—$CH(OH)$—CH_3	78.0
CH_2—$CH{=}CH(OH)CH_2$—CH_3	65.0
CH_3—$CH(OH)$—CH_2—$CH(OH)$—CH_3	2.6
Cyclic secondary alcohol	
Cyclohexanol	12.2
Aldehyde	
CH_3—CHO	7.8
Linear ketones	
CH_3—CO—CH_3	10.4
CH_3—CO—CH_2—CH_3	7.6
CH_3CO—CH_2—CH_2—CH_3	4.2
Cyclopropyl methyl ketone	0.5
Cyclic ketones	
Cyclopentanone	6.0
Cyclohexanone	6.2
2-Cyclohexanone	4.9
2-Methylcyclohexanone	6.0
3-Methylcyclohexanone	2.0
4-Methylcyclohexanone	0.5
Cycloheptanone	3.0
Cyclo-octanone	<0.1
4-Norbornanone	4.9

From: Lamed, R. J., and J. G. Zeikus. 1981. Novel NADP-linked alcohol-aldehyde/ketone oxidoreductase in thermophilic ethanologenic bacteria. Biochem. J. **195**: 183–190.

ficity of the enzyme which makes it of interest as an industrial ketone/aldehyde/alcohol oxidoreductase. Figure 13.5 illustrates the temperature–activity relationship of the enzyme when oxidizing propanol versus butanol. The optimum temperature for activity depends on substrate source and concentration. The molecular basis for the biphasic

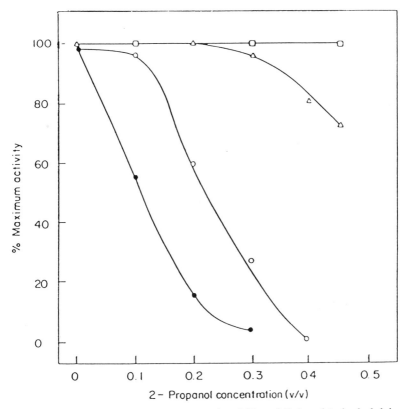

Figure 13.6. Temperature and 2-propanol stability of *T. brockii* alcohol dehydrogenase. Experimental conditions: purified enzyme was dissolved (16 μg/ml) in 50 mM potassium phosphate buffer (pH 7.1), 2 mM dithiothreitol, 0.05 mM NADP, and the 2-propanol concentrations indicated; the solution was heated aerobically in sealed tubes. At the indicated times the solution was cooled and assayed for activity at 40°C: (△), 15 min, 52°C; (○), 20 h, 52°C; (●), 15 min, 86°C; □, control. One hundred percent activity (unheated control) represents μmol/min per mg of protein of secondary alcohol dehydrogenase at 40°C. [From: Lamed, R. J., E. Keinan, and J. G. Zeikus. 1981. Potential applications of an alcohol-aldehyde/ketone oxidoreductase from thermophilic bacteria. *Enzyme Microb. Tech.* **3:** 144–148. Reprinted with permission of the copyright holder, Butterworth Publishers, a division of Reed Publishing (USA) Inc.]

Table 13.5. Typical chiral alcohols obtained by *T. brockii* alcohol dehydrogenase.

Secondary alcohols	OH (structure)	OH (structure)
Chloro-secondary alcohols	OH ⋯Cl (structure)	OH ⋯Cl (structure)
Olephinic secondary alcohols	OH (structure)	OH (structure)
Acetylenic secondary alcohols	OH ≡— (structure)	OH ≡ (structure)
Nitrilo secondary alcohols	OH ⋯CN (structure)	OH ⋯CN (structure)
Ether-containing secondary alcohols	OH ⋯O⋯ (structure)	
Phenyl secondary alcohols	OH ⋯Ph (structure)	OH ⋯Ph (structure)
Carboxylic ester secondary alcohols	OH ⋯CO_2CH_3 (structure)	OH ⋯CO_2CH_3 (structure)

From: Lamed, R. J., A. Bayer, B. C. Saha, and J. G. Zeikus. 1988. Biotechnological potential of enzymes from unique thermophiles. In: *Proc. 8th Intl. Biotechnology Symposium* (G. Durand, L. Bobichon, and J. Florent, eds.), pp. 371–383. Paris: Société Française de Microbiologie.

Arrhenius plot is not known but Q_{10} values were >2 below and above the 50°C deflection point. This enzyme has extraordinary thermoactivity and thermostability for an alcohol dehydrogenase.

Figure 13.6 illustrates the effect of temperature and 2-propanol concentration on the activity of *T. brockii* alcohol dehydrogenase. The enzyme displays extraordinary activity at both moderate temperature and high substrate concentration or at moderate substrate concentration and high temperature. The alcohol dehydrogenase of *T. brockii* is of interest to biotechnology because of its ability to form specialty chiral chemical products from a wide variety of substrates (Table 13.5) and because of its ability to function with high activity at moderate temperatures and high substrate concentration (Lamed, Keinan, and Zeikus 1981; Lamed et al. 1988).

AMYLASES

In moderately thermophilic (i.e., 50–65°C) bacterial algal mats found in Yellowstone National Park, starch is produced by thermophilic phototrophs and degraded by a food chain that includes thermoanaerobes

(Zeikus 1979, 1983). The amylases produced by thermoanaerobic species isolated from these environments possess novel biocatalytic properties (Hyun and Zeikus 1985a,b; Shen et al. 1988; Saha, Mathupala, and Zeikus 1988).

In the following list we describe the biochemical properties of β-amylase purified from *C. thermosulfurogenes*.

Specific activity (U/mg of protein)	4215
Chemical compositions	Tetramer
Molecular weight	210,000
	(Subunit, 51,000)
Isoelectric point (pH)	5.1
Optimum pH	6.0
pH stability	3.5–7.0
Optimum temperature (°C)	75
Thermostability (up to °C)	80
K_m for soluble starch (mg/ml)	1.68 (at 75°C)
K_{cat}	440,000/min (at 75°C)
Specificity	Hydrolyzes α-1,4 linkage forming maltose (exoacting)
Metal ion requirement	
for stability	Calcium
for activity	None
Inhibitors	*p*-Chloromercuribenzoate, cyclodextrins
Polyclonal antibody reaction	Does not cross-react with β-amylase from some other sources

As expected of an enzyme from a thermophile, it possesses high thermostability and thermoactivity. Notably, it also possesses high catalytic efficiency for a β-amylase when its K_m and K_{cat} are evaluated at 75°C. Future studies are required to demonstrate the amino acid sequence specific properties that account for its thermophilicity (i.e., thermoactivity and thermostability).

Now we describe the properties of a new class of amylase activity first discovered in and homogeneously purified from *Clostridium thermohydrosulfuricum*, namely, amylopullulanase (Hyun and Zeikus 1985b; Saha, Mathupala, and Zeikus 1988).

Specific activity (units/mg protein): 481
Chemical composition   Monomeric, glycoprotein
Molecular weight   136,500
Isoelectric point (pH)   5.9
pH optimum   5.0–5.5
pH stability   3.5–5.0
Temperature optimum (°C)   90
Thermostability (up to °C, 30 min)   90
K_m for pullulan (mg/ml)   0.675
K_{cat} 16,240/min
Specificity   Hydrolyzes both α-1,4 and α-1,6 (endoacting)
Metal ion requirement for activity   None
Inhibitors   Cyclodextrins, EDTA, N-bromosuccinimide

The enzyme is a monomeric glycoprotein with an apparent molecular weight of 136,500. It is called amylopullulanase because it has higher affinity for pullulan than for starch, it hydrolyzes α-1,4 and α-1,6 linkages in starch (unlike either α-amylase or pullulanase), and detailed kinetic analysis supports a single active site that cleaves both α-1,4 and α-1,6 glycosidic bonds (Saha, Mathupala, and Zeikus 1988; unpublished results Mathupala, Saha, and Zeikus; Daniels, this volume).

Amylopullulanase from *C. thermohydrosulfuricum* strain 39E is also extremely thermostable and thermoactive (see Fig. 13.7). The temperature–activity profile is normal of an enzyme from an extreme thermophile (Langworthy et al. 1979). Namely, it lacks significant activity at moderate temperature (40°C) and has high activity at extreme temperature (90°C). We have found this enzyme in a wide variety of both aerobic and anaerobic (Saha, Shen, Lamed, Mathupala, and Zeikus unpublished findings) thermophiles. The enzyme has very unique efficiency as an amylase because it can cleave both α-1,4 and α-1,6 linkages in starch. Thus, it may have commercial use in specialty corn syrup or saccharide production.

Future studies are required to demonstrate the amino acid specific properties and perhaps glycoconjugate specific properties that account for thermophilicity (i.e., activity and stability) of amylopullulanase. The molecular basis for enzyme thermoactivity is generally ascribed to tertiary properties of an amino acid sequence. Thus, thermophilic en-

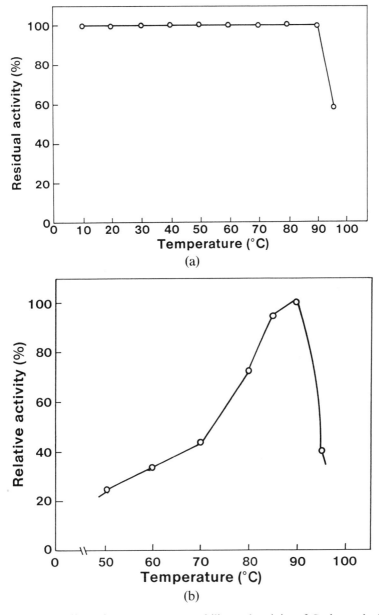

Figure 13.7. Effect of temperature on stability and activity of *C. thermohydrosulfuricum* pullulanase. (a) Thermal stability. The enzyme was placed in acetate buffer (50 mM, pH 6.0) with 5 mM CaCl$_2$ and preincubated at various temperatures for 30 min, and then residual pullulanase activities were assayed. (b) Effect of heat on activity. The enzyme activity was assayed at various temperatures by the standard assay method (30-min incubation). (From: Saha, B. C., S. Mathupala, J. G. Zeikus. 1988. Purification and characterization of a novel highly thermostable pullulanase from *Clostridium thermohydrosulfuricum*. *Biochem. J.* **252**: 343–348.)

zymes are thought to have high intramolecular bond interactions especially in the active site, at moderate temperature and these forces are reduced by high temperature (i.e., thermophilic enzymes are flexible or breathe at high temperature but are not very active at low temperature). Thermostability is generally ascribed to secondary properties of an amino acid sequence that accounts for resistance to irreversibly thermal denaturation; however, binding to a matrix (e.g., glycan immobilization) may also aid in thermostability.

XYLOSE (GLUCOSE) ISOMERASE

Table 13.6 compares the physiobiochemical properties of glucose isomerase from selected Yellowstone thermoanaerobes. Two different types of enzyme activities were found based on pH optima for activity of 7.0 versus 8.5. The former activity was purified and shown to be a tetrameric protein with a native molecular weight of 200,000.

The gene encoding for the glucose isomerase (GI) of *C. thermosulfurogenes* was cloned into *E. coli* and then into a *B. subtilis* host (see Table 13.7). It is of interest to note that the wild-type GI was regulated by induction and catabolite repression, whereas the cloned GI putatively lost its regulatory element but not its promoter in the cloning procedure since the enzyme was produced constitutively in the mesophilic hosts. The thermophilic GI was produced at higher levels in *B. subtilis* than the wild-type organism.

Table 13.8 illustrates the utility of heat treatment in purifying the cloned thermostable GI in mesophiles. After a simple heat treatment process at 85°C for 15 min, the enzyme obtained was of industrial purity (>80% homogeneous). Notably, the *B. subtilis* host–vector system is also of practical value for production of food-safe thermostable enzymes because host proteases were destroyed by the heat treatment process used in purification.

Detailed studies in progress on amino acid sequence analysis of the thermoanaerobe GI indicates high sequence homology to the thermolabile GI of *B. subtilis* but low homology to thermostable GI of *Streptomyces* (industrial strain). This suggests that multiple mechanisms for enzyme thermophilicity exist among glucose isomerases.

The saccharidases of thermoanaerobes have potential utility as industrial biocatalysts because of their novel biocatalytic mechanisms, thermostability, and thermoactivity. The genes for their enzymes can

Table 13.6. Properties of glucose isomerases from thermoanaerobes.

	C. thermosulfurogenes STRAIN 4B	*Thermoanaerobacter* STRAIN B6A	*C. thermohydrosulfuricum* STRAIN 39E
Regulation of GI synthesis	Induced by xylose	Induced by xylose, xylan	Induced by xylose
Optimum pH	7.0	7.0	8.5
Optimum temperature	75°C	75°C	80°C
Isoelectric pH	4.8	4.9	ND
K_m (mM)			
Xylose	20	16	ND
Glucose	140	120	ND
Fructose	60	50	ND
Molecular weight	200,000	200,000	ND
Subunits	4	4	ND

ND, not determined.
Source: C. Lee, doctoral dissertation, 1989.

Table 13.7. Comparison of growth media carbohydrates on synthesis of cloned glucose isomerase in *E. coli* and *B. subtilis* versus *C. thermosulfurogenes*.

| | | SPECIFIC ACTIVITY (U/MG OF PROTEIN) | | |
| | | CARBOHYDRATES ADDED | | |
STRAIN	PLASMID	XYLOSE	GLUCOSE	NONE
C. thermosulfurogenes (Control)		0.19	0.00	ND[a]
E. coli W595	pHSG262	0.00	0.00	0.00
E. coli W595	pCGI38	0.29	0.15	0.27
B. subtilis NA1	pTB523	0.00	0.00	0.00
B. subtilis NA1	pMLG1	1.03	0.81	0.98

E. coli and *B. subtilis* were grown in LB media with 1% carbohydrate as indicated and *C. thermosulfurogenes* was grown in TYE media with 1% carbohydrate prior to assay of activity at 65°C.
[a] ND, not determined.
Source: C. Lee, doctoral dissertation, 1989.

Table 13.8. Purification scheme of cloned thermostable glucose isomerase from *E. coli* and *B. subtilis*.

| | TOTAL PROTEIN (MG) | | TOTAL ACTIVITY (UNIT) | | SPECIFIC ACTIVITY (UNIT/MG) | | PURIFI-CATION (FOLD) | |
STEP	E.C.	B.S.	E.C.	B.S.	E.C.	B.S.	E.C.	B.S.
Crude cell extract	800	122	228	101	0.3	0.8	1	1
Heat treatment (85°C, 15 min)	144	37	216	75	1.5	1.9	5	2.5
DEAE–Sepharose anion exchange	88	28	176	61	2.0	2.2	7	2.8
FLPC-Superose 12 gel filtration	39	15	85	35	2.2	2.3	7.7	2.9

E.C., *E. coli* W595 containing the plasmid pCGI38; B.S., *B. subtilis* NA1 containing the plasmid pMLG1.
Source: C. Lee, doctoral dissertation, 1989.

be cloned and overexpressed in food-safe aerobes. These thermoactive saccharides have application in the production of specialty and commodity corn syrups, sweeteners, and saccharide-based chemicals and for use in the development of in situ sweeteners and flavor agents in processed foods.

SUMMARY

Anaerobic bacteria that have adapted to extreme environmental conditions appear to have evolved novel enzymes that can function under extreme conditions with high activity and stability. Their enzymes can serve as a source of robust catalysts for development of process biotechnologies requiring heat, acid or salt stability. Future studies at the molecular level are aimed at understanding the unique amino acid sequence specific properties of the enzymes from these bacteria that account for activity and stability at extreme environmental conditions versus normal conditions. It is hypothesized that these enzymes are not active at normal environmental conditions because they are not flexible but require extreme conditions to be active. As the genes for these enzymes are cloned and overproduced in conventional microbial hosts, it is hoped the biotechnologists and chemists will use them as models for biocatalysis research.

REFERENCES

Goodwin, S., and J. G. Zeikus. 1987a. Ecophysiological adaptations of anaerobic bacteria to low pH: Analysis of anaerobic digestion in acidic bog sediments. *Appl. Environ. Microbiol.* **53:** 57–64.

Goodwin, S., and J. G. Zeikus. 1987b. Physiological adaptations of anaerobic bacteria to low pH: metabolic control of proton motive force in *Sarcina ventriculi. J. Bacteriol.* **169:** 2150–2157.

Hyun, H. H., and J. G. Zeikus. 1985a. General biochemical characterization of thermostable extracellular β-amylase from *Clostridium thermosulfurogenes. Appl. Environ. Microbiol.* **49:** 1162–1167.

Hyun, H. H., and J. G. Zeikus. 1985b. General biochemical characterization of thermostable pullulanase and glucoamylase from *Clostridium thermohydrosulfuricum. Appl. Environ. Microbiol.* **49:** 1168–1173.

Lamed, R. J., and J. G. Zeikus. 1981. Novel NADP-linked alcohol-aldehyde/ketone oxidoreductase in thermophilic ethanologenic bacteria. *Biochem. J.* **195:** 183–190.

Lamed, R. J., E. Keinan, and J. G. Zeikus. 1981. Potential applications of an alcohol-aldehyde/ketone oxidoreductase from thermophilic bacteria. *Enzyme Microb. Technol.* **3**: 144–148.

Lamed, R. J., E. A. Bayer, B. C. Saha, and J. G. Zeikus. 1988. Biotechnological potential of enzymes from unique thermophiles. In: *Proc. 8th Intl. Biotechnology Symposium*, (G. Durand, L. Bobichon, and J. Florent, eds.), pp. 371–383. Paris: Société Française de Microbiologie.

Langworthy, T. A., T. D. Brock, R. W. Castenholtz, A. F. Esser, E. J. Johnson, T. Oshima, M. Tsuboi, J. G. Zeikus, and H. Zuber. 1979. Life at high temperature. In *Strategies of Microbial Life in Extreme Environments* (M. Shilo, ed.), pp. 489–502. Berlin: Dahlem Konferenzen.

Lowe, S. E., H. S. Pankratz, and J. G. Zeikus. 1989. Influence of pH extremes on sporulation and ultrastructure of *Sarcina ventriculi*. *J. Bacteriol.* **140**: 3775–3781.

Rengpipat, S., T. A. Langworthy, and J. G. Zeikus. 1988. *Halobacteroides acetoethylicus* sp. nov. a new obligately anaerobic halophile isolated from deep subsurface hypersaline environments. *Syst. Appl. Microbiol.* **11**: 28–35.

Rengpipat, S., S. E. Lowe, and J. G. Zeikus. 1988. Effect of extreme salt concentrations on the physiology and biochemistry of *Halobacteroides acetoethylicus*. *J. Bacteriol.* **170**: 3065–3071.

Saha, B. C., S. Mathupala, and J. G. Zeikus. 1988. Purification and characterization of a novel highly thermostable pullulanase from *Clostridium thermohydrosulfuricum*. *Biochem. J.* **252**: 343–348.

Schink, B., and J. G. Zeikus. 1983. *Clostridium thermosulfurogenes* sp. nov., a new thermophile that produces elemental sulphur from thiosulphate. *J. Gen. Microbiol.* **129**: 1149–1158.

Shen, G.-J., B. C. Saha, L. Bhatnagar, Y.-E. Lee, and J. G. Zeikus. 1988. Purification and characterization of a novel thermostable β-amylase from *Clostridium thermosulfurogenes*. *Biochem. J.* **254**: 835–840.

Zeikus, J. G. 1979. Thermophilic bacteria: ecology, physiology, and technology. *Enzyme Microb. Technol.* **1**: 243–252.

Zeikus, J. G. 1983. Metabolic communication between biodegradative populations in nature. *Microbes Natur. Environ.* **34**: 423–462.

Zeikus, J. G., A. Ben-Bassat, and P. W. Hegge. 1980. Microbiology of methanogenesis in thermal, volcanic environments. *J. Bacteriol.* **143**: 432–440.

Zeikus, J. G., P. W. Hegge, and M. A. Anderson. 1979. *Thermoanaerobium brockii* gen. nov. and sp. nov., a new chemoorganotrophic, caldoactive anaerobic bacterium. *Arch. Microbiol.* **122**: 41–48.

14
Large-Scale Bioconversion of Nitriles into Useful Amides and Acids

Toru Nagasawa and Hideaki Yamada

The application of the bioconversion process has been generally restricted to the production of fine chemicals that are difficult to obtain through conventional chemical methods. Recently, however, the industrial production of acrylamide, an important commodity chemical, using bacterial nitrile hydratase, was started in Japan. We describe here the first successful example, the enzymatic production of acrylamide, and the recent progress in the microbial transformation of nitriles. The cofactors and reaction mechanism of nitrile hydratase are also discussed.

BIOTECHNOLOGICAL POTENTIAL OF NITRILE METABOLISM

Many microorganisms can use nitriles as sources of carbon and/or nitrogen for growth (Robinson and Hook 1964; Mimura, Kawamoto, and Yamaga 1969; DiGeronimo and Antoine 1976; Firman and Gray 1976; Arnaud, Galzy, and Jallageas 1977; Harper 1977a,b; Kuwahara et al. 1980; Yamada et al. 1980; Yamada, Asano, and Tani 1980; Bandyopadhyay et al. 1986; Linton and Knowles 1986; Nagasawa, Kobayashi, and Yamada 1988). The general metabolic scheme for organisms growing on aromatic, heterocyclic, and aliphatic nitriles can be summarized as follows: benzonitrile and related aromatic nitriles, and heterocyclic nitriles are catabolized directly to the corresponding acids and ammonia by a "nitrilase" enzyme (EC 3.5.5.1, nitrile aminohydrolase) [Eq. (1)].

$$RCN + 2H_2O \rightarrow RCOOH + NH_3 \qquad (1)$$
(R can be phenyl or α,β-alkenyl)

On the other hand, aliphatic nitriles are catabolized in two stages, that is, conversion to the corresponding amide and then to the acid plus ammonia (Asano, Tani, and Yamada 1980; Linton and Knowles 1986). The enzyme that catalyzes the first step is clearly distinguishable from the nitrilase in the mode of degradation of nitriles. The enzyme that catalyzes the hydration of nitriles to amides was termed "nitrile hydratase" [Eq. (2)]. The second step is catalyzed by an "amidase" [Eq. (3)]

$$RCN + H_2O \rightarrow RCONH_2 \qquad (2)$$

$$RCONH_2 + H_2O \rightarrow RCOOH + NH_3 \qquad (3)$$
(R can be an alkyl)

Nitriles are widely used in organic synthesis to produce chemical compounds such as amides and organic acids. However, the conversion of nitriles in the traditional "chemical" way has several disadvantages: (1) reactions proceed only in either strongly acidic or basic media; (2) heat-energy consumption is high; and (3) byproducts (toxic substances such as HCN or a large amount of salt) are formed. Attention is currently focused on the application of enzymes to organic chemical processing. The newly developed "biological" procedure, in which microorganisms are used as catalysts, is more attractive, because the pH and temperature conditions are less severe, and because very pure products are formed without the production of secondary byproducts. In addition, bioconversions can be stereo- and regiospecific. Thus, nitrile-converting enzymes are expected to have great potential as catalysts for converting nitriles to the corresponding higher value amides or acids in organic chemical processing.

PRODUCTION OF ACRYLAMIDE WITH *PSEUDOMONAS CHLORORAPHIS* B23 NITRILE HYDRATASE

Acrylamide is one of the most important commodity chemicals and it is in great demand as a starting material for the production of various polymers for use as flocculants, stock additives, and polymers for petroleum recovery. About 200,000 tons of acrylamide are produced per year worldwide. Conventional synthesis involves the hydration of nitriles with the use of copper salts as catalysts. This method, however, suffers from various problems due to the complexity of the preparation of the catalyst, the difficulty in regenerating the used catalyst, and the complexity of the separation and purification of the acrylamide formed. Furthermore, it is desirable to produce acrylamide under moderate reaction conditions, because acrylamides are readily polymerizable.

It was particularly from the viewpoint of the bioindustry that the discovery of nitrile hydratase interested us. This was because nitrile hydratase seemed to be promising for the enzymatic transformation of nitriles into amides. Galzy and colleagues (Commeyras et al. 1973, 1977; Jallageas, Arnaud, and Galzy 1980; Bui, Arnaud, and Galzy 1982) in France mentioned the potential use of *Brevibacterium* R312 in the industrial field in patent applications and papers describing the production of amides and carboxylic acids. Our Kyoto University group (Asano et al. 1982) and Nitto Chemical Industry Company, Ltd. (Watanabe, Satoh, and Kouno 1979; Watanabe and Okumura 1986; Watanabe, Satoh, and Enomoto 1987) proposed an enzymatic production process for acrylamide involving nitrile hydratases from *P. chlororaphis* B23 and *Corynebacterium* N-774, selected as favorable strains, respectively.

At first, we isolated various nitrile-assimilating bacteria and cultivated them in medium containing different nitriles. Using resting cells, we examined the bioconversion of acrylonitrile into acrylamide (Asano et al. 1982). *P. chlororaphis* B23 was isolated from soil as an isobutyronitrile-utilizing bacterium, and it exhibited high acrylamide-producing activity. When resting cells of *P. chlororaphis* B23 were added as a catalyst to the reaction mixture in a 7.5-h reaction at 10°C, more than 400 g of acrylamide per liter accumulated (Fig. 14.1). The substrate, acrylonitrile, was added in small portions, because a high concentration of it is inhibitory toward nitrile hydratase activity. There was >99% conversion of the acrylonitrile into acrylamide, without the formation of acrylic acid as a byproduct. Such high productivity and such a high conversion yield suggested that this biocatalysis could be applied for the industrial production of acrylamide.

To develop this enzymatic reaction into an industrial process, it was necessary to improve the culture conditions. We tried to optimize the culture conditions (Yamada et al. 1986). The formation of the *P.*

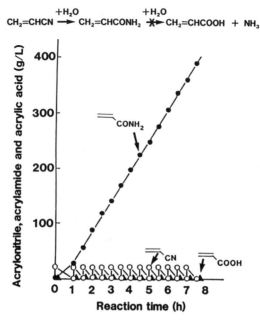

Figure 14.1. Time course of acrylamide production with *P. chlororaphis* B23 cells. Cells were grown at 28°C for 3 days on the culture medium containing 0.15% isobutyronitrile and 0.5% (wt/vol) dextrin. The reaction mixture contained 2 g (as dry weight) of washed cells, 10 mmol of potassium phosphate buffer, pH 7.0, and 560 mmol of acrylonitrile, which was added in portions of 40 mmol at 30-min intervals, in a total volume of 100 ml, and the reaction was carried out at 0–4°C with stirring.

chlororaphis B23 nitrile hydratase is induced by various nitriles and amides. Among the latter, methacrylamide showed the greatest induction (Table 14.1). Therefore, we chose methacrylamide as an inducer for that reason. The addition of ferrous or ferric ions to the medium greatly increased the formation of nitrile hydratase (Table 14.2). No other metal ions, including cobalt ions, can replace ferrous or ferric ions. This reflects the fact that nitrile hydratase contains ferric ions as a cofactor. We optimized the culture medium such that a medium containing 1 g of sucrose, 0.5 g of methacrylamide, 0.2 g of L-cysteine, 0.2 g of L-glutamate (Na), 0.2 g of L-proline, 50 mg of KH_2PO_4, 50 mg of K_2HPO_4, 50 mg of $MgSO_4 \cdot 7H_2O$, and 1 mg of $FeSO_4 \cdot 7H_2O$ per 100 ml

Table 14.1. Effects of nitriles, amides and acids on the formation of nitrile hydratase by *P. chlororaphis* B23.

NITRILES, AMIDE AND ACIDS	ENZYME ACTIVITY (units/ml CULTURE BROTH)	SPECIFIC ACTIVITY (units/mg DRY CELL WEIGHT)
Isobutyronitrile	160	26
Acetonitrile	3.4	0.8
Propionitrile	26	4.6
n-Butyronitrile	0	0
Methacrylonitrile	0	0
Crotononitrile	0	0
Acrylonitrile	no growth	
Isobutyramide	160	28
Formamide	0	0
Acrylamide	18	6.9
Crotonoamide	150	32
Methacrylamide	360	65
n-Butyramide	45	6
n-Propionamide	92	14
Succinamide	0	0
Lactamide	0	0
Acetic acid	0	0
Propionic acid	0	0
n-Butyric acid	0	0
Isobutyric acid	54	11
n-Valeric acid	0	0
Isovaleric acid	0	0
Methacrylic acid	0	0

Various nitrile compounds (0.5 ml), amides (0.5 g), and acids (0.5 ml) were added to 100 ml of the basal medium, which consisted of 1 g of sucrose, 0.2 g of L-cysteine, 0.2 g of L-glutamate (Na), 0.2 g of L-proline, 50 mg of KH_2PO_4, 50 mg of K_2HPO_4, 50 mg $MgSO_4 \cdot 7H_2O$, and 1 mg of $FeSO_4 \cdot 7H_2O$ per 100 ml of tap water. Cultivation was carried out until the early stationary phase (19–32 h) of growth.

of tap water was most suitable for the preparation of cells with high nitrile hydratase activity. Instead of amino acids, a cheap soybean hydrolysate can be used. When *P. chlororaphis* B23 was cultivated for 26 h at 25°C in this optimum medium, with the pH controlled from pH 7.5 to 7.8, the enzyme productivity was great. The enormous amount of nitrile hydratase is formed in the cells, which corresponds to >20% of the total soluble protein (Fig. 14.2). The increase in enzyme activity

Table 14.2. Effects of inorganic compounds on the formation of nitrile hydratase by *P. chlororaphis* B23.

INORGANIC COMPOUND (mg/100 ml)		ENZYME ACTIVITY[a]	SPECIFIC ACTIVITY[b]
None		7.8	6.5
$CaCl_2 \cdot 2H_2O$	1	2.8	2.0
$BaSO_4 \cdot 7H_2O$	1	7.0	5.0
LiCl	1	11	6.9
$NiSO_4 \cdot 7H_2O$	1	23	13
$ZnSO_4 \cdot 7H_2O$	1	20	13
$CuSO_4 \cdot 5H_2O$	1	10	6.7
$CoSO_4 \cdot 7H_2O$	1	9.4	7.8
$MnSO_4 \cdot 7H_2O$	1	5.6	4.0
$Na_2MoO_4 \cdot 2H_2O$	1	8.4	6.0
$PbCl_2$	1	2.9	2.9
$FeSO_4 \cdot 7H_2O$	0.5	35	27
	1	38	24
	2	35	27
$Fe_2(SO_4)_3$	1	37	26

Various mineral compounds were added at the indicated concentrations to a medium consisting of 1 g of sucrose, 0.38 g of isobutyronitrile, 50 mg of KH_2PO_4, 50 mg of K_2HPO_4, and 50 mg of $MgSO_4 \cdot 7H_2O$ per 100 ml of tap water. During the cultivation, the pH of each medium was controlled at around 7.2–7.8 by the addition of 4 M NaOH or H_2SO_4 at times.
[a] Units/ml culture broth.
[b] Units/mg dry cell weight.

was 900-fold and that in specific activity, respectively, 90-fold the initial level. Thus, as a first step for the enzymatic production of acrylamide on an industrial scale, we established suitable culture conditions for *P. chlororaphis* B23.

The next problem we had to overcome was that when *P. chlororaphis* B23 is grown on a medium containing sucrose, mucilage polysaccharides, which resemble levan, are produced. Therefore, it is not easy to harvest the cells by brief centrifugation due to the high viscosity of the culture medium. This is an important technical disadvantage that cannot be overlooked. The viscosity of the culture me-

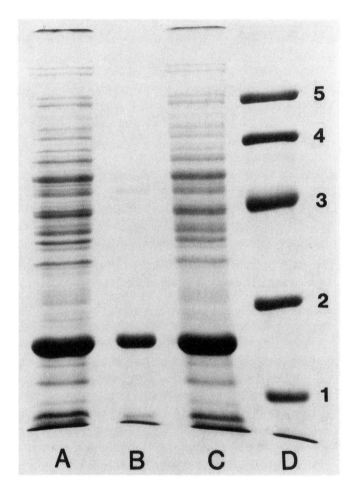

Figure 14.2. Protein profiles of cells cultivated under the optimum conditions for the production of nitrile hydratase on SDS-polyacrylamide gel electrophoresis. Lanes A and C were loaded with 56 μg of protein of a cell-free extract of *P. chlororaphis* B23 cells under the optimal conditions. Lane B was loaded with the purified *P. chlororaphis* B23 nitrile hydratase (19 μg). Lane D was loaded with the following molecular weight standards: 1, soybean trypsion inhibitor (20, 100); 2, carbonic anhydrase (30,000); 3, ovalbumin (43,000); 4, bovine serum albumin (67,000); and 5, phosphorylase b (94,000).

dium also adversely affects aeration. In other words, the accumulation of mucilage polysaccharides results in a waste of the energy derived from the added sucrose. Thus, we attempted to isolate mucilage polysaccharide nonproducing mutants (Ryuno, Nagasawa, and Yamada 1988). Cells were mutagenized with N-methyl-N'-nitro-N-nitroso-guanidine (MNNG) and then spread onto agar plates, with appropriate dilution. The parent strain mainly formed large, translucent, swollen colonies, but colonies with a different appearance also occurred at a frequency of 1%. The latter were smaller, clearer shaped, and less sticky than those of the parent strain. They were picked up and then sedimentation testing of each culture broth was carried out by centrifugation at 8,000 $\times$ g for 10 min (Fig. 14.3). About 40% of the strains picked were precipitated completely or partially on this brief centrifugation, because their culture fluids were not mucilaginous. Those that were completely precipitated of the brief centrifugation were designated as Am strains. The parent strain, in contrast with the Am strains,

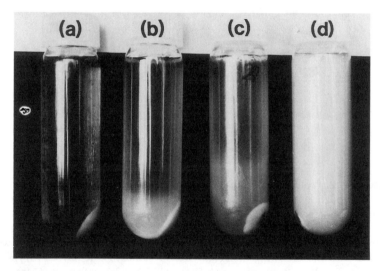

Figure 14.3. Sedimentation testing of culture broth of the parent strain and mutants. Sedimentation testing was carried out by centrifugation at 8,000 $\times$ g for 10 min. (a), (b), (c) Mutants derived from *P. chlororaphis* B23, which formed small and clear-shaped colonies on the medium containing 1.0% (wt/vol) sucrose, 0.3% yeast extract, 0.3% malt extract (Difco, Detroit) and Polypepton (Daigo, Osaka), pH 7.2. (d) The parent strain.

did not precipitate on centrifugation, even for 1 h at 12,000 × g. Strain Am-3 exhibited 1.8-fold higher nitrile hydratase activity than the parent strain, and so was selected and was the highest nitrile hydratase producer. To isolate high nitrile hydratase-producing mutants, strain Am-3 was mutagenized again with MNNG. We picked mutants that grew rapidly on isobutyronitrile medium. One mutant, Am-324, derived from Am-3, exhibited higher nitrile hydratase activity than that of the parent strain. Moreover, when 0.2% (wt/vol) methacrylamide was fed after 42 h of cultivation, growth and the specific activity of nitrile hydratase was greatly enhanced (Fig. 14.4). Thus, an advantageous mutant, strain Am-324, which can be easily precipitated on brief centrifugation and which exhibits high nitrile hydratase activity, was bred from the parent strain, *P. chlororaphis* B23. As shown in Table 14.3, with the improvement of the culture medium and the isolation of mutants, the specific activity increased 3,000 times compared with the initial observed level. Thus, we established the basis for the industrial

Table 14.3.   Genealogies of the nitrile hydratase mutants.

STRAIN	SPECIFIC ACTIVITY (U/mg OF DRY CELLS)	TOTAL ACTIVITY (U/ml)
Parent (medium A) ↓	0.72	0.40
Parent (medium R) ↓ MNNG treatment	66	363
Am 3 ↓ MNNG treatment	65	465
Am 324 | Feeding of ↓ methacrylamide	125	952
Am 324	141	1260

Medium A consisted of 0.5 g of dextrin, 0.2 g of K_2HPO_4, 0.1 g of NaCl, 0.02 g of $MgSO_4 \cdot 7H_2O$, and 0.15 ml of isobutyronitrile per 100 ml of tap water (pH 7.0). Medium B consisted of 1.0 g of sucrose, 0.05 g of KH_2PO_4, 0.05 g of K_2HPO_4, 0.05 g of $MgSO_4 \cdot 7H_2O$, 0.001 g of $FeSO_4 \cdot 7H_2O$, 1.5 g of soybean hydrolysate, and 0.8 g of methacrylamide per 100 ml of tap water (pH 7.8). Nitrile hydratase activity of the mutants was assayed in cells grown in medium R.

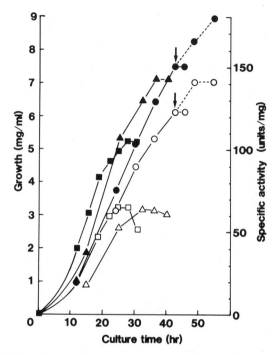

Figure 14.4. Growth and nitrile hydratase activity of the parent, and strains Am-3 and Am-324 in the presence of a high amount of methacrylamide. Cultivations were carried out in 2-L jar fermentors containing 1 L of medium containing 1.0% (wt/vol) sucrose, 0.05% KH_2PO_4, 0.05% K_2HPO_4, 0.05% $MgSO_4 \cdot 7H_2O$, 0.001% $FeSO_4 \cdot 7H_2O$, 0.2% L-cysteine, 0.2% L-glutamate (Na), and 0.2% L-proline, pH 7.8, supplemented with 0.8% (wt/vol) methacrylamide. After 43 h cultivation (➡), 0.2% (wt/vol) methacrylamide was fed. (■), (▲), (●), Growth of the parent strain, and strains Am-3 and Am-324, respectively. (□), (△), (○), Specific activity of the parent strain, and strains Am-3 and Am-324, respectively.

produciton of acrylamide by means of the enzymatic hydration process.

We cooperated with Nitto Chemical Industry Company, Ltd., to establish this process for industrial use. A new bioreactor containing immobilized cells, entrapped on a cationic acrylamide-based polymer gel, was designed (Watanabe 1987), and a very compact and efficient commercial plant was constructed by Nitto Chemical Industry Company, Ltd. (Nakai et al. 1988). When the enzymatic hydration process is compared with the conventional process involving a copper salt catalyst for the produciton of acrylamide (Fig. 14.5), it can be seen that the step of recovery of the remaining acrylonitrile can be omitted due to the high conversion yield (almost 100%) of the enzymatic hydration, and the step of removal of copper ions from the product can also be omitted (Nakai et al. 1988). Overall, the enzymatic process is simpler and more economical. It can be carried out below 10°C under mild reaction conditions and requires no special equipment as a source of energy. The immobilized cells can be used repeatedly and a highly pure product is formed. The enzymatic process involving *P. chlororaphis* B23 cells is already in use and at present about 6,000 tons of acrylamide is produced per year by Nitto Chemical Industry Company, Ltd. Japan (Vandamme 1987).

Why can *P. chlororaphis* B23 accumulate so much acrylamide? The first explanation is that *P. chlororaphis* B23 exhibits enormously high nitrile hydratase activity, compared to its amidase activity (Nagasawa, Ryuno, and Yamada 1989). The nitrile hydratase activity toward acrylonitrile is at least 4,000 times higher than the amidase activity toward acrylamide in the cells. Next, acrylonitrile, a powerful nucleophilic reagent, inactivates the active thiol residue of the typical thiol-enzyme, amidase (Nagasawa, Ryuno, and Yamada 1989) (Fig. 14.6). On the other hand, nitrile hydratase is not as sensitive to acrylonitrile. Thus, the accumulation of acrylamide and the nonformation of acrylic acid are possible. In addition, the nitrile hydratase purified from *P. chlororaphis* B23 exhibits high resistance to high concentrations of acrylamide. The *P. chlororaphis* B23 enzyme acts on acrylonitrile even in the presence of 17.5% acrylamide. On the other hand, the nitrile hydratase from *Brevibacterium* R312 is not so tolerant of high concentrations of acrylamide and cannot act on acrylonitrile at all in the presence of 17.5% acrylamide (Fig. 14.7) (Nagasawa, Ryuno, and Yamada 1989). Therefore, the *P. chlororaphis* B23 enzyme is a more suitable catalyst from the viewpoint of acrylamide accumulation.

CHARACTERIZATION AND COFACTORS OF IRON-CONTAINING NITRILE HYDRATASE

Nitrile hydratase has been purified and crystallized from *P. chlororaphis* B23 (Nagasawa et al. 1987) and *Brevibacterium* R312 (Na-

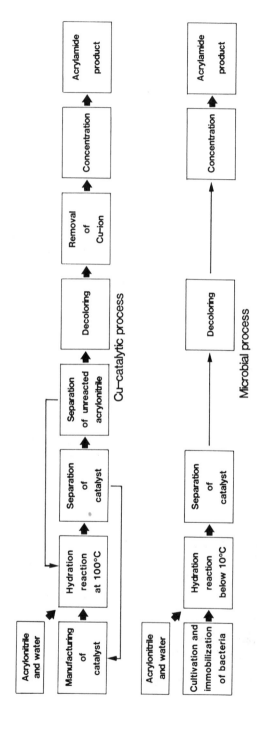

Figure 14.5. Comparative flowsheet for the microbial and conventional processes.

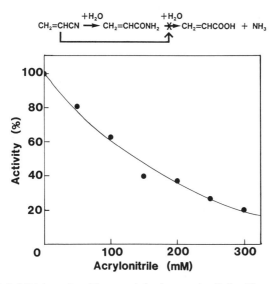

Figure 14.6. Inhibition of amidase activity by acrylonitrile. The amidase activity of *P. chlororaphis* B23 was assayed in the presence of various concentrations of acrylonitrile.

Figure 14.7. Effect of acrylamide on the activity of nitrile hydratases from *P. chlororaphis* B23 and *Brevibacterium* R312. The nitrile hydration reaction was carried out at 5°C in the reaction mixture containing 15 U of nitrile hydratase of *P. chlororaphis* B23 (○) or *Brevibacterium* R312 (△), 17.5% (wt/vol) acrylamide, 2.5% (wt/vol) acrylonitrile, 0.2% (wt/vol) *n*-butyric acid, and 10 m*M* potassium phosphate buffer (pH 7.5).

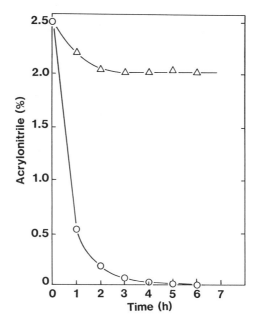

gasawa, Ryuno, and Yamada 1986) (Fig. 14.8). The crystalline enzymes appeared to be homogeneous on analysis by polyacrylamide gel electrophoresis, analytical ultracentrifugation, and double diffusion in agarose gel. Both enzymes are active toward various aliphatic nitriles, particularly, nitriles with three to six carbon atoms, e.g., propionitrile, n-butyronitrile, acrylonitrile, and cyclopropyl cyanide, which are the most suitable substrates (Table 14.4). The P. chlororaphis B23 enzyme stoichiometrically catalyzed the hydration of nitriles to amides, and no formation of acid or ammonia was detected. The P. chlororaphis B23 enzyme has a molecular mass of about 100 kilodaltons (kDa) and consists of four subunits identical in molecular mass (approximately 25 kDa). The Brevibacterium enzyme has a molecular mass of about 85 kDa and is composed of two subunits (26 kDa and 27.5 kDa). These nitrile hydratases contain iron, and significant concentrations of other transition metals were not detected on inductively coupled radiofrequency plasma spectrophotometric analysis. The iron atoms are tightly bound to the protein. Concentrated solutions of the two highly purified enzymes have a green color and exhibit a broad absorption in the visible range with an absorption maximum at 720 nm (Fig. 14.9). A loss of enzyme activity occurs in parallel with the disappearance of the absorption in the visible range under a variety of conditions. The physicochemical properties of these nitrile hydratases are summarized in Table 14.5. The two enzymes displayed different immunological properties as well.

Figure 14.10(A) shows the electron spin resonance (ESR) spectrum of the native nitrile hydratase isolated from Brevibacterium R312 at 77K (Nagasawa et al. 1987). Of special interest is that the ESR features ($g_{max} = 2.284$, $g_{mid} = 2.140$, and $g_{min} = 1.971$) of the present enzyme are characteristic of the rhombic low-spin Fe(III) ($S = 1/2$) type. The enzyme is a nonheme iron enzyme, and not a hemoprotein or ferredoxin, because of the lack of iron-porphyrin and acid-labile sulfur (Nagasawa, Ryuno, and Yamada 1986; Nagasawa et al. 1987)]. To the best of our knowledge, this nitrile hydratase is the first known nonheme iron enzyme containing a typical low-spin Fe(III) site (Nagasawa et al. 1987). There is a unique example of such a low-spin Fe(III) type in the iron (III) complex of bleomycin, a glycopeptide antitumor antibiotic (Sugiura 1980). A spectral blue shift from 720 to 690 nm observed on the addition of propionitrile, suggests the binding of the substrate to the

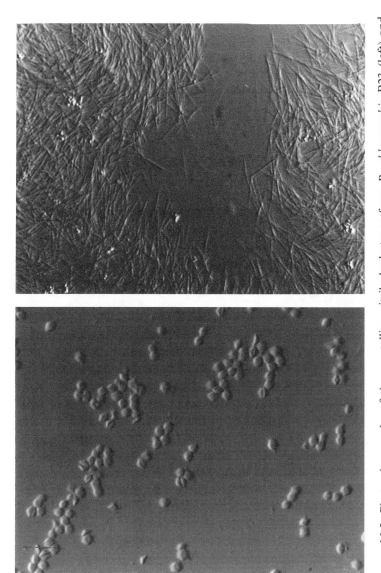

Figure 14.8. Photomicrographs of the crystalline nitrile hydratases from *P. chlororaphis* B23 (left) and *Brevibacterium* R312 (right).

Table 14.4. Substrate specificity of the nitrile hydratase from *P. chlororaphis* B23.

SUBSTRATE	RELATIVE ACTIVITY (%)	K_m (mM)
Propionitrile	100[a]	29.4
Potassium cyanide	0	
Acetonitrile	1.78	
n-Butyronitrile	76.5	1.03
Isobutyronitrile	0.08	0.0035 (K_i)
n-Valeronitrile	3.00	2.33
Isovaleronitrile	0	
n-Capronitrile	45.0	
Pivalonitrile	2.00	
Acrylonitrile	81.2	34.6
Methacrylonitrile	14.9	3.80
Crotononitrile	0	
2-Methyl-2-butenenitrile	0	
3-Pentenenitrile	6.00	
2-Methyl-3-butenenitrile	0	
Cyclopropyl cyanide	97.0	
Lactonitrile	0	
Benzonitrile	0	
Benzylcyanide	0	
p-Hydroxybenzonitrile	0	
4-Cyanopyridine	0	
Ethylene cyanhydrine	8.56	
β-Cyano-L-alanine	0	
Chloroacetonitrile	30.7	22.2
Cyanoacetamide	1.86	
β-Hydroxyacetonitrile	1.92	
Methoxyacetonitrile	1.86	
Cyanoacetic acid	0	
Cyanoacetic acid ethylester	0.40	

The reaction was carried out in a reaction mixture (2 ml) containing 0.2 mmol of a substrate, 60 μmol of potassium phosphate buffer (pH 7.0), and an appropriate amount of the enzyme at 20°C for 30 min.
[a] 1,840 μmol/min/mg protein.

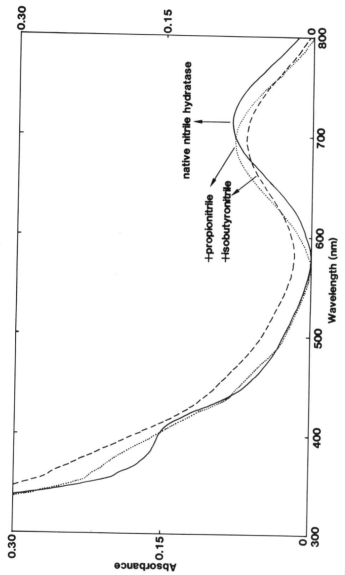

Figure 14.9. Absorption spectrum of the native nitrile hydratase and the shift of the absorption maximum on the addition of the substrate, propionitrile, and the inhibitor, isobutyronitrile.

293

Table 14.5. Summary of the physicochemical properties of the nitrile hydratases from *P. chlororaphis* B23 and *Brevibacterium* R312.

	P. chlororaphis B23	*Brevibacterium* R312
Molecular mass		
Sephadex G 150	105,000	77,000
HPLC (Toyosoda G 3000 SW)	100,000	87,000
Cellulofine GCL-2000sf	90,000	90,000
Sedimentation equilibrium	100,000	85,000
Sedimentation velocity		79,400
Subunit molecular mass		
SDS gel electrophoresis	25,000	α 26,000
		β 27,500
Number of subunits	4	4
$S_{20,w}^0$	5.52	3.56
$E_{1\%}^{1\ cm}$ at 280 nm	16.4	12.2
Absorption maxima (nm)	240, 280, 720	240, 280, 720
720/280	0.014	0.014
Metal (Fe) (mol/mol)	4	4
Optimum temperature (°C)	20	25
Heat stability (°C)	20	20
Optimum pH	7.5	7.8
pH stability	6–7.5	6.5–8.5

Fe(III) site of the enzyme (Fig. 14.9). Figure 14.10(B) shows the ESR features of the native enzyme induced on the addition of propionitrile, a preferred substrate for this enzyme reaction, which reverted to the original ESR spectrum after several minutes [Fig. 14.10(C)]. In contrast, after the large ESR change occurring on the addition of isobutyronitrile, no alteration was observed for several hours [Fig. 14.10(E)]. It is known that isobutyronitrile is barely hydrated, in spite of its extraordinarily high affinity for the enzyme ($K_i = 5.4\ \mu M$). On the other hand, the addition of propionamide, the enzymatic reaction product, did not affect the ESR spectrum of the native nitrile hydratase [Fig. 14.10(D)]. The native *P. chlororaphis* B23 enzyme showed a similar ESR spectrum to that of the native *Brevibacterium* R312 enzyme. Table 14.6 summarizes the ESR parameters of nitrile hydratase in comparison with those of low-spin iron(III) complexes of porphyrin (Tang

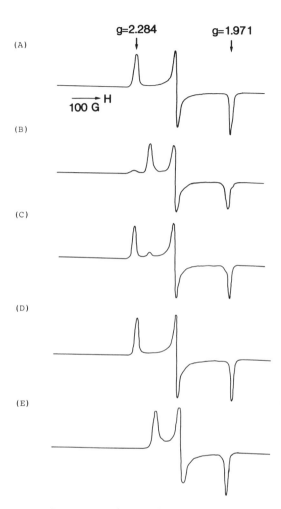

Figure 14.10. (A) ESR spectra of the native nitrile hydratase (0.05 mM) from *Brevibacterium* R312; (B) and the enzyme sample obtained immediately after mixing with propionitrile (10 m*M*); (C) and the time courses (5 min) for sample B; (D) the enzyme plus propionamide (10 m*M*); (E) and the enzyme plus isobutyronitrile (10 m*M*) in 0.01 *M* HEPES-KOH buffer (pH 7.2), at 77 K.

Table 14.6. ESR Characteristics of nitrile hydratase in comparison with low-spin iron(III)–bleomycin complexes.

	AXIAL LIGTN	g_{max}	g_{mid}	g_{min}	$(g_{max} - g_{min})$
Native R312 enzyme		2.284	2.140	1.971	0.313
Propionitrile-bound R312 enzyme		2.230	2.163	1.982	0.248
Isobutyronitrile-bound R312 enzyme		2.207	2.124	1.984	0.223
Mercury-treated native B23 enzyme		2.286	2.141	1.969	0.317
Propionitrile-bound B23 enzyme		2.227	2.161	1.984	0.243
Fe(PPIXDME)(THF)(SCH$_2$Ph)	O—Fe—S	2.29	2.22	1.97	0.32
Fe(PPIXDME)(N-MeIm)(SC$_6$H$_4$NO$_2$)	N—Fe—S	2.42	2.26	1.91	0.51
Fe(PPIXDME)(N-MeIm)(OC$_6$H$_4$NO$_2$)	N—Fe—O	2.61	2.21	1.84	0.77
Fe(PPIXDME)(N-MeIm)(N-MeIm)	N—Fe—N	2.90	2.29	1.57	1.33
Fe(bleomycin)(SH)	N—Fe—S	2.223	2.148	1.999	0.224
Fe(bleomycin)(OH)	N—Fe—O	2.431	2.185	1.893	0.538
Fe(bleomycin)(NH$_3$)	N—Fe—N	2.545	2.178	1.837	0.708
Fe(bleomycin)(CH$_3$(CH$_2$)$_3$NH$_2$)	N—Fe—N	2.537	2.179	1.850	0.687
Fe(bleomycin)(CH$_3$)$_3$CNH$_2$	N—Fe—N	2.497	2.181	1.866	0.631

et al. 1976) and bleomycin (Sugiura 1980). In general, the g value splitting of low-spin Fe(III) complexes decreases in the order of the N—Fe—N > N—Fe—O > N—Fe—S axial ligation modes. The small ($g_{max} - g_{min}$) value of nitrile hydratase may suggest that one of the axial ligands involves the thiolate donor, probably the cysteine thiol group. Indeed, the ESR features of the present enzyme are very similar to those of the oxidized cytochrome P-450 state with g values near 2.4, 2.2, and 1.9 (Tang et al. 1976). The recently determined 2.6-Å crystal structure of *P. putida* cytochrome P-450 clarified that the heme iron atom is coordinated with the axial sulfur ligand by a cysteine residue (Poulos et al. 1985). Most five-coordinated Fe(III) complexes are high-spin, and accordingly the most probable other axial ligand of the native nitrile hydratase may involve water, because of the easy replacement by propionitrile or isobutyronitrile.

To determine whether or not H_2O is coordinated to the iron(III) site of the native R312 enzyme, we prepared a sample in $H_2^{17}O$. On the substitution of $H_2^{16}O$ by water enriched in ^{17}O, which has a nuclear spin of 5/2, 2.5-G broadening (half-peak width) of the $g_{min} = 1.971$ line was clearly observed. The $g_{mid} = 2.140$ and $g_{max} = 2.284$ were also broadened to 18.2G (peak to trough) and 22.2G (half peak width) from 15.4G and 21.6G, respectively, as seen for the native enzyme in $H_2^{16}O$. The broadenings are attributable to the transferred hyperfine interaction, demonstrating that ^{17}O derived from $H_2^{17}O$ is coordinated to the iron(III)-active center of nitrile hydratase. On the basis of the broadening of about 3G at the $g = 9.67$ resonance on $H_2^{16}O \rightarrow H_2^{17}O$ replacement, one ligand sphere of the high-spin Fe(III) center in protocatechuate 3,4-dioxygenase was shown to contain water (Lipscomb 1982). Similar ESR line broadening due to $H_2^{17}O$ has been observed for reduced activated aconitase (Emtage et al. 1983) and metmyoglobin (Vuk-Pavlovic and Siderer 1977). The present ESR study showed the following: (1) nitrile hydratase is the first known nonheme iron enzyme with a typical low-spin Fe(III) coordination environment, (2) the axial position of the iron(III) site in the native enzyme may be occupied by thiolate and aquo groups, and (3) aliphatic nitrile substrates directly bind to the iron(III)-active center through water substrate replacement.

It was recently discovered that in addition to iron, nitrile hydratase involves an active carbonyl cofactor as a second prosthetic group, probably pyrroloquinoline quinone (PQQ) (Nagasawa and Yamada

1987). Nitrile hydratase is strongly inhibited by various carbonyl re-agents, phenylhydrazine being the most effective inhibitor. The color of a nitrile hydratase solution changes from green to yellow on the addition of phenylhydrazine. Absorption band appeared at around 390 and 450 nm, and the absorbance values at these wavelengths increased when the concentration of phenylhydrazine was increased. The inhibition of the nitrile hydratase reaction by phenylhydrazine was irreversible, i.e., the activity of the phenylhydrazine-treated enzyme was not recovered on dialysis or gel filtration. These results suggest that there are carbonyl groups that are reactive with phenylhydrazine in the nitrile hydratase molecule, and that these carbonyl groups are essential for the enzyme activity. Because the carbonyl cofactor was attached to the enzyme through a covalent linkage, the chromophores were isolated by acid hydrolysis, protease digestion, and successive chromatographic separation. The isolated chromophores showed similar spectroscopic characteristics to those obtained from the amine oxidase of *Aspergillus niger*, to which PQQ is covalently linked (Ameyama et al. 1985). The isolated chromophores are potent activators of apo-D-glucose dehydrogenase (EC 1.1.99.17), supporting the presence of PQQ or a PQQ-like compound in nitrile hydratase. Our nitrile hydratase is the first example of a quinoprotein that is not an oxidoreductase. So far, three redox forms have been found to participate in biological oxidations; namely PQQ, PQQH and $PQQH_2$ (Fig. 14.11). Dekker et al. (1982) showed that PQQ in aqueous solutions is in part covalently hydrated, namely, $PQQ-H_2O$ or $PQQ-2H_2O$ (Fig. 14.11), and that only the hydrated forms are fluorescent, emitting green light. A higher amount of a form of hydrated PQQ is formed at low temperatures. Interestingly, the optimal temperature for hydration by the enzyme is between 20° and 25°C, which is very low compared to that for other enzymes. The occurrence of PQQ or a compound closely resembling it in nitrile hydratase is significant. We can assume therefore a new function of PQQ in the hydration reaction catalyzed by nitrile hydratase. We assume that a hydrated form of PQQ, $PQQ-H_2O$, or $PQQ-2H_2O$ probably participates in the activation of H_2O and adds H_2O to the $C\equiv N$ bond of the nitrile group.

Recently, the first evidence for an important interaction between the low-spin ferric site and PQQ was provided by the results of ESR experiments (Sugiura et al. 1988). Figure 14.12 compares the ESR spectra of

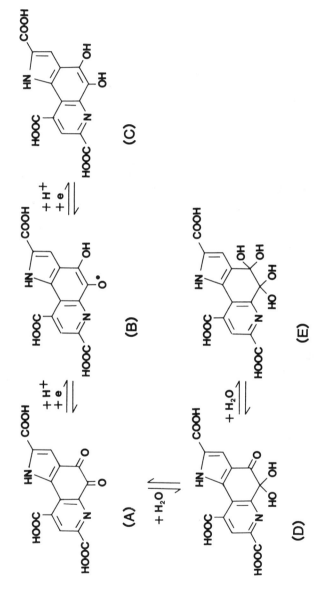

Figure 14.11. Structures of (A) PQQ, (B) PQQH, (C) PQQH₂, (D) PQQ-H₂O, and (E) PQQ-2H₂O.

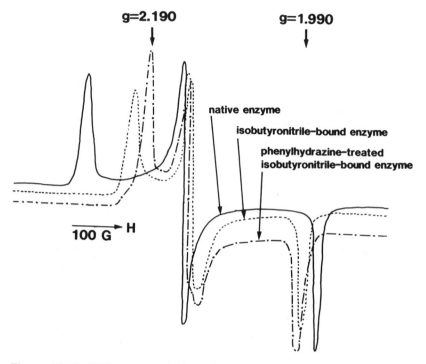

Figure 14.12. ESR spectra of the native *Brevibacterium* R312 enzyme, the isobutyronitrile-bound enzyme, and the phenylhydrazine-treated isobutyronitrile-bound enzyme at pH 7.2. The enzyme (0.05 m*M*) was treated with isobutyronitrile (1.0 m*M*) and/or phenylhydrazine (0.5 m*M*) in 0.01 *M* HEPES-KOH buffer (pH 7.2).

the native and phenylhydrazine-treated enzyme from *Brevibacterium* R312. Evidently, phenylhydrazine induced a significant change in the ESR features of the native nitrile hydratase. In contrast, benzylamine and benzonitrile, which are structurally close to phenylhydrazine, caused no alteration of the ESR spectrum or the enzymatic activity of the native enzyme. It is known definitely that this enzyme hydrates aliphatic nitriles, but not aromatic nitriles. As clearly shown in Figure 14.12, the isobutyronitrile (inhibitor)-bound enzyme was also remarkably affected by the addition of phenylhydrazine, and the transformed ESR features were distinct from those caused by phenylhydrazine on the low-spin Fe(III)-ESR spectrum of the nitrile hydratase obtained from *Brevibacterium* R312. Similar ESR spectral changes on addition of phenylhydrazine were also observed for the *P. chlororaphis* B23 nitrile hydratase. The present results strongly indicate the direct binding of phenylhydrazine to an organic cofactor rather than the iron(III) site of the enzyme. In addition, the isobutyronitrile-bound enzyme treated with sodium dithionite and exposed to air gave a transient ESR signal near $g = 2.00$ (Fig. 14.13). The ESR absorption at $g = 2.00$ appears to be characteristic of the semiquinone radical and the signal disappeared quickly on air oxidation. Nitrile hydratase includes a low-spin Fe(III) active site and a PQQ prosthetic group, which reacts strongly with phenylhydrazine. Therefore, the present low-spin ferric ESR change due to carbonyl reagents implies the occurrence of a dis-

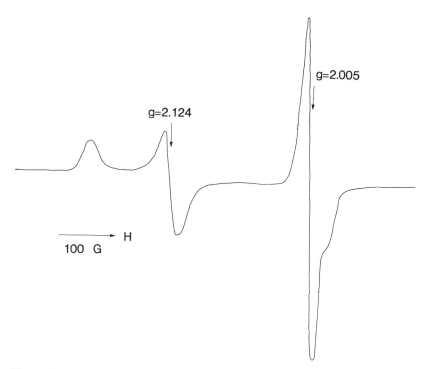

Figure 14.13. ESR signals of the low-spin Fe(III) site and semiquinone radical in the isobutyronitrile-bound enzyme. After the isobutyronitrile-bound *Brevibacterium* R312 enzyme (0.05 m*M*) had been treated with sodium dithionite (0.5 m*M*) at pH 7.2, the sample was exposed to air and then frozen.

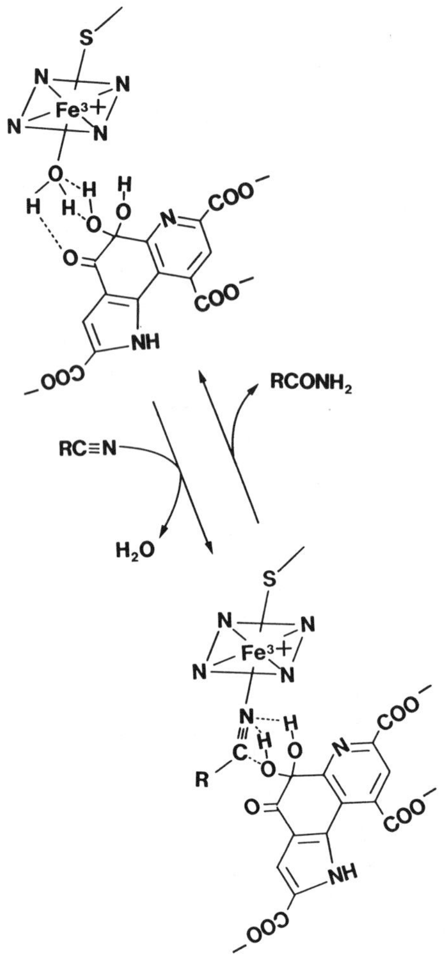

Figure 14.14. Proposed mechanism for the nitrile hydration catalyzed by nitrile hydratase.

tinct interaction between the Fe(III) center and PQQ in the active site of the native enzyme. The present results constitute the first evidence for an important interaction between the metal site and PQQ in nitrile hydratase.

This interaction between the low-spin Fe(III) site and PQQ allows one to propose a mechanism involving their individual roles in the catalysis. The active iron(III) cofactor is known to provide the binding site for the nitrile substrate. In the active center of nitrile hydratase, there may be an interaction, such as a hydrogen bond, between the Fe(III) site and a form of hydrated PQQ (Fig. 14.14) (Nagasawa and Yamada 1989). PQQ has been shown to form a PQQ-H_2O adduct. Phenylhydrazine alters the active site of the enzyme and inactivates the enzyme. A nitrile and the corresponding amide and acid are all at the same oxidation level, and thus the overall reaction in each case is a hydrolysis. Thus it would appear that these reactions proceed via a simple hydrolytic mechanism. However, the discovery of ferric ions and PQQ as cofactors of nitrile hydratase seems not to exclude the possibility that a biological oxidation/reaction mechanism is involved. A well known chemical analogy is the conversion of nitriles to amides by the hydroperoxide anion under mildly alkaline conditions. Further studies on the reaction mechanism of nitrile hydratase are in progress.

OCCURRENCE OF A COBALT-INDUCED AND COBALT-CONTAINING NITRILE HYDRATASE IN *RHODOCOCCUS RHODOCHROUS* J1

The substrate specificity of the *P. chlororaphis* B23 nitrile hydratase is shown in Table 14.4. Aliphatic nitriles are very good substrates for the *Pseudomonas* B23 enzyme; however, this enzyme does not act on aromatic nitriles, such as benzonitrile or 3-cyanopyridine (Nagasawa et al. 1987). The *Brevibacterium* R312 nitrile hydratase exhibits a similar substrate specificity (Nagasawa, Ryuno, and Yamada 1986). Therefore, we tried to find a nitrile hydratase which acts on aromatic nitriles. We isolated *R. rhodochrous* J1 through screening procedures from soil and found a nitrile hydratase that acts on aromatic nitriles (Nagasawa et al. 1988).

In Figure 14.15, the time course of nitrile hydratase formation in *R. rhodochrous* J1 is shown. For the formation of nitrile hydratase, both crotonamide and cobalt ions are required (Nagasawa, Takeuchi, and Yamada 1988). Ferric or ferrous ions cannot replace cobalt ions. On the other hand, the formation of nitrile hydratase in *P. chlororaphis* B23 and *Brevibacterium* R312 requires ferric or ferrous ions. The nitrile hydratase protein of *R. rhodochrous* J1 is not synthesized without the addition of cobalt ions, even when crotonamide is added (Nagasawa, Takeuchi, and Yamada 1988). It is interesting that cobalt ions control the synthesis of the nitrile hydratase protein.

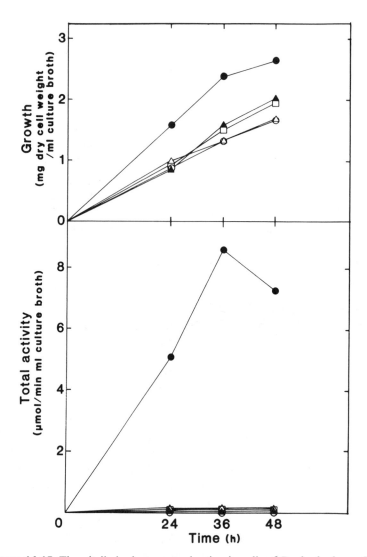

Figure 14.15. The nitrile hydratase production in cells of *R. rhodochrous* J1 in the presence and absence of CoCl₂ and crotonamide. *R. rhodochrous* J1 was cultivated at 28°C with shaking in medium consisting of 10 g of glucose, 0.5 g of K_2HPO_4, 0.5 g of $MgSO_4 \cdot 7H_2O$, 0.2 g of NaCl, 2.0 g of NH_4NO_3, and 2 g of casamino acids (Difco, Detroit) per liter of distilled water, pH 7.2, supplemented as follows: (○), absence of crotonamide and CoCl₂; (●), presence of 0.0001% (wt/vol) $CoCl_2 \cdot 6H_2O$ and 0.2% (wt/vol) crotonamide; (△), presence of 0.0001% (wt/vol) $CoCl_2 \cdot 6H_2O$; (□), presence of 0.2% (wt/vol) crotonamide; and (▲), presence of 0.001% (wt/vol) or 0.0001% (wt/vol) $FeSO_4 \cdot 7H_2O$ and 0.2% (wt/vol) crotonamide.

We have purified and crystallized the *R. rhodochrous* J1 nitrile hydratase (Nagasawa, Takeuchi, and Yamada 1988). The enzyme contains cobalt atoms, but not any other transition metals. The cobalt atoms are tightly bound to the protein. The enzyme shows a red color with an absorption maximum at 415 nm. Its molecular mass is 505 kDa and it is composed of two subunits (29 kDa and 26 kDa). This cobalt-containing nitrile hydratase is more heat-stable than the ferric ion-containing nitrile hydratases (Nagasawa, Takeuchi, and Yamada 1989a). The role of the cobalt atoms and the content of PQQ are currently being examined.

We optimized the culture medium as described elsewhere (Nagasawa, Takeuchi, and Yamada 1989b). The nitrile hydratase found in the cells of *R. rhodochrous* J1 grown on the optimum medium corresponds to more than 50% of all soluble protein. Thus, a very active catalyst for aromatic nitriles has been discovered.

PRODUCTION OF NICOTINAMIDE WITH *R. RHODOCHROUS* J1 NITRILE HYDRATASE

The vitamin nicotinamide is used as an animal feed supplement. 3-Cyanopyridine is chemically synthesized by means of ammoxidation reaction of picoline and is converted to nicotinamide with alkali. The yield of the chemical synthesis of nicotinamide is low due to the formation of nicotinic acid as a byproduct. Thus, we attempted the nitrile hydratase-catalyzed production of nicotinamide from 3-cyanopyridine using *R. rhodochrous* J1 cells (Nagasawa et al. 1988). *R. rhodochrous* J1 nitrile hydratase exhibits high activity toward 3-cyanopyridine. As shown in Figure 14.16, 100% of the added 12 M 3-cyanopyridine was converted to nicotinamide, without the formation of nicotinic acid, the highest yield being 1,465 g of nicotinamide per liter of reaction mixture containing resting cells (1.17 g as dry cell weight) in 9 h. When 0.553 g of cells was added, 100% of the added 9 M 3-cyanopyridine was converted to nicotinamide in 22 h (Nagasawa et al. 1988). The nicotinamide formed is stable, and further conversion of the nicotinamide to nicotinic acid at long times was due to the low activity of nicotinamide as a substrate for the amidase(s) present in this organism. Due to the high yield, this process is promising for the industrial production of nicotinamide. In Figure 14.17, the progress of nicotinamide

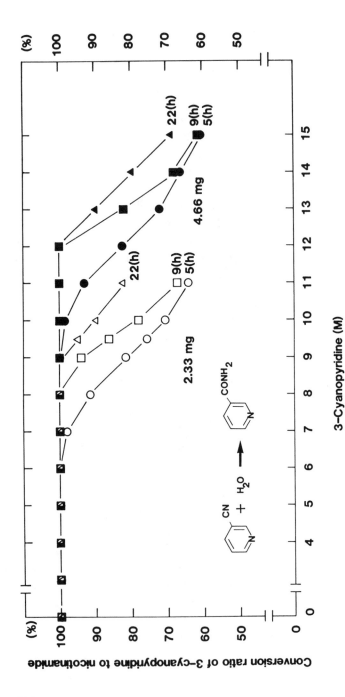

Figure 14.16. Enzymatic conversion of various amounts of 3-cyanopyridine to nicotinamide. The reaction mixture (4 ml) contained 10 mM potassium phosphate buffer (pH 7.0), various amounts of 3-cyanopyridine, and resting cells [2.33 mg (dry weight)] from 4 ml of culture broth or resting cells [4.66 mg (dry weight)] from 8 ml of culture broth. The incubation was carried out at 25°C with shaking for 5 h (○, ●), 9 h (□, ■), or 22 h (△, ▲).

Figure 14.17. Appearance of nicotinamide crystals during the enzymatic synthesis. Photographs (from left) were taken after 0, 1, 6, and 18 h incubation.

production is shown. In the case of the enzymatic production of acrylamide, acrylonitrile is added in portions due to its high inhibitory effect. On the other hand, in this case, a high amount of 3-cyanopyridine can be added at one time. 3-Cyanopyridine is probably not so inhibitory toward the nitrile hydratase in the *R. rhodochrous* J1 cells. In the course of the reaction, the solid 3-cyanopyridine dissolved completely, and crystals of nicotinamide appeared gradually after 6 h of incubation and the reaction mixture solidified after 18 h of incubation. Thus, this may be an example of a pseudocrystal fermentation (that is, crystalline substrate → solution of substrate → solution of product → crystalline product).

The substrate specificity of the *R. rhodochrous* J1 nitrile hydratase for various aromatic nitriles was also examined (Mauger, Nagasawa, and Yamada 1989a,b). The amides were accumulated at each concentration, as shown in Table 14.7, with molar conversion yields of 100% from the corresponding nitrile, without formation of the corresponding acids. Isonicotinamide is used for the industrial production of isonicotinic acid hydrazide, a tuberculostatic, and pyrazinamide is also a useful tuberculostatic (Mauger, Nagasawa, and Yamada 1989a). 2,6-Difluorobenzamide is useful for the synthesis of agricultural chemicals.

Table 14.7. Formation of various amides from nitriles by the *R. rhodochrous* J1 nitrile hydratase.

Structure	Yield	Structure	Yield
3-pyridyl-CONH₂	1465 g/L	2,6-difluorophenyl-CONH₂	393 g/L
4-pyridyl-CONH₂	1099 g/L	thienyl-CONH₂	254 g/L
2-pyridyl-CONH₂	977 g/L	indolyl-CONH₂	697 g/L
pyrazinyl-CONH₂	985 g/L	furyl-CONH₂	888 g/L
		phenyl-CONH₂	848 g/L

Various kinds of aromatic heterocyclic amides can be produced, as shown in Table 14.8 (Mauger, Nagasawa, and Yamada 1989b). Due to the high yields of this process, the easy cultivation of *R. rhodochrous* J1 cells and the stability of resting cells, which are stable at −20°C for more than 2 years, the use of this enzymatic hydration process seems to be promising for the industrial production of various aromatic and heterocyclic amides.

In addition, the *R. rhodochrous* J1 nitrile hydratase acts on aliphatic nitriles, too. Above all, we achieved very high accumulation of acrylamide. Therefore, the *R. rhodochrous* J1 nitrile hydratase will probably be useful for the industrial production of acrylamide as well (Nagasawa et al. 1989).

PRODUCTION OF NICOTINIC ACID WITH *R. RHODOCHROUS* J1 NITRILASE

When *R. rhodochrous* J1 was cultivated in the nutrient medium containing isovaleronitrile as an inducer in the absence of cobalt ions, nitrilase was also abundantly produced in the cells, without the formation of nitrile hydratase (Nagasawa, Kobayashi, and Yamada 1988). We also attempted the nitrilase-catalyzed production of nicotinic acid, a useful vitamin, from 3-cyanopyridine using resting cells of *R. rhodochrous* J1 (Mathew et al. 1988). Under the optimum conditions, 100% of the added 3-cyanopyridine was converted to nicotinic acid, the highest yield being 172 g of nicotinic acid per liter of reaction mixture containing 2.89 g (dry weight) of cells in 26 h (Fig. 14.18), and needle shaped crystals of ammonium nicotinate appeared during the enzymatic synthesis (Fig. 14.19). As ammonium nicotinate can be easily converted into nicotinic acid under suitable conditions, without treatment with ion-exchange resins, this enzymatic process seems to be economical. In a similar manner, *p*-aminobenzoic acid (110 g/L), a useful vitamin, can be produced (Kobayashi et al. 1989). Thus, *R. rhodochrous* J1 nitrilase is also promising for the synthesis of useful

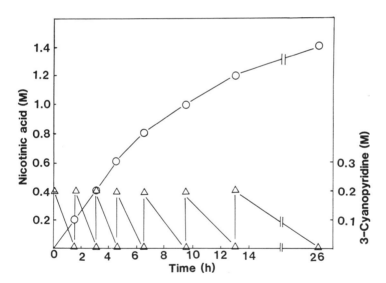

Figure 14.18. Accumulation of nicotinic acid by resting *R. rhodochrous* J1 cells. The reaction mixture (50 ml) consisted of 200 m*M* 3-cyanopyridine, 0.01 *M* potassium phosphate buffer (pH 8.0), and resting cells [145 mg (dry weight) of cells] harvested from 50 ml of culture broth. The incubation was carried out at 25°C with shaking. During the course of the reaction, 200 m*M* 3-cyanopyridine was added. (△), 3-Cyanopyridine; (○), nicotinic acid.

Table 14.8. Substrate specificity of the *R. rhodochrous* J1 nitrilase.

SUBSTRATE	RELATIVE ACTIVITY (%)	K_m (mM)	SUBSTRATE	RELATIVE ACTIVITY (%)	K_m (mM)
Benzonitrile	100 *	2.10	3-Quinolinecarbonitrile	8.9	
3-Hydroxybenzonitrile	22.2	4.55	2-Furonitrile	138.8	28.6
4-Hydroxybenzonitrile	7.7		2-Thiophenecarbonitrile	108.5	13.2
3-Cyanobenzaldehyde	4.1		5-Norbornene-2-carbonitrile	1.0	
4-Cyanobenzaldehyde	10.2				
3-Aminobenzonitrile	74.8	2.15	Cyclopentanecarbonitrile	1.6	
4-Aminobenzonitrile	23.0	1.80	Cycloheptanecarbonitrile	2.3	
3-Tolunitrile	99.6	1.90	4-Cyano-1-cyclohexene	3.4	
4-Tolunitrile	116	2.90	1,4-Piperazinedicarbonitrile	1.2	
3-Nitrobenzonitrile	74.2	5.71			
4-Nitrobenzonitrile	4.3		Phenylacetonitrile	1.8	
Isophthalonitrile	46.8	7.81	2-Thiopheneacetonitrile	10.7	
Terephthalonitrile	3.6		3-Thiopheneacetonitrile	3.7	
3-Chlorobenzonitrile	137	7.41			
4-Chlorobenzonitrile	114	10.0	Cinnamonitrile	9.9	

Compound			Compound	
2-Methoxybenzonitrile	1.2		2-Furanacrylonitrile	12.6
3-Methoxybenzonitrile	61.9	2.00	Crotononitrile	1.7
4-Methoxybenzonitrile	29.4	5.81	2-Methyl-3-butenenitrile	1.2
4-Acetylbenzonitrile	5.9		3-Pentenenitrile	8.2
4-Cyanothiophenol	6.2		Fumaronitrile	1.3
α-Bromo-3-tolunitrile	25.4	6.67	Diaminomaleonitrile	1.1
α-Bromo-4-tolunitrile	3.2		2-Chlorocyanoethylene	2.0
α,α,α-Trifluoro-3-tolunitrile	39.5	3.94	3-Ethoxyacrylonitrile	8.9
4-Fluorobenzonitrile	26.2			
2-Cyanopyridine	7.7		Valeronitrile	2.3
3-Cyanopyridine	22.1	12.5	Isocapronitrile	2.4
4-Cyanopyridine	25.6	20.0	Chloroacetonitrile	1.9
Cyanopyrazine	37.3	8.93	Methoxyacetonitrile	5.2
2-Cyanonaphthalene	15.8		Cyanoacetic acid ethylester	5.5
5-Cyanoindole	27.3	12.8	(S)-(±)-2-Methylbutyronitrile	1.1
Piperonylonitrile	84.3	7.58	(±)-2-Phenylglycinonitrile	1.4

The reaction was carried out at 25°C for 10 min in a reaction mixture (2 ml) containing 20 μmol of potassium phosphate buffer (pH 8.0), 12 μmol of a nitrile, 1 μmol of dithiothreitol, 10% (vol/vol) methanol, and an appropriate amount of enzyme.

[a] 15.9 μmol/min/mg protein.

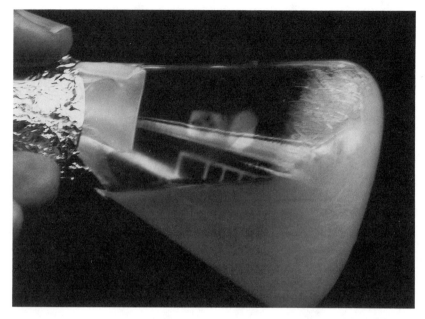

Figure 14.19. Appearance of ammonium nicotinate during the enzymatic synthesis. The photograph was taken after 26 h of incubation.

acids under mild reaction conditions (Table 14.8). Various aromatic and heterocyclic nitrile compounds are also suitable substrates for this nitrilase. This nitrilase also exhibits regiospecificity for dicyanobenzenes (Kobayashi, Nagasawa, and Yamada 1988). Only one nitrile group of dicyanobenzene is attacked, namely, 3- and 4-cyanobenzoic acids are synthesized from isophthalonitrile and terephthalonitrile, respectively, with conversion ratios of >99% (Fig. 14.20).

We have purified and crystallized the nitrilase from an extract of isovaleronitrile-induced cells of *R. rhodochrous* J1 (Kobayashi, Nagasawa, and Yamada 1989). The enzyme has a molecular mass of about 78 kDa and consists of two subunits identical in molecular mass. The purified enzyme exhibits a pH optimum of 7.6 and a temperature optimum of 45°C. The enzyme stoichiometrically catalyzes the hydrolysis of benzonitrile to benzoic acid and ammonia, and no amide formation is detected. The enzyme requires a thiol compound, such as dithiothreitol, L-cysteine, and reduced glutathione, to exhibit maxi-

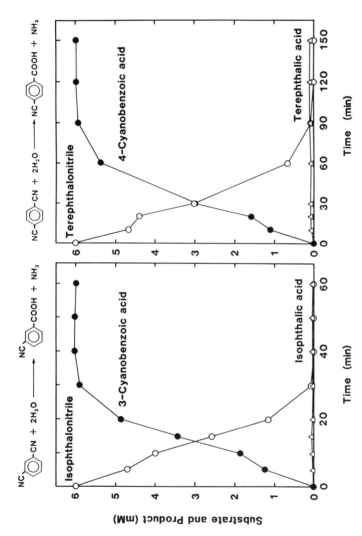

Figure 14.20. Time courses of the 3-cyanobenzoic acid and 4-cyanobenzoic acid syntheses. The reaction was carried out at 25°C in a reaction mixture (500 ml) containing 6 mM dinitrile compound, 1 mM dithiothreitol, 0.1 M potassium phosphate buffer (pH 7.5), 10% (vol/vol) methanol, and 267 U of the *R. rhodochrous* J1 nitrilase. (○), Dinitrile compound; (●), cyanobenzoic acid.

313

mum activity. The enzyme does not contain metals or PQQ, and it is entirely different from nitrile hydratase (Kobayashi, Nagasawa, and Yamada 1989).

In the course of these studies, we were often surprised at the variety of enzymes and new phenomena involved in nitrile metabolism in bacteria. It seems that microorganisms have great potential as biocatalysts for industrial use. The industrial production of acrylamide, a typical commodity chemical, was catalyzed by the enzyme nitrile hydratase. In the near future, the enzymatic production of nicotinamide from 3-cyanopyridine on an industrial scale is also expected. These facts greatly encourage us and confirm that the further application of biocatalysis is promising for not only fine chemicals but also commodity chemicals, too, and that it will become more commonly used for the industrial production of useful compounds.

ACKNOWLEDGMENTS

We are grateful to Dr. Y. Sugiura, Professor at Chemical Institute of Kyoto University, and Dr. O. Adachi, Professor at Department of Agricultural Chemistry of Yamaguchi University for their supports and helpful advices. We wish to thank Dr. M. Kobayashi, Dr. J. Mauger, Dr. C. D. Mathew, Dr. Y. Asano, Mr. K. Ryuno, Mr. H. Shimizu, Mr. K. Takeuchi, Mr. Mihara, Mr. N. Yanaka, Mr. A. Miura, and Mr. H. Nanba for their invaluable technical discussion and assistances during the course of this work.

REFERENCES

Ameyama, M., E. Shinagawa, K. Matsushita, K. Takimoto, K. Nakashima, and O. Adachi. 1985. Mammalian choline dehydrogenase is a quinoprotein. *Agric. Biol. Chem.* **49:** 3623–3626.

Arnaud, A., P. Galzy, and J. L. Jallageas. 1977. Etude de L'Acetonitrilase D'une Souche de *Brevibacterium*. *Agric. Biol. Chem.* **41:** 2183–2191.

Asano, Y., Y. Tani, and H. Yamada. 1980. A new enzyme, nitrile hydratase, which degrades acetonitrile in combination with amidase. *Agric. Biol. Chem.* **44:** 2251–2252.

Asano, Y., T. Yasuda, Y. Tani, and H. Yamada. 1982. Microbial degradation of nitrile compounds—a new enzymatic method of acrylamide production. *Agric. Biol. Chem.* **46:** 1183–1189.

Baydyopadhyay, A. K., T. Nagasawa, Y. Asano, K. Fujishiro, Y. Tani, and H. Yamada. 1986. Purification and characterization of benzonitrilases from *Arthrobacter* sp. strain J-1. *Appl. Environ. Microbiol.* **51**: 302–306.

Bui, K., A. Arnaud, and P. Galzy. 1982. A new method to prepare amides by bioconversion of corresponding nitriles. *Enzyme Microbiol. Technol.* **4**: 195–197.

Commeyras, A., A. Arnaud, P. Galzy, and J. L. Jallageas. 1973. Demande de Brevet D'Invention, No. 73, 33613.

Commeyras, A., A. Arnaud, P. Galzy, and J. L. Jallageas. 1977. U.S Patent 4,000,081.

Dekker, R. H., J. A. Duine, J. Frank, J. P. E. H. Verwiel, and J. Westerling. 1982. Covalent addition of H_2O, enzyme substrates and activators to pyrrolo-quinoline quinone, the coenzyme of quinoproteins. *Eur. J. Biochem.* **125**: 69–73.

DiGeronimo, M. T., and A. D. Antoine. 1976. Metabolism of acetonitrile and propionitrile by *Nocardia rhodochrous* LL100–21. *Appl. Environ. Microbiol.* **31**: 900–906.

Emtage, M. H., T. A. Kent, M. C. Kennedy, H. Beinert, and E. Munck. 1983. Mossbauer and EPR studies of activated aconitase: development of a localized valence state at a subsite of the [4Fe-4S] cluster on binding of citrate. *Proc. Natl. Acad. Sci. USA* **80**: 4674–4678.

Firman, J. L., and D. O. Gray. 1976. The biochemical pathway for the breakdown of methyl cyanide(acetonitrile) in bacteria. *Biochem. J.* **158**: 223–229.

Harper, D. B. 1977a. Microbial metabolism of aromatic nitriles. *Biochem. J.* **165**: 309–319.

Harper, D. B. 1977b. Fungal degradation of aromatic nitriles. *Biochem. J.* **167**: 685–692.

Jallageas, J. C., A. Arnaud, and P. Galzy. 1980. Bioconversions of nitriles and their applications. In *Advances in Biochemical Engineering, Vol. 14* (A. Fiechter, ed.). pp. 1–32. Berlin-Heidelberg-New York: Springer-Verlag.

Kobayashi, M., T. Nagasawa, and H. Yamada. 1988. Regiospecific hydrolysis of dinitrile compounds by nitrilase from *Rhodococcus rhodochrous* J1. *Appl. Microbiol. Biotechnol.* **29**: 231–233.

Kobayashi, M., T. Nagasawa, and H. Yamada. 1989. Nitrilase of *Rhodococcus rhodochrous* J1—purification and characterization. *Eur. J. Biochem.* **182**: 349–356.

Kobayashi, M., T. Nagasawa, N. Yanaka, and H. Yamada. 1989. Nitrilase-catalyzed production of p-aminobenzoic acid from p-aminobenzonitrile with *Rhodococcus rhodochrous* J1. *Biotechnol. Lett.* **11**: 27–30.

Kuwahara, M., H. Yanase, K. Kikuchi, and K. Okuzumi. 1980. Metabolism of succinonitrile in *Aeromonas* sp. Hakkokogaku Kaishi **58**: 441–447.

Linton, E. A., and C. J. Knowles. 1986. Utilization of aliphatic amides and nitriles by *Nocardia rhodochrous* LL100-21. *J. Gen. Microbiol.* **132:** 1493–1501.

Lipscomb, J. D. 1982. In *Oxygenases and Oxygen Metabolism.* (M. Nozaki, S. Yamamoto, Y. Ishimura, M. J. Coon, L. Ernster, and R. W. Estabrook, eds.). pp. 27–38. Academic Press: New York.

Mathew, C. D., T. Nagasawa, M. Kobayashi, and H. Yamada. 1988. Nitrilase-catalyzed production of nicotinic acid from 3-cyanopyridine in *Rhodococcus rhodochrous* J1. *Appl. Environ. Microbiol.* **54:** 1030–1032.

Mauger, J., T. Nagasawa, and H. Yamada. 1989a. Nitrile hydratase-catalyzed production of isonicotinamide, picolinamide and pyrazinamide from 4-cyanopyridine, 2-cyanopyridine and cyanopyrazine in *Rhodococcus rhodochrous* J1. *J. Biotechnol.* **8:** 87–96.

Mauger, J., T. Nagasawa, and H. Yamada. 1989b. Synthesis of various aromatic amide derivatives using nitrile hydratase of *Rhodococcus rhodochrous* J1. *Tetrahedron* **45:** 1347–1354.

Mimura, A., T. Kawamoto, and K. Yamaga. 1969. Application of microorganisms to the petrochemical industry—assimilation of nitrile compounds by microorganisms. *J. Ferment. Technol.* **47:** 631–638.

Nagasawa, T., K. Ryuno, and H. Yamada. 1986. Nitrile hydratase of *Brevibacterium* R312—purification and characterization. *Biochem. Biophys. Res. Commun.* **139:** 1305–1312.

Nagasawa, T., H. Nanba, K. Ryuno, K. Takeuchi, and H. Yamada. 1987. Nitrile hydratase of *Pseudomonas chlororaphis* B23—purification and characterization. *Eur. J. Biochem.* **162:** 691–698.

Nagasawa, T., H. Yamada, Y. Sugiura, and J. Kuwahara. 1987. Nitrile hydratase: the first non-heme iron enzyme with a typical low-spin Fe(III)-active center. *J. Am. Chem. Soc.* **109:** 5848–5850.

Nagasawa, T., and H. Yamada. 1987. Nitrile hydratase is a quinoprotein—a possible new function of pyrroloquinoline quinone: activation of H_2O in an enzymatic hydration reaction. *Biochem. Biophys. Res. Commun.* **147:** 701–709.

Nagasawa, T., M. Kobayashi, and H. Yamada. 1988. Optimum culture conditions for the production of benzonitrilase by *Rhodococcus rhodochrous* J1. *Arch. Microbiol.* **150:** 89–94.

Nagasawa, T., C. D. Mathew, J. Mauger, and H. Yamada. 1988. Nitrile hydratase-catalyzed production of nicotinamide from 3-cyanopyridine in *Rhodococcus rhodochrous* J1. *Appl. Environ. Microbiol.* **54:** 1766–1769.

Nagasawa, T., K. Takeuchi, and H. Yamada. 1988. Occurrence of a cobalt-induced and cobalt-containing nitrile hydratase in *Rhodococcus rhodochrous* J1. *Biochem. Biophys. Res. Commun.* **155:** 1008–1016.

Nagasawa, T., J. Mauger, H. Shimizu, and H. Yamada. 1989. Enzymatic production of acrylamide by cobalt-containing nitrile hydratase of *Rhodococcus rhodochrous* J1. *Appl. Microbiol. Biotechnol.* (in press).

Nagasawa, T., K. Ryuno, and H. Yamada. 1989. Evaluation of *Pseudomonas chlororaphis* B23 as a catalyst for the enzymatic production of acrylamide. *Experientia* **45:** 1066–1070.

Nagasawa, T., K. Takeuchi, and H. Yamada. 1989a. Purification and characterization of nitrile hydratase from *Rhodococcus rhodochrous* J1. *Eur. J. Biochem.* (in press).

Nagasawa, T., K. Takeuchi, and H. Yamada. 1989b. Optimum culture conditions for the production of a cobalt-containing nitrile hydratase by *Rhodococcus rhodochrous* J1. *Arch. Microbiol.* (in press).

Nagasawa, T., and H. Yamada. 1989. Microbial transformations of nitriles. *Trends Biotechnol.* **7:** 153–158.

Nakai, K., I. Watanabe, Y. Sato, and K. Enomoto. 1988. Development of an acrylamide manufacturing process using microorganisms. *Nippon Nogeikagaku Kaishi* **62:** 1443–1450.

Poulos, T. L., B. C. Finzel, I. C. Gunsalus, G. C. Wagner, and J. Kraut. 1985. The 2.6-A crystal structure of *Pseudomonas putida* cytochrome P-450. *J. Biol. Chem.* **260:** 16122–16130.

Robinson, W. G., and K. Hook. 1964. Ricinine nitrilase-reaction product and substrate specificity. *J. Biol. Chem.* **239:** 4257–4267.

Ryuno, K., T. Nagasawa, and H. Yamada. 1988. Isolation of advantageous mutants of *Pseudomonas chlororaphis* B23 for the enzymatic production of acrylamide. *Agric. Biol. Chem.* **52:** 1813–1816.

Sugiura, Y. 1980. Bleomycin-iron complexes. Electron spin resonance study, ligand effect, and implication for action mechanism. *J. Am. Chem. Soc.* **102:** 5208–5215.

Sugiura, Y., J. Kuwahara, T. Nagasawa, and H. Yamada. 1988. Significant interaction between the low-spin iron(III) site and pyrroloquinoline quinone in the active center of nitrile hydratase. *Biochem. Biophys. Res. Commun.* **154:** 522–528.

Tang, S. C., S. Koch, G.C. Papaefthymiou, S. Foner, R. B. Frankel, J. A. Ibers, and P. H. Holm. 1976. Axial ligation modes in iron(III) porphyrins. Models for the oxidized reaction states of cytochrome P-450 enzymes and the molecular structure of iron(III) protoporphyrin IX dimethyl ester *p*-nitrobenzenethiolate. *J. Am. Chem. Soc.* **98:** 2414–2434.

Vandamme, E. 1987. *Downstream Processing in Biotechnology* (R. de Bruyne and A. Huygheboert, eds.). Antwerp: The Royal Society of Engineers.

Vuk-Pavlovic, S., and Y. Siderer. 1977. Probing axial ligands in ferric hemoproteins: an ESR study of myoglobin and horse radish peroxidase in $H_2^{17}O$. *Biochem. Biophys. Res. Commun.* **79:** 885–889.

Watanabe, I., Y. Satoh, and T. Kouno. 1979. Jap. Patent 129,190.

Watanabe, I., and M. Okumura. 1986. Jap. Patent 162,193.

Watanabe, I. 1987. Acrylamide production method using immobilized nitrilase-containing microbial cells. In *Methods in Enzymology, Vol. 136* (K. Mosbach, ed.), pp. 523–530. New York: Academic Press.

Watanabe, I., Y. Satoh, and K. Enomoto. 1987. Screening, isolation and taxonomical properties of microorganisms having acrylonitrile-hydrating activity. *Agric. Biol. Chem.* **51:** 3193–3199.

Yamada, H., Y. Asano, T. Hino, and Y. Tani. 1980. Microbial utilization of acrylonitrile. *J. Ferment. Technol.* **57:** 8–14.

Yamada, H., Y. Asano, and Y. Tani. 1980. Microbial utilization of glutaronitrile. *J. Ferment. Technol.* **58:** 495–500.

Yamada, H., K. Ryuno, T. Nagasawa, K. Enomoto, and I. Watanabe. 1986. Optimum culture conditions for production by *Pseudomonas chlororaphis* B23 of nitrile hydratase. *Agric. Biol. Chem.* **50:** 2859–2865.

15
Aldolases in Organic Synthesis

C.-H. Wong

Asymmetric carbon–carbon bond formation in aldol addition reactions is one of the most useful and efficient methods in synthetic organic chemistry. Most of the organic aldol reactions developed recently are stoichiometric and require the use of a metal or metal-like enolate complex to achieve stereoselectivity. Because of the instability of the enolate complex in aqueous solutions, the aldol reactions must be carried out in organic solvents at low temperature. This requirement limits the application of organic aldol reactions to synthesis of molecules soluble in organic solvents. For those compounds containing polyfunctional groups, such as carbohydrates, the synthesis still relies on other methods. The aldol addition reactions catalyzed by aldolases, however, are performed in aqueous solution at neutral pH without protection of the substrate functional groups. It is obvious that both chemical and enzymatic aldol reactions are complementary and the combination of these two approaches will extend the synthetic utility of aldol reactions.

More than 15 different aldolases have been reported, each of which catalyzes a distinct aldol reaction with high stereoselectivity. Table 15.1 summarizes some aldolase-catalyzed stereospecific reactions associated with carbohydrate synthesis (Whitesides and Wong 1985).

FRUCTOSE-1,6-DIPHOSPHATE ALDOLASE (FDP ALDOLASE)

The enzyme FDP aldolase can be isolated from animal cells or microorganisms. The former is a Schiff base-forming and the latter is a zinc-enolate-forming enzyme with regard to the activation of substrate during the catalysis. The enzyme from rabbit muscle is commercially available and is the one that has been used most often as a synthetic

Table 15.1. Types of aldolase-catalyzed reactions.

DONOR	ACCEPTOR	ENZYME[a]	PRODUCT
		FDP Aldolase	
		Fuc-1-P Aldolase	
		Rha-1-P Aldolase	
		Sialic Acid Aldolase	
		KDO Aldolase	
		KDPG Aldolase	
		KDF Aldolase	
		KDG Aldolase	
		KHG Aldolase	
		MHKG Aldolase	
		KDHA Aldolase	
		KDO-8-P Synthase	
		Deoxyribose Aldolase	

[a] See Whitesides and Wong (1985) for details.

[a] Abbreviations: FDP, D-fructose-1,6-diphosphate; Fuc-1-P, L-fuculose-1-phosphate; Rha-1-P, L-rhamnulose-1-phosphate; KDPG, 2-keto-3-deoxy-6-phosphogluconate; KDF, 2-keto-3-deoxy-D-fuconate; KDG, 2-keto-3-deoxy-D-glucarate; KHG, 2-keto-4-hydroxyglutarate; MHKG, 4-methyl-4-hydroxy-2-ketoglutarate; KDHA, 2-keto-3-deoxyheptonate; KDO, 3-deoxy-D-mammo-octulosonate.

catalyst (Bednarski et al. 1989; Turner and Whitesides 1989; Durrwachter and Wong 1988; Drueckhammer et al. 1989). Figure 15.1 illustrates the reaction mechanism and the substrate specificity of this enzyme. It is interesting to note that the enzyme is highly specific for dihydroxyacetone phosphate (DHAP) but will accept a variety of aldehydes as acceptors. As shown in the figure, a small change of the DHAP moiety results in a substantial decrease of activity. The aldehyde components, however, can be varied. Many aldehydes have been shown to be good substrates for the enzyme in the synthesis of common and uncommon sugars and their isotopically labeled counterparts.

The zinc FDP aldolase from microorganisms such as yeast or *E. coli* is not commercially available. We have recently constructed a 300-fold overproducing *E. coli* strain which produces the *E. coli* FDP aldolase. Figure 15.2 illustrates the construction of a 5.8-kb plasmid for expression of the enzyme (Von der Osten et al. 1989). The zinc aldolase was shown to be more stable than the rabbit aldolase. The half-life at room temperature in the presence of 0.3 mM zinc chloride is 60 days compared to 2 days for the rabbit enzyme. Based on limited examples, both enzymes, however, have the same substrate specificity and stereoselectivity. With regard to the mechanism, the role of zinc ion is still not well defined. Based on the nuclear magnetic resonance (NMR) study, it was proposed that Zn^{2+} participates in the activation of DHAP (Smith and Mildvan 1981). The study with Fourier Transform-Infrared [(FT-IR) Spectroscopy], however, showed that Zn^{2+} was involved in the polarization of the carbonyl group of the aldehyde substrate (Belasco and Knowles 1983). In any case, both enzymes are now readily available and useful for synthetic aldol reactions.

One problem associated with the FDP aldolase-catalyzed reactions is the need for the preparation of DHAP. This molecule is not stable in solution ($t \sim 20$ h, pH 7), and the synthesis is not trivial. It may be generated in situ from FDP (Bednarski et al. 1989; Whitesides and Wong 1985), or prepared separately by chemical (Effenberger and Straug 1987; Colbran et al. 1967) or enzymatic methods (Whitesides and Wong 1985). An alternative solution is to use a mixture of dihydroxyacetone and a small amount of inorganic arsenate to replace DHAP (Drueckhammer et al. 1989). Mechanistic studies indicate that in aqueous solution, dihydroxyacetone reacts with inorganic arsenate spontaneously to form dihydroxyacetone arsenate, which is a mimic of

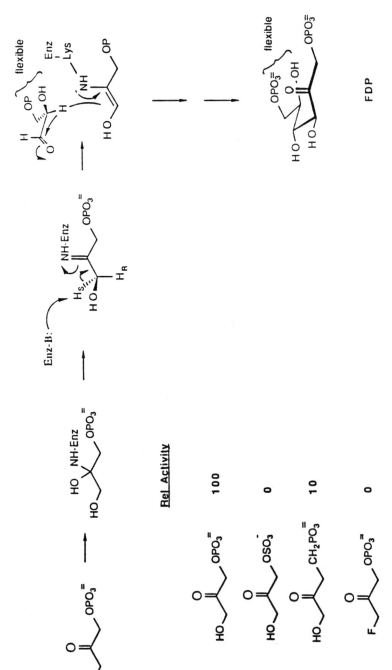

Figure 15.1. Mechanism and substrate specificity of FDP aldolase from rabbit muscle.

Rel Activity

100

0

10

0

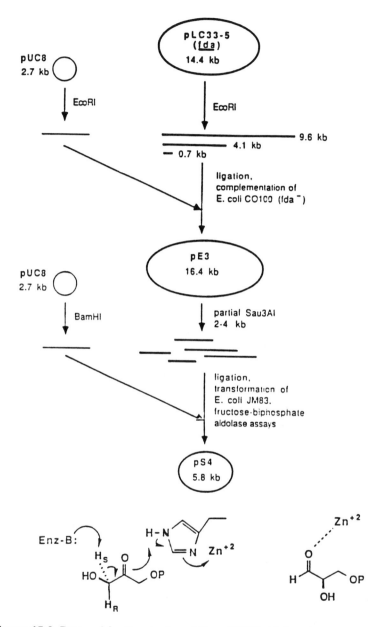

Figure 15.2. Protocol for the cloning of *E. coli* FDP aldolase for the construction of a 300-fold overproduction strain. The lower part indicates possible roles of zinc ion in catalysis.

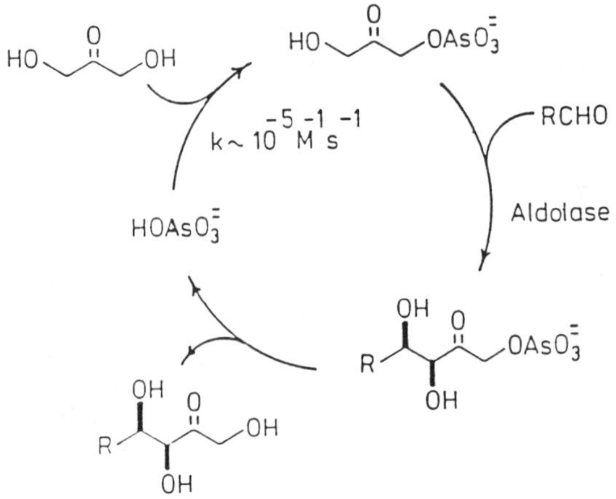

Figure 15.3. Replacement of dihydroxyacetone phosphate with dihydroxyacetone and arsenate in FDP aldolase reactions.

DHAP and thus is accepted by FDP aldolase as a substrate. After enzymatic aldol condensation, the arsenate moiety dissociates from the aldol product and reacts repeatedly with dihydroxyacetone (Fig. 15.3). This eliminates the need for the preparation of DHAP and for the removal of the phosphate group from the aldol product. The strategy has been successfully extended to many other reactions using organic phosphate-requiring enzymes.

SYNTHESIS

A recent development in enzymatic aldol reactions is the synthesis of piperidines and pyrrolidines structurally related to monosaccharides (Fig. 15.4) (Pederson, Kim, and Wong 1988). These molecules are useful slow-binding glycosidase inhibitors. Protonation of the N group of piperidines results in a tight binding to the enzyme. The combined enzymatic aldol reactions and catalytic reductive amination as illustrated in Figure 15.4 is a very efficient approach to the synthesis of these molecules. The *N*-butyl derivative of 1-deoxynojirimycin **3** has been shown to be a potent anti-AIDS agent with no cytotoxicity (Karpas et al. 1988). This compound is easily synthesized by hydrogenolysis (H$_2$/Pd) of **3** with *n*-butanal. Figure 15.5 illustrates the procedures

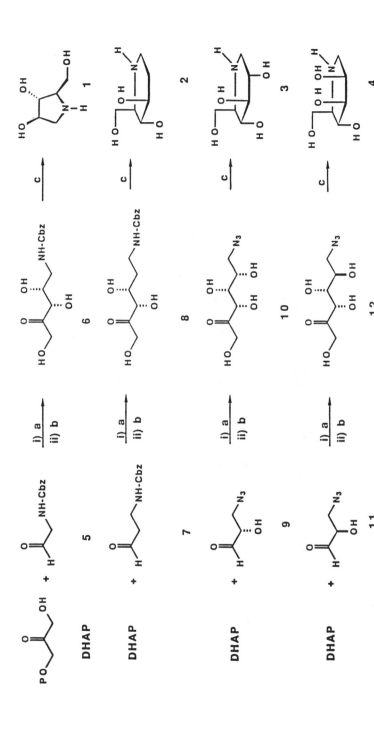

Figure 15.4. A combined enzymatic aldol condensation and reductive amination as an efficient route to 1,4-dideoxy-1,4-imino-D-arabinitol (1), fagomine (2), 1-deoxynojirimycin (3), and 1-deoxymannojirimycin (4). (a) FDP aldolase; (b) acid phosphatase; (c) H₂/Pd.

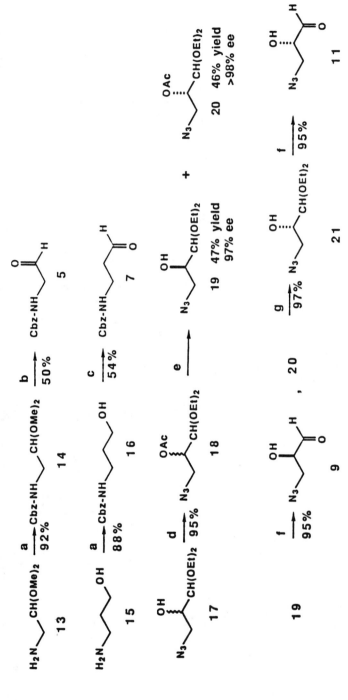

Figure 15.5. Synthesis of the aldehydes used in aldol reactions. (a) Cbz-Cl, aq acetone, NaHCO₃; (b) THF/H₂ (COOH)₂, reflux 4 days; (c) pyridinium chlorochromate, PCC; (d) Ac₂O, pyridine; (e) *Pseudomonas* lipoprotein lipase, 51% conversion; (f) HCl, H₂O, 65°C, 12 h; (g) 1 *M* NaOH.

used for the preparation of substrates for the aldol reactions. It is worth noting that (R, S)-3-azido-2-hydroxypropanoldiethylacetal acetate can be easily resolved by lipase-catalyzed hydrolysis (Von der Osten et al. 1989). This resolution process provides both 9 and 11 with very high enantiomeric excess for the synthesis of 3 and 4.

THERMODYNAMICALLY CONTROLLED C–C BOND FORMATION

Despite the fact that enzymatic aldol reactions are becoming useful in synthetic carbohydrate chemistry, preparation of the aldehyde substrates, particularly those containing chiral centers, remains a difficult

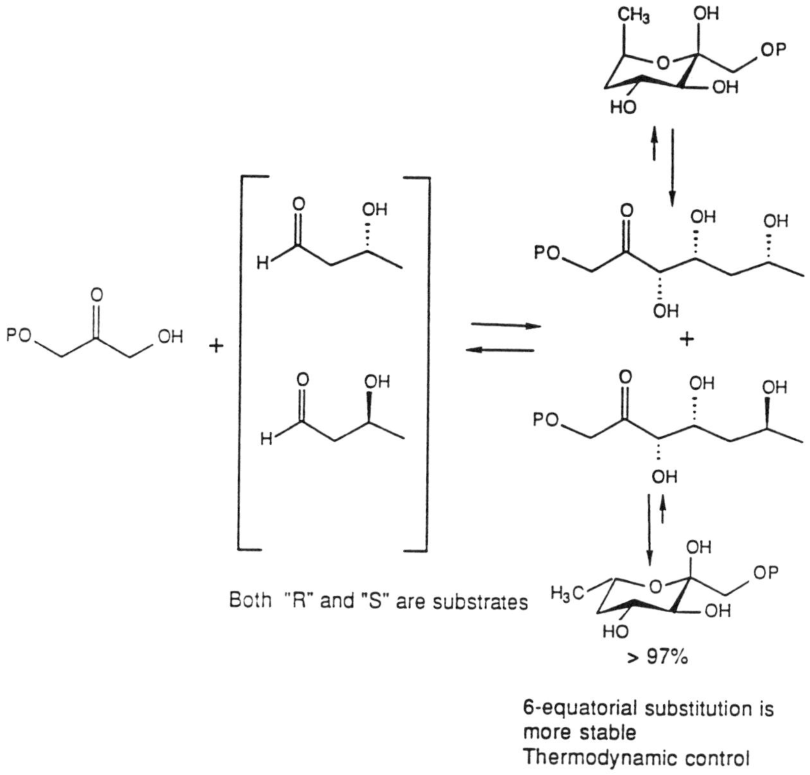

Both "R" and "S" are substrates

> 97%

6-equatorial substitution is more stable
Thermodynamic control

Figure 15.6. Thermodynamically controlled aldol reaction to 5-deoxy-6-methylfructose.

problem. Many interesting α-substituted aldehydes are not chirally stable in aqueous solution. Racemization may occur and result in the production of a mixture of diastereomers in the aldol reaction. In some cases, the stereoselective aldol condensation can be accomplished with the use of racemic aldehyde substrates in a kinetically controlled process; that is, the reaction can be stopped before it reaches equilibrium to obtain a single diastereomer if one of the enantiomeric aldehyde substrates reacts faster than the other (Durrwachter and Wong 1988).

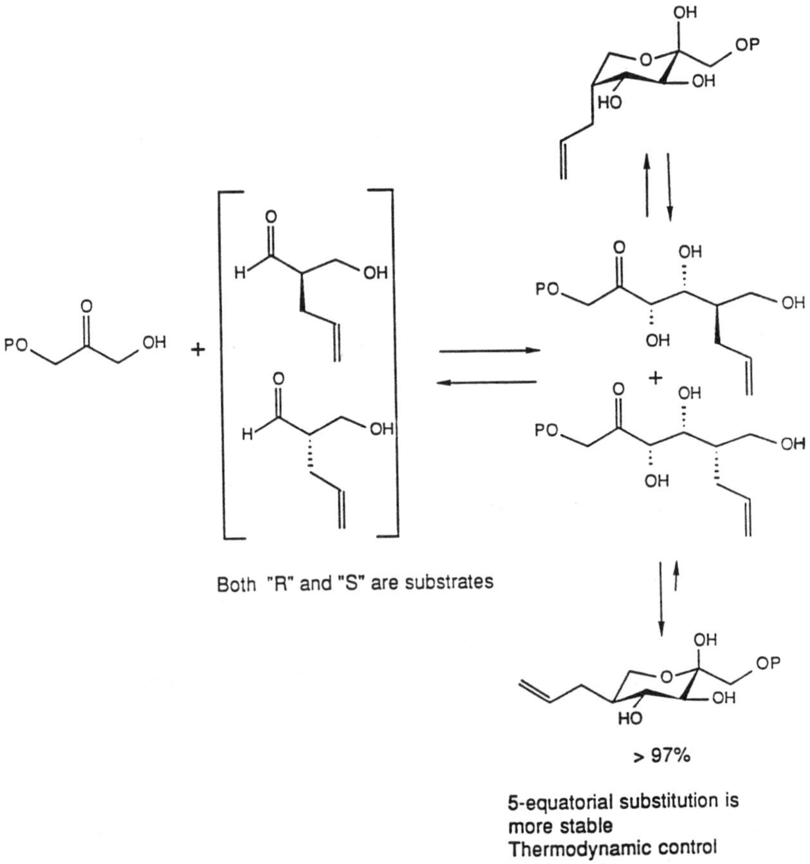

Figure 15.7. Thermodynamically controlled aldol reaction to C-allyl sugars.

One can also utilize the thermodynamically controlled process to prepare one single diastereomeric product if that product is more stable than the other diastereomer (Durrwachter and Wong 1988). For example, in the aldol reaction with racemic 3-hydroxy butanal, two diastereomeric products could be produced. Because of the ring formation of the aldol product and because of the reversible nature of the aldol reaction, only the more stable product with the 6-equatorial methyl substitution is produced when the reaction reaches equilibrium (Fig. 15.6). In another example, the aldol reaction with racemic 2-allyl-3-hydroxypropanal generates a single product with the 5-allyl substituent at the equatorial position (Fig. 15.7).

N-ACETYLNEURAMINIC ACID ALDOLASE (Neu5Ac ALDOLASE)

Sialic acids are a family of amino sugars that are derivatives of Neu5Ac. They play many important roles in biochemical recognition. Development of practical and efficient procedures for the synthesis of sialic acids will facilitate the study of sialic acid-associated biochemical problems. The enzyme Neu5Ac aldolase catalyzes the synthesis of Neu5Ac from pyruvate and *N*-acetyl mannosamine (Auge, David, and Gautheron 1984; Simon, Bednarski, and Whitesides 1988). We have recently exploited the synthetic utility of this enzyme from a *Clostridia* species which is available from Toyobo (Kim et al. 1988). As indicated in Table 15.2, the enzyme accepts many aldoses containing five or six carbons as substrates, but only pyruvate can be accepted as a donor. Figure 15.8 indicates the enzyme-catalyzed formation of Neu5Ac and the structures of substrates that are accepted by the enzyme.

With Neu5Ac in hand, we have undertaken the synthesis of certain derivatives of Neu5Ac which could be used as inhibitors of sialidases and cytidine monophosphate-*N*-acetylneuraminic acid synthetase. The procedures are described in Figure 15.9. The methyl ester of Neu5Ac was treated with acetyl chloride at room temperature to give the β-chloride. Catalytic hydrogenolysis of the C–Cl bond occurred with inversion to give the α-2-deoxy derivative which was subsequently hydrolyzed to give the free sugar. Elimination of hydrogen chloride from the sugar chloride in the presence of triethylamine gave the 2,3-

Table 15.2. The kinetic parameters (K_m, M; V_{max}, U mg^{-1}) of Neu5Ac aldolase for several substrates in the aldol condensation in 0.1 M phosphate (pH 7.5 and 25°C).

NO.	SUBSTRATE	K_m	V_{max}	
Acceptor + pyruvate				
1	N-Acetyl-D-mannosamine (ManNAc)	0.7	25	
2	N-Acetyl-D-glucosamine	—	0	
3	6-O-Acetyl-ManNAc	0.5	3	
4	D-Mannose	2.8	50	
5	D-Glucose	2.3	1.8	
6	6-O-Acetyl-D-mannose	2.0	10	
7	L-Glucose	—	0	
8	D-Allose	—	0.1	
9	2-Deoxy-D-glucose	1.8	31	
10	2-Deoxy-D-galactose	1.3	4.5	
11	L-Fucose	—	0.9	
12	D-Lyxose	1.7	3.3	
13	D-Arabinose	—	0.8	
14	L-Xylose	—	0.3	
15	2-Deoxy-D-ribose	—	0.6	
16	D-Glyceraldehyde	—	0	
17	L-Glyceraldehyde	—	0	
18	Glycolaldehyde	—	0	
Donor + ManNAc				
1	Pyruvate	0.01	25	
2	Acetylphosphonate	—	0	
3	3-Fluoropyruvate	—	0	a
4	3-Bromopyruvate	—	0	a
5	3-Hydroxypyruvate	—	0	a
6	Acetopyruvate	—	0	a
7	Acetoacetate	—	0	a
8	2-Oxobutyrate	—	0	a
9	Phosphoenolpyruvate	—	0	a

[a] Uchida, Y., Y. Tsukada, T. Sugimori. 1984. *J. Biochem.* **96:** 507.

dehydro derivative which on hydrogenation and hydrolysis gave the β-form of the 2-deoxy sugar.

Another sialic acid of interest to us is 9-O-acetyl-Neu5Ac, the receptor of influenza A glycoprotein. The aldol acceptor used in the synthesis was 6-O-acetyl-N-acetylmannosamine which was prepared via pro-

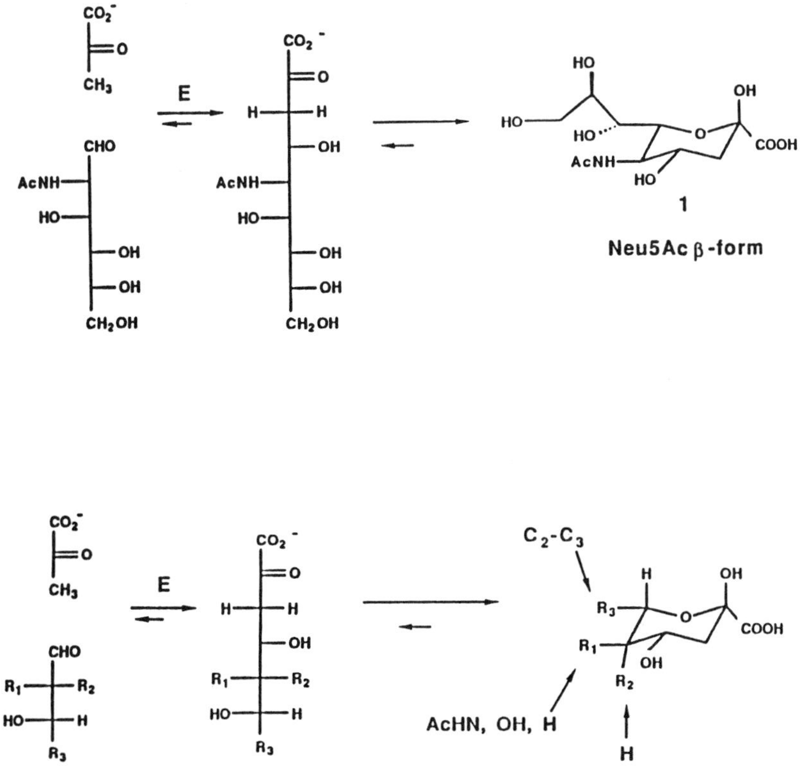

Figure 15.8. (Top) N-Acetylneuraminic acid aldolase catalyzed reaction. (Bottom) Structures of aldoses accepted by the enzyme. E = neuraminic acid aldolase.

tease N-catalyzed transesterification in anhydrous dimethylformamide [subtilisin was used by others in this solvent (Riva et al. 1988)] using N-acetylmannosamine and isopropenyl acetate as substrates (Wang et al. 1988).

The 6-O-acetyl sugar was prepared in ~90% yield and used in the subsequent aldol reaction in aqueous solution at pH 7.5. With 3 equivalents of pyruvate, Neu5Ac was prepared in 84% yield (Fig. 15.10).

It is worth noting that the use of enol esters as irreversible transesterification reagents not only facilitates the reaction but also improves

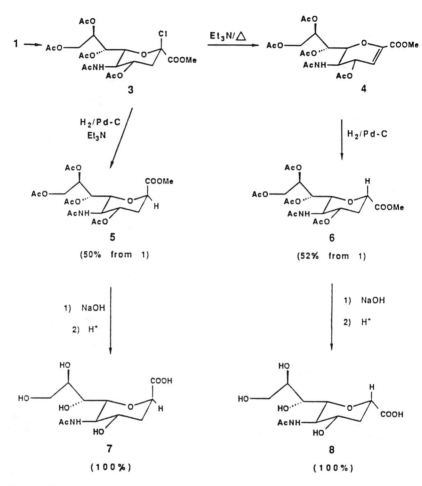

Figure 15.9. Preparation of α- and β-2-deoxy-N-acetylneuraminic acid.

the selectivity. Esters of this type have been very useful in many other enzymatic transesterification processes (Wang et al. 1988), making the process very attractive for enantioselective transformation. Figure 15.11 illustrates the use of an enol ester in the synthesis of a chiral glycerol derivative for use in the synthesis of platelet-activating factor. Because enzymes usually possess the same enantiotopic group selec-

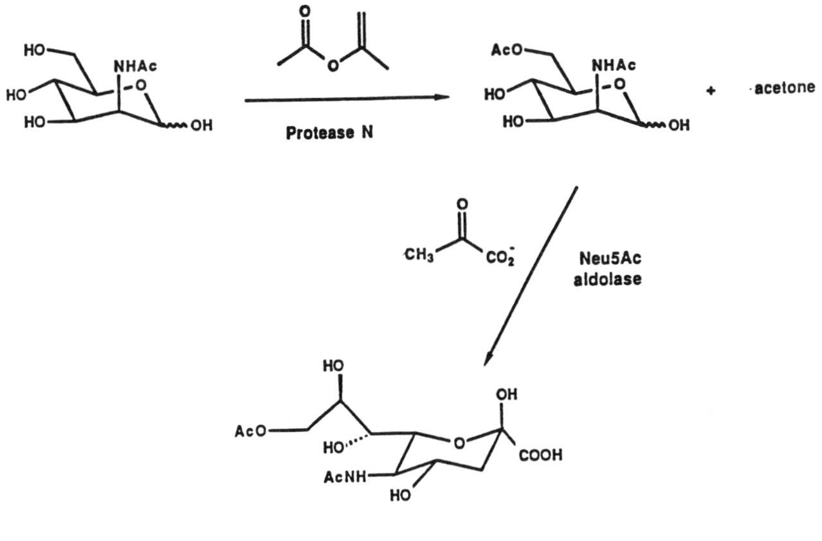

Figure 15.10. Two-step enzymatic synthesis of 5,9-diacetylneuraminic acid.

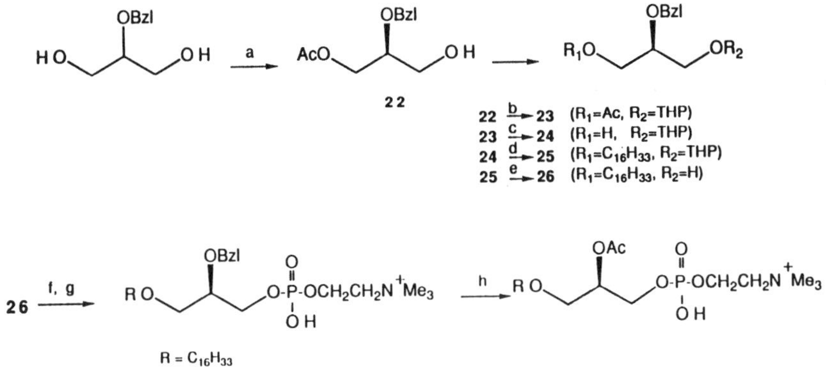

Figure 15.11. Reagents: (a) vinyl acetate, *Pseudomonas* lipase; (b) DHP, pyridinium *p*-toluenesulfonate; (c) CH₃ONa, CH₃OH; (d) NaH then C₁₆H₃₃ Br; (e) CH₃COOH/THF/H₂O (4:2:1); (f) Cl₂P(O)OCH₂CH₂Br, then hydrolysis; (g) Me₃N, then AcOAg; (h) H₂, Pd-C, then Ac₂O in pyridine.

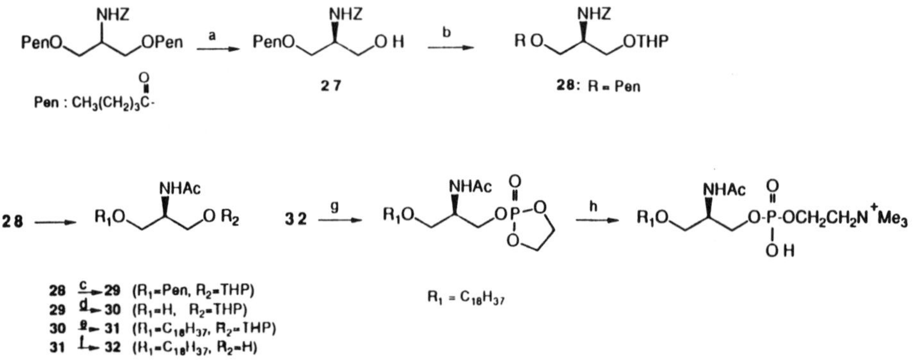

Figure 15.12. Reagents: (a) pancreatic lipase hydrolysis; (b) DHP, pyridinium *p*-toluenesulfonate; (c) H₂, Pd-C, then AcCl, pyridine; (d) LiBH₄; (e) NaH, then C₁₈H₃₇Br; (f) CH₃COOH/THF/H₂O (4:2:1); (g) ClP(O) (—OCH₂CH₂O—); (h) Me₃N.

tivity in hydrolysis and in transesterification, the enantiomeric isomer of the monoester prepared in the irreversible transesterification process can be obtained via the enzyme-catalyzed hydrolysis of 2-*O*-benzyl glycerol diacetate. Similarly, (*S*)-*N*-benzyloxycarbonyl serinol monopentanoate was prepared via lipase-catalyzed hydrolysis of *N*-benzyloxycarbonyl-serinol dipentanoate, and the (*R*)-enantiomer was prepared via the enzyme-catalyzed acylation of *N*-benzyloxycarbonyl-serinol with the corresponding isopropenyl acylate. In this case a complete reversal of enantioselectivity was observed as compared to the transformation of glycerol derivatives. Compound **27** was used in the synthesis of a phospholipase A₂ inhibitor, the 2-acetamido analog of platelet-activating factor (Fig. 15.12).

ACKNOWLEDGMENT

This research was supported by the NSF and the NIH.

REFERENCES

Auge, C., S. David, and C. Gautheron. 1984. *Tetrahedron Lett.* **25:** 4663.

Bednarski, M. D., E. S. Simon, N. Bischofberger, W.-D. Fessner, M.-J. Kim, W. Lees, T. Saito, H. Waldmann, and G. M. Whitesides. 1989. *J. Am. Chem. Soc.* **111:** 627.

Belasco, J. G., and J. R. Knowles. 1983. *Biochemistry* **22:** 122.

Colbran, R. L., J. K. N. Jones, N. K. Matheson, and I. Rozema. 1967. *Carbohyd. Res.* **4:** 355.

Drueckhammer, D. G., J. R. Durrwachter, R. L. Pederson, D. C. Crans, L. Daniels, and C.-H. Wong. 1989. *J. Org. Chem.* **54:** 70.

Durrwachter, J. R., and C.-H. Wong. 1988. *J. Org. Chem.* **53:** 4175.

Effenberger, O. F., and A. Straub. 1987. *Tetrahedron Lett.* **28:** 1641.

Karpas, A., G. W. J. Fleet, R. A. Dwek, S. Petursson, S. K. Namgoong, N. G. Ramsden, G. S. Jacob, and T. W. Rademacher. 1988. *Proc. Natl. Acad. Sci. USA* **85:** 9229.

Kim, M. J., W. J. Hennen, H. M. Sweers, and C.-H. Wong. 1988. *J. Am. Chem. Soc.* **110:** 6481.

Pederson, R. L., M. J. Kim, and C.-H. Wong. 1988. *Tetrahedron Lett.* **29:** 4645.

Riva, S., J. Chopineau, A. P. G. Kieboom, and A. M. Klibanov. 1988. *J. Am. Chem. Soc.* **110:** 584.

Simon, E. S., M. D. Bednarski, and G. M. Whitesides. 1988. *J. Am. Chem. Soc.* **110:** 7159.

Smith, G. M., and A. S. Mildvan. 1981. *Biochemistry* **20:** 4340.

Turner, N. J., and G. M. Whitesides. 1989. *J. Am. Chem. Soc.* **111:** 624.

Von der Osten, C. H., A. J. Sinskey, C. F. Barbas, III, R. L. Pederson, Y. F. Wang, and C.-H. Wong. 1989. *J. Am. Chem. Soc.* **111:** 3924.

Wang, Y. F., J. J. Lalonde, M. Momongan, D. E. Bergbreiter, and C.-H. Wong. 1988. *J. Am. Chem. Soc.* **110:** 7200.

Whitesides, G. M., and C.-H. Wong. 1985. *Angew Chem. Int. Ed. Engin.* **24:** 617.

16
Two-Liquid Phase Biocatalysis
Reactor Design

J. M. WOODLEY

Biocatalysis offers some significant advantages over traditional chemical catalysis from the viewpoint of both the organic chemist and the engineer, for example, stereospecific, regiospecific, and reaction-specific catalysis under mild operating conditions. However, the process engineering required to transfer this technology from laboratory to industrial scale is not as mature as that in the chemical industry. This is particularly true in the field of biocatalytic reactions which involve organic compounds of low water solubility. Biocatalysis in the presence of organic solvents is one strategy that has been proposed to mediate these reactions (Tramper, van der Plas, and Linko, 1985; Laane, Tramper, and Lilly, 1987). Dependent on the solubilities of the reactant(s) and product(s), varying amounts of organic solvent may be added to the otherwise aqueous reaction medium (Lilly et al. 1987). In the extreme it is possible to perform enzymic reactions in essentially dry organic solvent (Klibanov 1986). Many reactions involve water-soluble as well as poorly water-soluble organic reactant(s) and/or product(s) (Lilly and Woodley 1985); for these cases the use of a heterogeneous liquid–liquid reaction medium is particularly attractive (Lilly, 1982, 1983; Lilly and Woodley, 1985). The merits of operating biocatalytic reactions in the presence of organic solvents are illustrated by examples throughout this volume and those specific to two-liquid phase systems have been well documented previously (Lilly and Woodley, 1985; Brink et al. 1988). In this chapter the engineering of two-liquid phase biocatalytic reactions is discussed and in particular the considerations for reactor design are addressed. These considerations enable a preliminary evaluation to be made of the feasibility of a potential process, prior to scale-up from the laboratory.

Optimal reactor design and operation for a two-liquid phase biocatalytic reaction will be primarily dependent on the nature of the organic phase, which may be a reservoir of reactant(s) or of product(s) or a combination of reaction components. Specifically, the methodology presented here applies to an organic phase rich in reactant alone. In many cases the poorly water-soluble organic reactant will be a liquid at reaction temperature (and pressure), forming the organic phase itself. In other reactions the poorly water-soluble organic reactant will be a solid, and hence dissolved in an added essentially water-immiscible organic solvent to form the second liquid phase. Presented here is a method for reactor design based on the biocatalysis of a poorly water-soluble organic liquid reactant to a water-soluble product. Although this is only one of a series of possible component distributions, several industrially important reactions fall in this class, for example, stereo-

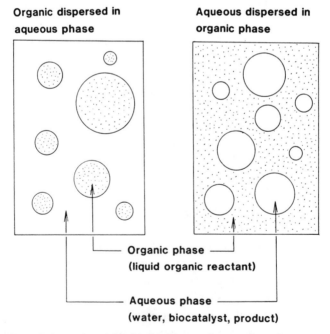

Figure 16.1. Schematic of liquid–liquid reaction medium for conversion of poorly water-soluble organic liquid reactant to water-soluble product. (▧), Organic phase; (□), aqueous phase.

specific ester hydrolysis (Brookes, Lilly, and Drozd, 1986; Williams et al. 1987), benzene and toluene oxidations to their corresponding *cis*-glycols (Ballard et al. 1983; Brazier, Lilly, and Herbert, in press). Product solubility is a further consideration in reactor design, but beyond the scope of this work. Consequently the analysis presented here is applicable only to reactions producing a water-soluble product. In each of the previously cited examples at least one product is water-soluble.

Schematically illustrated in Figure 16.1 is the reaction medium. The organic phase (organic liquid reactant) and the aqueous phase (containing water, biocatalyst, and ultimately product) are mixed, dispersing one phase within the other, so as to create a large liquid–liquid interfacial area suitable for facile mass transfer of the reactant from the organic to the aqueous phase. The organic phase may be dispersed within the aqueous phase or the system inverted with aqueous phase droplets dispersed within a continuous organic phase. At the conclusion of the reaction the liquid phases are separated to yield a product-rich aqueous phase and an organic phase, suitable for recycle if required.

NOMENCLATURE

[Ba]	$(g\ L^{-1})$	Aqueous phase biocatalyst concentration
[Ea]	$(g\ L^{-1})$	Aqueous phase enzyme concentration
$K_L A$	(min^{-1})	Reactant mass transfer coefficient
n	(min^{-1})	Reactor stirrer speed
Ra	$(mmol\ min^{-1}\ L^{-1})$	Reaction rate (aqueous phase basis)
Ra (max)	$(mmol\ min^{-1}\ L^{-1})$	Maximum reaction rate (aqueous phase basis)
Rs	$(mmol\ min^{-1}\ g^{-1})$	Specific reaction rate
$[Sa*]$	(mM)	Aqueous phase reactant saturation concentration
$[Sab]$	(mM)	Aqueous phase reactant concentration
ϕ	$(—)$	Phase ratio

EXPERIMENTAL DETERMINATION OF REACTOR DESIGN PARAMETERS

Three sets of laboratory measurements are required to determine the fundamental parameters of such a previously described reaction:

1. The saturation concentration of reactant within the aqueous phase, [Sa*].
2. The biocatalytic reaction rate as a function of the aqueous phase reactant concentration. This is dependent on the catalyst concentration, [Ba].
3. The reactant transfer rate from organic to aqueous phase as a function of the aqueous phase reactant concentration. This is dependent on the saturation concentration of reactant within the aqueous phase, [Sa*], and the reactant transfer coefficient, $K_L A$.

These measurements, if used correctly, will yield information about the location of catalysis within the reaction medium. This may be categorized into three reaction zones (see Fig. 16.2). Biocatalyst may con-

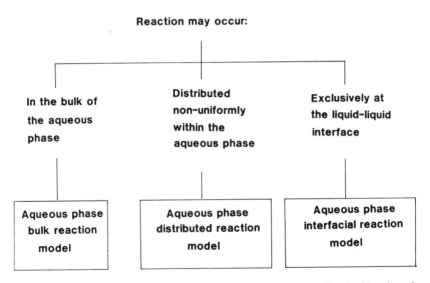

Figure 16.2. A classification of reaction zone models to describe the kinetics of a two-liquid phase biocatalytic reaction.

vert reactant to product in the bulk of the aqueous phase solution, exclusively at the liquid–liquid interface or distributed in a nonuniform manner between the interface and the bulk of the aqueous phase. The presence of biocatalyst, preferentially, in the organic phase is highly unlikely in view of its predominantly hydrophilic nature. The location of the reaction zone is critical in determining the reaction kinetics and hence ultimately the reactor design.

Obtaining the necessary data to determine the reactor design parameters (listed previously) and hence the location of the reaction zone require three sets of experiments to be carried out, schematically represented in Figure 16.3.

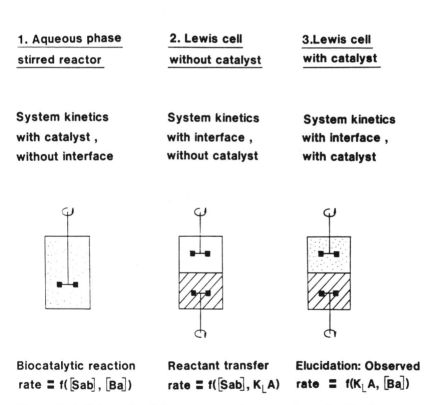

1. Aqueous phase stirred reactor

2. Lewis cell without catalyst

3.Lewis cell with catalyst

System kinetics with catalyst , without interface

System kinetics with interface , without catalyst

System kinetics with interface , with catalyst

Biocatalytic reaction rate ≡ f([Sab], [Ba])

Reactant transfer rate ≡ f([Sab], $K_L A$)

Elucidation: Observed rate ≡ f($K_L A$, [Ba])

Figure 16.3. Schematic of laboratory experiments to determine kinetic parameters of the system. (▨), Organic phase; (□), aqueous phase; (▦), aqueous phase with biocatalyst.

1. Biocatalytic reaction rate may be determined in an all aqueous phase stirred vessel in which known amounts of reactant are dissolved (reactant concentrations up to saturation concentraiton). Known concentrations of biocatalyst are added to the solution and subsequently the initial reaction rates may be measured in the normal manner.

2. Substrate transfer rate may be determined in a modified Lewis cell (Lewis 1954) in which reactant (organic phase) is placed alongside (either above or below, dependent on density) aqueous phase (excluding catalyst). The phase boundary is defined by a known liquid–liquid interfacial area, specified by the size of the Lewis cell. Both phases are stirred sufficiently to keep them well mixed but without disturbance of the flat liquid–liquid interface. The rate of transfer of reactant to aqueous phase is evaluated by measuring reactant concentrations in the aqueous phase as a function of time.

3. Elucidation of catalytic reaction zone is determined by observing the reaction rate obtained, with biocatalyst, in a modified Lewis cell. Comparison of this reaction rate with that obtained by using the results of experiments (*1*) and (*2*) to predict the reaction rate yield information about the location of the reaction zone.

Results from experiment (*1*) are likely to fall into three classes of reaction rate–reactant concentration relationship, regardless of whether they obey Michaelis–Menten kinetics or otherwise. Characteristic reaction rate–reactant concentration profiles are illustrated in Figure 16.4 for each of these classes, labeled A, B, and C. In class C the maximum reaction rate cannot be achieved in a two-liquid phase system. The maximum rate can be achieved in classes A and B, although with a lower reactant transfer requirement in A than B. Strategies to lower these mass transfer requirements and overcome the problems of class C reactions are being addressed by current research.

The second set of experiments, (*2*), requires the measurement of the reactant transfer rate as a function of the aqueous phase reactant concentration, and for solvent transfer this process may be described by the following equation:

$$\text{Reactant transfer rate} = K_L A([Sa*] - [Sab])$$

Measurements of the transient aqueous phase reactant concentration are made in a modified Lewis cell. Integration of this transient deter-

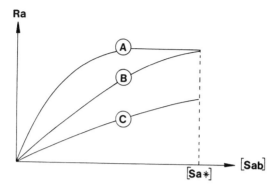

Figure 16.4. Three classes of reaction rate aqueous phase reactant concentration relationship. (A) $Ra = Ra$ (max) at $[Sab] < [Sa*]$; (B) $Ra = Ra$ (max) at $[Sab] = [Sa*]$; (C) $Ra = Ra$ (max) at $[Sab] > [Sa*]$.

mines the value for the reactant transfer coefficient, $K_L A$. This value gives the gradient of the linear plot of transfer rate as a function of aqueous phase reactant concentration, illustrated in Figure 16.5.

Superimposing Figure 16.5 on Figure 16.4 results in two rate-reactant concentration profiles as shown in Figure 16.6, for a class A reaction. For reactions catalyzed at the liquid–liquid interface then the

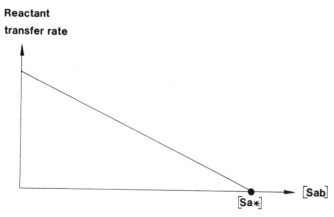

Figure 16.5. Reactant transfer rate (organic phase solvent) as a function of aqueous phase reactant concentration. Solvent transfer rate = $K_L A ([Sa^*] - [Sab])$.

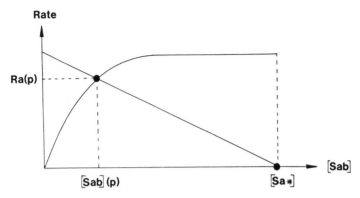

Figure 16.6. Prediction of reaction rates and aqueous phase reactant concentrations at steady state. Ra (p) and [Sab] (p) are the aqueous phase bulk reaction model predictions of reaction rate and aqueous phase reactant concentration, respectively.

reaction rate is that given by the concentration at the interface—for the reaction illustrated in Figure 16.6 it will be the maximum reaction rate. It is not clear whether reactions that are catalyzed exclusively at the liquid–liquid interface in a heterogeneous reaction medium can yield any products in an aqueous phase medium such as that used in experiment (*1*). However the rationale for the measurements made in experiment (*1*) is that they can provide the basis for prediction (together with the results of experiment (*2*) for comparison with observed measurements in experiment (*3*). Thus the reaction zone can still be elucidated with this method even under such extreme conditions as exclusively interfacial catalysis, regardless of the medium.

For reactions occurring evenly throughout the bulk of the aqueous phase then both the rate and aqueous phase reactant concentration are defined by the intersection of the two rate-reactant concentration profiles, [*Ra*(p) and [Sab](p) in Figure 16.6]. This intersection defines a steady state, when the reactant transfer rate is equal to biocatalytic reaction rate. For the aqueous phase distributed reaction model then rate and concentration predictions will lie between the interfacial and bulk predictions. The experimental procedure for elucidation of the reaction zone is diagrammatically represented in Figure 16.7.

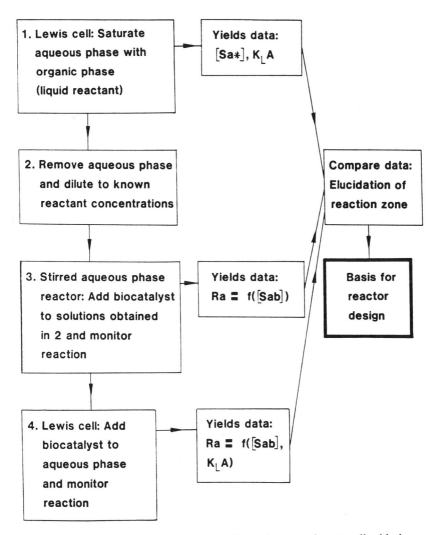

Figure 16.7. Procedure for elucidation of reaction zone in a two-liquid phase biocatalytic reaction.

EXAMPLE: BENZYL ACETATE HYDROLYSIS BY PIG LIVER ESTERASE

The previously described technique for elucidation of the reaction zone is well illustrated by the hydrolysis of benzyl acetate (a poorly water-soluble organic liquid) to benzyl alcohol (a poorly water-soluble organic liquid) and acetic acid (a predominantly water-soluble liquid) (Woodley, Brazier, and Lilly, submitted). Benzyl acetate is strongly chromophoric and has an aqueous phase saturation concentration of 13.5 mM. Consequently, dissolving and measuring the poorly water-soluble organic reactant in the aqueous phase is not difficult. Figure 16.8 illustrates the results of a set of experiments to determine the reaction rate (plotted as specific enzyme activity) as a function of the aqueous phase reactant concentration, [Sab]. Although these results do not obey Michaelis–Menten behavior, zero-order kinetics are observed above 4 mM (30% of saturation), hence making this a class A reaction.

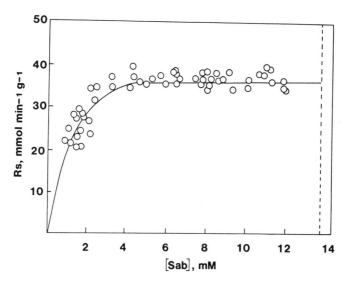

Figure 16.8. Specific enzyme activity as a function of aqueous phase reactant concentration for the hydrolysis of benzyl acetate by pig liver esterase. Experimental details given in Woodley, Brazier, and Lilly.

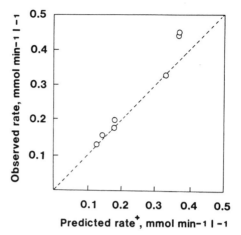

Figure 16.9. Comparison of observed with predicted reaction rates in a Lewis cell for the hydrolysis of benzyl acetate by pig liver esterase. [+]Predictions based on the aqueous phase bulk reaction model.

Figure 16.9 is a plot of the observed Lewis cell reaction rate results (for different sizes of interface and catalyst concentration) against those predicted by the aqueous phase bulk reaction model, by the method illustrated in Figure 16.6. Within the bounds of experimental error the results fall on the parity line. A similar result was observed for the measured and predicted aqueous phase substrate concentrations. This implies that the benzyl acetate hydrolysis by pig liver esterase is occurring in the bulk of the aqueous phase. The implications of this result for reactor design are discussed in the following section.

OPTIMAL REACTOR OPERATION

Knowledge of the reaction zone (i.e., the particular location and volume of reaction liquor in which reaction is occurring) enables optimization to focus on that area. Therefore, for reactions whose kinetics may be described by the aqueous phase bulk reaction model (e.g., the benzyl acetate hydrolysis by pig liver esterase), it is instructive to carry out a substrate mass balance on the aqueous phase:

$$\frac{d[\text{Sab}]}{dt} = \left(\begin{array}{c}\text{Rate of substrate}\\\text{transfer}\end{array}\right) - \left(\begin{array}{c}\text{Rate of biocatalytic}\\\text{reaction}\end{array}\right)$$

$$\frac{d[\text{Sab}]}{dt} = (K_L A([\text{Sa}*] - [\text{Sab}])) - (Ra)$$

At steady state,

$$\frac{d[\text{Sab}]}{dt} = 0$$

Hence,

$$K_\text{L}A([\text{Sa}*] - [\text{Sab}]) = Ra$$

where

$$Ra = f([\text{Sab}], [\text{Ba}])$$

Hence the aqueous phase reactant concentration, [Sab], may be defined if the reactant mass transfer coefficient, $K_\text{L}A$, and the aqueous phase biocatalyst concentration, [Ba], are known, together with the data obtained from the previously described sets of experiments (1), (2), and (3). Such a mass balance can also yield the fraction of aqueous phase reactant saturation for particular values of reaction rate (at specified catalyst concentration) and mass transfer coefficient (Harbron, Narendranathan, and Lilly, 1984).

Optimal operation occurs when the ratio of mass transfer coefficient to catalyst concentration (i.e., $K_\text{L}A/[\text{Ba}]$) is such that both potential mass transfer duty (characterized by $K_\text{L}A$) and catalytic activity (characterized by [Ba]) are fully used. Figure 16.10 shows rate–reactant concentration profiles for the optimum use of both $K_\text{L}A$ and [Ba] as defined by the intersection of the two profiles. At steady state the rate of reactant transfer and biocatalytic reaction are equal, but the aqueous phase reactant concentration that they define may be greater or less than the optimum concentration, β (shown in Fig. 16.10) if $K_\text{L}A$ and/or [Ba] are incorrectly chosen. Reactions operating with aqueous phase reactant concentrations greater than β are in the kinetically controlled regimen. This makes poor use of the available reactant mass transfer duty and either $K_\text{L}A$ needs to be reduced or [Ba] increased to operate optimally. Reactions operating with aqueous phase reactant concentrations less than β are in the reactant mass transfer controlled regimen. This makes poor use of the available catalytic activity and either $K_\text{L}A$ needs to be increased or [Ba] reduced to operate optimally.

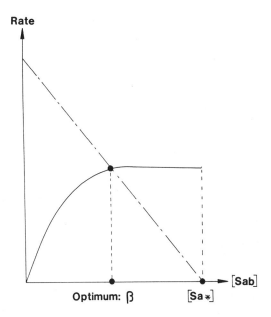

Figure 16.10. Plot to show optimum aqueous phase reactant concentration. Characteristic of optimum: (1) maximum rate of biocatalytic reaction per unit mass of biocatalyst in the aqueous phase; (2) minimum aqueous phase reactant concentration required to achieve that rate. (—·—), Optimum $K_L A$.

The optimum reactant concentration, β, with which to operate in the aqueous phase, is dependent on the shape of the biocatalytic reaction rate–reactant concentration profile, but for a given profile it is defined by the ratio ($K_L A/$[Ba]). Consequently it is possible to construct a plot of $K_L A$ against [Ba] for different aqueous phase reactant concentrations. Each of the plotted lines in Figure 16.11 is characterized by a constant and particular aqueous phase reactant concentration. Steeper gradient lines are characterized by lower aqueous phase reactant concentrations. The full line in Figure 16.11 represents that defined by the optimum aqueous phase reactant concentration. Hence, the required mass transfer duty, $K_L A$, to optimally use a given biocatalyst concentration, [Ba], can be evaluated directly from the design chart. The catalyst concentration is set by the required reactor productivity.

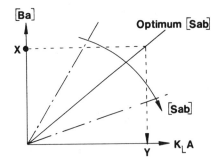

Figure 16.11. General design chart ($K_L A$ against [Ba]). Y is the $K_L A$ required to use catalyst concentration, [Ba], of X optimally.

REACTOR DESIGN AND SCALE-UP

This simple design method gives an estimate of the reactant mass transfer coefficient, $K_L A$, required for a given reactor productivity. This then enables a preliminary selection of reactor type. Liquid–liquid contacting equipment, suitable for biocatalytic reactions, may be characterized by the range of mass transfer duties (i.e., minimum and maximum $K_L A$ values) achievable in that particular design (e.g., Doraiswamy and Sharma, 1984). Knowledge of the $K_L A$ required will therefore eliminate some of these possibilities.

Scale-up will also require examination of several other considerations. Figure 16.12 is a design chart (plot of $K_L A$ against [Ba]) for the optimal hydrolysis of benzyl acetate by pig liver esterase in a two-liquid phase system. The plot is drawn as log $K_L A$ against log [Ea] for simplicity. At a high aqueous phase enzyme concentration, [Ea], the proportion of reactant bound to the enzyme, as reactant–enzyme complex, becomes significant, causing a loss of specific activity with increasing catalyst concentration. This situation occurs because of the low concentration of reactant in the reaction zone, characteristic of reactions with poorly water-soluble organic reactants catalyzed in the bulk of the aqueous phase. Another operational limit occurs at a high fraction of aqueous phase reactant saturation. Above this concentration (defined by a particular ratio of $K_L A/$[Ba]) a loss of enzyme activity is observed over a period of time. This loss of catalyst stability may be explained by the high liquid–liquid interfacial area (relative to catalyst concentration) or by a time-dependent effect of a relatively high aqueous phase reactant concentration, both of which occur at a high

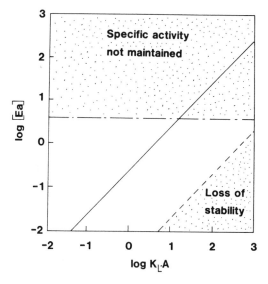

Figure 16.12. Design chart for benzyl acetate hydrolysis by pig liver esterase. (———), Optimum operation; (—·—), limit to catalyst concentration; and (-----), aqueous phase reactant concentration.

operational $K_L A/$[Ba] ratio. These bounds therefore define an operating window. In scale-up, further bounds will be introduced (e.g., limiting power input to create adequate mass transfer), thus reducing the size of the operating window as the process scale is increased.

This then is the real power of this technique—the ability it gives to take laboratory scale measurements of mass transfer and reaction kinetics and use them to draw conclusions about larger scale operation. The rationale behind this argument is that the design method presented here is based on fundamental mass transfer-reaction concepts, which are therefore independent of scale. An added complication is that measurements made over a flat liquid–liquid interface (laboratory reactor) may not accurately simulate and be applicable to those made over a dispersed liquid–liquid interface (likely industrial reactor). However, preliminary results indicate that this is not a problem (Woodley, Cunnah, and Lilly, submitted).

Using a design chart, such as Figure 16.12, as a basis for reactor scale-up also implies that $K_L A$ needs to be maintained constant, independent of scale. It is likely that this engineering challenge will be easier to meet for liquid–liquid systems than for gas–liquid systems (due to the lower interfacial tensions and phase density differences in liquid–liquid systems compared to gas–liquid systems). Nevertheless, the ability to scale-up on this basis is crticial to industrial scale implementation of these reactions.

PROCESS DESIGN

Figure 16.13 shows a schematic diagram of a two-liquid phase biocatalytic process for the conversion of a poorly water-soluble organic liquid

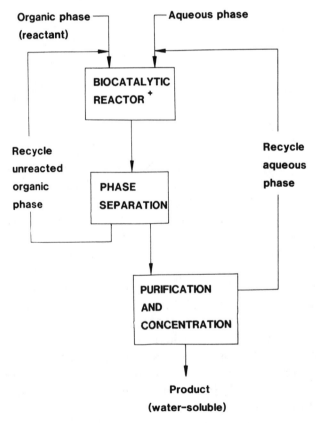

Figure 16.13. Schematic of a two-liquid phase biocatalytic process for conversion of a poorly water-soluble organic liquid reactant to water-soluble product. +Assumed catalyst remains in reactor,e.g., by biocatalyst immobilization.

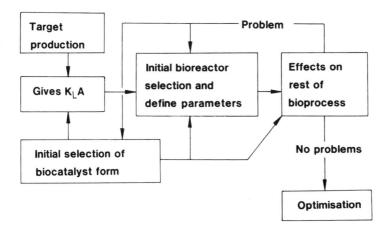

Figure 16.14. Integrated process design strategy.

reactant to a water-soluble product. While it is clear that the reactor is at the heart of the flowsheet, reactor design, and operation cannot be considered in isolation of the rest of the process. Reactor selection, design, and operation all have an impact on subsequent product recovery and reuse/recycle of catalyst, if required. It is important therefore that design methods incorporate such reasoning and Figure 16.14 shows a schematic integrated process design strategy. Target productivity, together with preliminary selection of biocatalyst form, define the required K_LA—the subject of this chapter. This then sets guidelines for initial reactor selection which subsequently set operating parameters (for example, in a stirred tank reactor, stirrer speed, n, and phase ratio, ϕ). The effects of these decisions on the rest of the process must then be considered and if there are problems for product recovery then either the catalyst selection or reactor selection, design, and operation, or some combination of these, may need to be reexamined. Figure 16.14 shows this iterative procedure.

CONCLUSIONS

The industrial use of two-liquid phase biocatalysis for carrying out reactions involving poorly water-soluble organic compounds will depend on the development of appropriate process technology. In this paper a powerful technique for using laboratory scale kinetic measurements as a basis for reactor design has been presented. Although only applied to one class of reaction, current research now embraces other classes, e.g., solute transfer problems, high (Km/[Sa∗]) ratio systems, etc. Nevertheless, the philosophy of building a design method on fundamental mass transfer–reaction concepts will be common to all these reaction classes and hence has wider application as a design aid.

ACKNOWLEDGMENTS

The author is grateful to the Biotechnology Directorate of the Science and Engineering Research Council for support of this program and to Professor M. D. Lilly for stimulating discussions.

REFERENCES

Ballard, D. G. H., A. Courtis, I. M. Shirley, and S. C. Taylor. 1983. A biotech route to polyphenylene. *J. Chem. Soc. Chem. Commun.* **634:** 954–955.

Brazier, A. J., M. D. Lilly, and A. B. Herbert. Toluene cis-glycol synthesis by *Pseudomonas putida;* kinetic data for reactor evaluation. *Enzyme Microb. Technol.* (in press).

Brink, L. E. S., J. Tramper, K. Ch. A. M. Luyben, and K. Van't Riet. 1988. Biocatalysis in organic media. *Enzyme Microb. Technol.* **10:** 736–743.

Brookes, I. K., M. D. Lilly, and J. W. Drozd. 1986. Stereospecific hydrolysis of *d, l*-menthyl acetate by *Bacillus subtilis:* mass transfer-reaction interactions in a liquid–liquid system. *Enzyme Microb. Technol.* **8:** 53–57

Doraiswamy, L. K., and M. M. Sharma. 1984. *Heterogeneous Reactions: Analysis, Examples and Reactor Design. Vol. 2: Fluid–Fluid–Solid Reactions.* New York: John Wiley & Sons.

Harbron, S., T. J. Narendranathan, and M. D. Lilly. 1984. The effects of mass transfer on the kinetic behaviour of two-liquid phase biocatalytic reactions. In *Proceedings of Third European Congress on Biotechnology, Vol. IV,* pp. 369–374. Weinham: Verlag Chemie.

Klibanov, A. M. 1986. Enzymes that work in organic solvents. *Chemtech.* **16:** 354–359.

Laane, C., J. Tramper, and M. D. Lilly (Eds.). 1987. *Biocatalysis in Organic Media.* Amsterdam: Elsevier.

Lewis, J. B. 1954. The mechanism of mass transfer of solutes across liquid–liquid interfaces. Part I. The determination of individual transfer coefficients for binary systems. *Chem. Engin. Sci.* **3:** 248–259.

Lilly, M. D. 1982. Two-liquid-phase biocatalytic reactions. *J. Chem. Tech. Biotechnol.* **32:** 162–169.

Lilly, M. D. 1983. Two-liquid phase biocatalytic reactors. *Philos. Trans. R. Soc. Lond. B* **300:** 391–398.

Lilly, M. D., and J. M. Woodley. 1985. Biocatalytic reactions involving water-insoluble organic compounds. In *Biocatalysts in Organic Syntheses*, pp. 179–192. (J. Tramper, H. C. van der Plas, and P. Linko, eds.). Amsterdam: Elsevier.

Lilly, M. D., A. J. Brazier, M. D. Hocknull, A. C. Williams, and J. M. Woodley. 1987. Biological conversions involving water-insoluble organic compounds. In *Biocatalysis in Organic Media*, pp. 3–17. (C. Laane, J. Tramper, and M. D. Lilly, eds.) Amsterdam: Elsevier.

Tramper, J., H. C. van der Plas, and P. Linko (Eds.). 1985. *Biocatalysts in Organic Syntheses*. Amsterdam: Elsevier.

Williams, A. C., J. M. Woodley, P. A. Ellis, and M. D. Lilly. 1987. Denaturation and inhibition studies in a two-liquid phase biocatalytic reaction: the hydrolysis of menthyl acetate by pig liver esterase. In *Biocatalysis in Organic Media*, pp. 399–404. (C. Laane, J. Tramper, and M. D. Lilly, eds.). Amsterdam: Elsevier.

Woodley, J. M., P. J. Cunnah, and M. D. Lilly. Stirred tank two-liquid phase bioreactor studies: kinetics evaluation and modelling of substrate mass transfer. *Biocatalysis* (submitted).

Woodley, J. M., A. J. Brazier, and M. D. Lilly. Lewis cell studies to determine reactor design data for two-liquid phase bacterial and enzymic reactions. *Biotechnol. Bioeng.* (submitted).

17
Enzymes That Do Not Work in Organic Solvents
Too Polar Substrates Give Too Tight Enzyme–Product Complexes

ANTONIUS P. G. KIEBOOM

"Enzymes in Organic Synthesis" was the title of the review article by which Suckling and Suckling (1974) advocated the use of biocatalysts as an additional and valuable tool in preparative organic chemistry. During the past decade, the upscaling of enzymatic reactions from "biochemical scale" (milligram) to "preparative scale" (multigram) amounts has been successful in several areas of organic chemistry. In particular, the work of Jones (1986), Whitesides (Whitesides and Wong 1985; Akiyama et al. 1988), and Klibanov (1986) provides many exciting examples of enzyme-assisted organic synthesis (Butt and Roberts 1986, 1987; Sonnet 1988; Yamada and Shimizu 1988; Roberts 1988).

In addition, two major breakthroughs concerning enzymatic reactions have been established. First, Klibanov (1986) demonstrated that enzymes work not only in aqueous solutions but also in organic solvents. The basic concept for this approach is that the enzyme does not interact with all the bulk solvent water but only the layer directly bound to it. Second, Dewar (1986) stated that adsorption of a proper substrate in the active site of the enzyme can occur only if all water is excluded from between them. Consequently, any subsequent reaction is thought to take place in the absence of solvent and may be compared with chemical conversions in the gas phase. It is evident that the use of enzymes in (nearly) water-free solvent systems is of practical importance in preparative organic chemistry using lipophilic substrates as well as to force hydrolytic enzymes toward synthesis.

Combination of these two novel concepts about enzyme reactions, one leaving out the outer-enzyme water and the other the inner-enzyme water, might suggest that there is no important role left for water other than as reactant or, in miniscule amounts, for retaining the correct conformation of the enzyme. This is not so because replacement of water by an apolar organic solvent will largely affect the association between enzyme and either substrate or product, in particular if they show strong hydrogen bonding in aqueous solution. Such an effect on K_m may dramatically affect the apparent reaction rate as will be exemplified below. A too tight enzyme–substrate and/or enzyme–product complex is thought to be responsible for the fact that carbohydrate-converting enzymes (hydrolases, transferases, isomerases, oxidoreductases) are not catalytically active in water-free solvent systems (Straathof et al. 1988a; unpublished results).

In the case of carbohydrate-converting enzymes, it is concluded that either the use of substrates containing fewer hydroxyl substituents ("naked sugars") or the enzymatic conversion of carbohydrates into more lipophilic products should be more feasible in organic solvents. This, indeed, has been confirmed experimentally for the conversion of glucose into octyl β-glucoside in octanol using the catalyst β-glucosidase.

DISCUSSION

Formation of an enzyme–substrate (the term "substrate" has been used for both substrate and product) complex in aqueous solution requires the disruption of a number of water hydrogen bonds with both the enzyme active site (the active site includes both the catalytically active site and the noncatalytic binding site of the enzyme) (n_E) and the substrate molecule (n_s). This has schematically been depicted in Figure 17.1. The equilibrium constants may be formulated as follows:

$$K_{ESW} = \frac{[ES]*[W](n_E + n_s)}{[EW]*[SW]} \tag{1}$$

as the apparent association constant of the enzyme–substrate complex in water, i.e., as reflected by $1/K_m$ and $1/K_i$ values from kinetic measurements;

$$K_{ESO} = \frac{[ES]}{[E]*[S]} \qquad (2)$$

either in the gas phase or in an apolar organic solvent;

$$K_{EW} = \frac{[EW]}{[E]*[W]^{n_E}} \qquad (3)$$

as a measure of the water affinity of the enzyme active site;

$$K_{SW} = \frac{[SW]}{[S]*[W]^{n_S}} \qquad (4)$$

as a measure of the substrate-water binding.

Combination of (i)–(iv) gives

$$K_{ESO} = K_{EW}*K_{SW}*K_{ESW} \qquad (5)$$

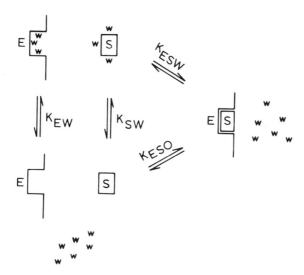

Figure 17.1. Mutual correlation of the equilibria between enzyme, substrate (or product) and water. EW, hydrated enzyme active site; SW, substrate (or product) solvated by waters; E, non-hydrated enzyme; S, non-hydrated substrate (or product); W, water.

from which it is directly seen that hydration of both the enzyme active site and the substrate result in higher K_{ESO} values. In this respect it has to be noted that the higher the hydrophilicity of substrate (or product), the higher will be the hydrophilicity of the enzyme active site. This correlation is a fundamental prerequisite to obtain sufficient enzyme–substrate (or product) complex formation in aqueous solution. Therefore, enzymes with carbohydrates as their natural substrates should have extremely high K_{ESO} values due to a high K_{EW} as well as a high K_{SW}.

Assuming enzyme and substrate to be water-free in the gas phase or in a nonhydrogen bonding forming organic solvent, the free energy of complex formation is given by

$$\Delta G(\text{org solv}) = \Delta G(\text{water}) - (n_E + n_s) * \Delta G(\text{H-bonding})$$

where $\Delta G(\text{org solv})$ and $\Delta G(\text{water})$ are the free energies of the equilibrium

$$\text{Enzyme} + \text{substrate} \rightleftharpoons \text{Enzyme–substrate complex}$$

in an organic solvent and water, respectively, and $\Delta G(\text{H-bonding})$ is the mean value for the cleavage of one water from either enzyme, substrate or product:

$$\text{E–W} \rightleftharpoons \text{E} + \text{W} \quad \text{or} \quad \text{S–W} \rightleftharpoons \text{S} + \text{W}$$

Thus, the increase in stability of the enzyme–substrate complex in terms of the free energy upon going from water to the organic solvent is given by

$$\Delta\Delta G(\text{water-org solv}) = -(n_E + n_s) * \Delta G(\text{H-bonding})$$

For a monosaccharide molecule strongly hydrogen-bonded to four waters (Franks 1983; Suggett 1976) ($n_s = 4$) and an enzyme active site involving also four hydrogen-bonded waters ($n_E = 4$) this points to a maximal gain in complexation energy, using 5 kcal/mol per hydrogen bond, of approximately 40 kcal/mol, i.e.

$$K_{ESO} \approx 10^{30} * K_{ESW}$$

(In aqueous solution, a maximal value of approximately 5 kcal/mol for the free energy of complexation is found.) Assuming $K_{ES} \approx 1/K_m \approx 1/K_i$, this means an extraordinarily low rate of dissociation of the enzyme–substrate and/or enzyme–product complex, even in the case of a diffusion-limited rate of association and K_m and K_i values as high as 1 M. (The fact that K_m and K_i values often contain contributions of reaction rate constants does not affect the reasoning principally.) It may be noted that a $\Delta\Delta G$(water-org solv) difference of only 5 kcal/mol, corresponding to the effect of one H-bonding difference between water and an organic solvent, will be enough to slow down enzymic carbohydrate conversions in organic solvents to immeasurably low rates. In other words, the use of organic solvents will result in severe substrate and product inhibition in the case of carbohydrate- or other highly-polar-compound-converting enzymes. As shown by the reaction profile in Figure 17.2, rapid enzymatic ES $\rightleftharpoons$ EP interconversion will still occur in apolar organic solvents but replenishment of either substrate or product molecules in the enzyme active site is prohibited by the high dissociation energy barriers. [The water effect on ΔG and $\Delta\Delta G$ mentioned will only be partly counteracted by the ΔS term of complexation. If we compare the carbohydrate–enzyme complexation, carbohydrate·(aq) + enzyme·(aq) $\rightleftharpoons$ complex + water, with the complexation of cations with crown ethers, cation·(aq) + crown ether·(aq) $\rightleftharpoons$ complex + water, this seems to be justified. In the latter process, ΔH varies ca. 6 kcal/mol going from water to ethanol, whereas ΔS only

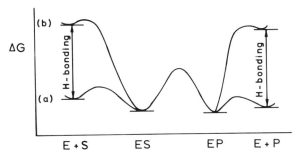

Figure 17.2. Reaction profiles for an enzymatic reaction in water (a) and in an organic solvent (b) for substrates and products that form strong hydrogen bonding with water.

varies a few e.u.'s (Christensen, Eatough, and Izatt 1974). In addition, it may be noted that the dissociation rates of alkaline earth cryptates, which might be considered models for highly polar substrate–enzyme complexes, decrease considerably upon going from water to propylene carbonate (Cox, van Truong, and Schneider 1984). Thus, not only the thermodynamics but also the kinetics of decomplexation will be in disfavor with respect to the high turnover numbers of the active site.]

These considerations suggest that either proper carbohydrate-like compounds without severe hydrogen-bonding properties (polydeoxy or naked sugars) are possible substrates for carbohydrate-converting enzymes in organic solvents or that the conversion of carbohydrates into more lipophilic products in such systems should be feasible. In addition, minor amounts of water present in the system and the hydrogen-bonding properties of the organic solvent will be of crucial importance for the successful application of such enzymic conversions.

Some preliminary experimental results have been obtained for the conversion of glucose to octyl β-glucopyranoside using β-glucosidase in octanol (Figure 17.3). Although a tight enzyme–substrate complex will occur in the organic solvent octanol, the enzyme–product complex is expected to be weaker and thus replenishment of the enzyme active site by new substrate might become possible. [Substrate and product

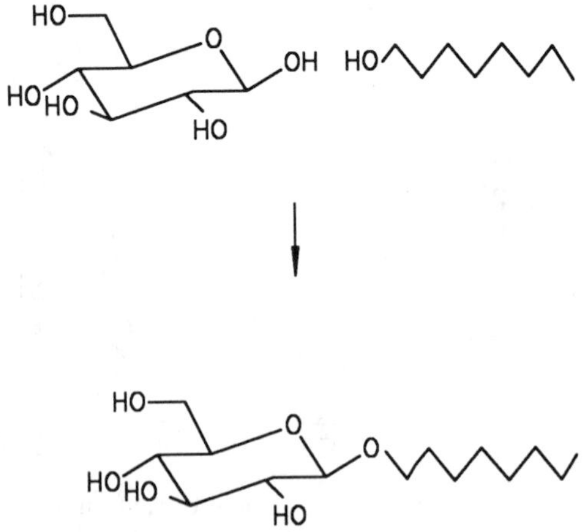

Figure 17.3. Octyl β-glucopyranoside formation.

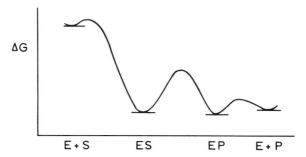

Figure 17.4. Reaction profile for the enzymic conversion of glucose (S) into octyl β-glucoside (P) in octanol according to Scheme 1.

hydrophobicity is directly reflected by the partition coefficient P in the octanol–water two-phase system. At 25°C, glucose is predominantly present in the water layer ($p < 0.01$) whereas the product octyl β-glucoside prefers the octanol phase ($p = 29$) (Straathof, van Bekkum, and Kieboom 1988b). In other words, the product is much more extractable from the active site by apolar organic solvents (e.g., octanol).] In contrast to the reaction profile given in Figure 17.2, where comparable polarities of substrate and product were assumed as in the case of glucose $\rightleftharpoons$ fructose isomerization, we now obtain a more advantageous down-hill energy profile from E + S up to E + P (Figure 17.4). Indeed, shaking a suspension of glucose (0.5 g), β-glucosidase (0.1 g) in dry octanol (10 ml) for 1 day at 37°C resulted in some octyl β-glucoside formation (ca. 2%), corresponding with a total turn over number of a few hundred glucoses per mole of enzyme (Nijkamp, Hoefnagel, and Kieboom, unpublished results). A first example which shows that, under certain circumstances, carbohydrate-converting enzymes might possess catalytic activity in organic solvents.

REFERENCES

Akiyama, A., M. Bednarski, M. J. Kim, E. S. Simon, H. Waldmann, and G. M. Whitesides. 1988. *CHEMTECH*: 627.

Butt, S., and S. M. Roberts. 1986. *Natural Prod. Rep.*: 489.

Butt, S., and S. M. Roberts. 1987. *Chem. Brit.*: 127.

Christensen, J. J., D. J. Eatough, and R. M. Izatt. 1974. *Chem. Rev.* **74**: 351.

Cox, B. G., N. van Truong, and H. Schneider. 1984. *J. Am. Chem. Soc.* **106:** 1273.

Dewar, M. J. S. 1986. *Enzyme* **36:** 8.

Franks, F. 1983. *Cryobiology* **20:** 335.

Jones, J. B. 1986. *Tetrahedron* **42:** 3351.

Klibanov, A. M. 1986. *CHEMTECH*: 354.

Roberts, S. M. 1988. *Chem. Ind. (Lond.)*: 384.

Sonnet, P. E. 1988. *CHEMTECH*: 94.

Straathof, A. J. J., J. P. Vrijenhoef, E. P. A. T. Sprangers, H. van Bekkum, and A. P. G. Kieboom. 1988a. *J. Carbohydr. Chem.* **7:** 223.

Straathof, A. J. J., H. van Bekkum, and A. P. G. Kieboom. 1988b. *Starch* **40:** 438.

Suckling, C. J., and K. E. Suckling. 1974. *Chem. Soc. Rev.* **3:** 398.

Suggett, A. 1976. *J. Solution Chem.* **5:** 33.

Whitesides, G. M., and C. H. Wong. 1985. *Angew. Chem.* **97:** 617.

Yamada, H., and S. Shimizu. 1988. *Angew. Chem.* **100:** 640.

Index